MW01536401

# FINITE ELEMENTS 1-2-3

# FINITE ELEMENTS 1-2-3

**A. J. Baker**
*Department of Engineering Science and Mechanics*
*University of Tennessee*

**D. W. Pepper**
*Department of Mechanical Engineering*
*California State University–Northridge*

**McGraw-Hill, Inc.**

New York St. Louis San Francisco Auckland Bogotá Caracas
Hamburg Lisbon London Madrid Mexico Milan Montreal New Delhi
Paris San Juan São Paulo Singapore Sydney Tokyo Toronto

This book was set in Times Roman.
The editors were B. J. Clark and David A. Damstra;
the production supervisor was Friederich W. Schulte.
The cover was designed by John Hite.
R. R. Donnelley & Sons Company was printer and binder.

**FINITE ELEMENTS 1-2-3**



1 2 3 4 5 6 7 8 9 0 DOC DOC 9 0 9 8 7 6 5 4 3 2 1

P/N 003467-2
PART OF
ISBN 0-07-909975-0

**Library of Congress Cataloging-in-Publication Data**

Baker, A. J., (date).
Finite elements 1-2-3 / A. J. Baker, D. W. Pepper.
p. cm.
Includes bibiliographical references and index.
ISBN 0-07-909 975-0 (set)
1. Finite element method. I. Pepper, D. W. (Darrell W.)
II. Title. III. Title: Finite elements one-two-three.
TA347.F5B34 1991
620'.001'51535—dc20 90-6626

# ABOUT THE AUTHORS

**A. J. Baker** received his B.S. in mechanical engineering from Union College (1958) and M.Sc. (1968) and Ph.D. (1970) degrees in engineering science from the State University of New York–Buffalo. He is currently professor of engineering science and mechanics at the University of Tennessee in Knoxville, with special interests in fluid mechanics, finite element analysis, and computational fluid dynamics (CFD). At UTK since 1975, he was designated a *Chancellor's Research Scholar* in 1987, the University's highest award recognizing academic research.

Dr. Baker is cofounder, principal scientist, and president of Computational Mechanics Corporation, Knoxville, a small business firm engaged in engineering analysis and computer code development in CFD. Prior professional employment includes principal research scientist at Textron/Bell Aerospace in Buffalo (1970–1975) and mechanical engineer with Union Carbide Corporation/Linde Division in Buffalo (1958–1964). Dr. Baker has authored over 160 technical publications and the textbook *Finite Element Computational Fluid Mechanics* (1983).

**D. W. Pepper** received his B.S. in mechanical engineering (1968), M.S. in aeronautical engineering (1970), and Ph.D. (1973) degrees from the University of Missouri–Rolla. From 1974 to 1987 he worked for E. I. Dupont at the Savannah River Laboratory in Aiken, South Carolina, where he held various technical and managerial positions. In 1987 he joined the Marquardt Company, an aerospace propulsion firm in Van Nuys, California, as chief scientist, with responsibility for computational fluid dynamics related projects associated with the National Aerospace Plane (NASP). He is the cofounder of Advanced Projects Research, Inc., a small business engineering research and development company located in Moorpark, California, which specializes in advanced numerical algorithms and parallel processing for fluid dynamics, heat transfer, and environmental problems. He recently became professor of mechanical engineering at California State University–Northridge.

# CONTENTS

# PREFACE

This book, and the accompanying PC-based *Computational Mechanics Laboratory*, has emerged from over a decade of learning how to teach the introductory level of finite element analysis to practicing engineers functioning in the real world. The *finite element method* is the engineer's implementation of a thoroughly formal mathematical theory for constructing *approximate solutions* to the partial differential equations occurring throughout engineering mechanics and physics. The intrinsic beauty of the method is that this theory completely accounts for each and *every critical decision* that must be made in the design of a numerical algorithm. No other approximation method in use today is so complete in giving firm and accurate guidance on algorithm construction, consistent boundary condition implementation, direct adjustment of spatial order of accuracy, and geometric flexibility. The absolutely general geometric capabilities of finite elements is common knowledge. However, the *true power* of the method lies in its *theoretical completeness*!

The potential reader of this book may well have a background familiarity with finite difference methods for solving differential equations. This method was first applied in the mid-1930s and has since enjoyed a continual expansion with fundamental advances coming from many recognized contributors such as Lax, Wendroff, Harlow, MacCormack, and Beam and Warming. A finite difference procedure directly replaces partial derivatives appearing in an equation with divided difference expressions established from a local Taylor series. Linear terms are very easy to handle, but conceptions for nonlinearities and other complications rests totally on the imagination of the algorithm designer. In these cases, a *finite difference scheme* is hypothesized and then tested numerically to validate the decision.

The transparent simplicity of a finite difference is its main attribute and ultimately its major detraction. One directly generates the algebraic replacement for any isolated spatial or temporal derivative. This algebraic (recursion relation) expression is immediately coded in Fortran, which was originally devised to precisely translate such formulas into executable syntax. However, as this is basically a one-dimensional concept, no theoretical guidance is given on multidimensional formulations in nonrectangular regions, applications in complex curvilinear geometries, or handling of complicated flux boundary conditions in any given situation.

Conversely, and in direct distinction, the theoretical construction supporting the finite element procedure shines in all areas. Of fundamental importance, the methodology employs *calculus* throughout the theory. Hence, the multidimensional formulation in absolutely arbitrary geometries is no more difficult than the simple one-dimensional development, since one simply replaces first and second ordinary derivatives with the gradient and divergence operators. The theory employs an *integral statement*, valid for the exact solution as well as *any* (!) approximation, and all formulational steps are carried out prior to defining a discretization. Hence, integration-by-parts and the divergence theorem are regularly utilized to *analytically* expose the correct expression for *any boundary condition* on any (curved) boundary surface! Nonlinear terms are always exactly handled, via the calculus; hence one is never required to "guess" at an equivalent approximation form.

After all formulational details are completed analytically, the finite element method then introduces a discretization of the solution domain into the familiar triangle-, quadrilateral-, etc., shaped cells that permit algorithm restatement into the nodal form amenable to digital computation. This terminal operation naturally yields a matrix algebra statement rather than the recursion relation form produced by differencing. These two statements are really synonymous, however, since the recursion corresponds to expression of the generic interior (i.e., non-boundary condition) row in a matrix statement. Factually, a recursion relationship could always be obtained by simply "assembling" terms over all finite elements sharing node point $j$, or point $i, j$ in two dimensions. The preference is to not do this, however, since it yields creation of a nonversatile restatement that becomes very cumbersome for nonuniform meshes, nonlinearities, and boundary condition alterations, for example. Instead, the finite element method directly expresses the terminal *global matrix statement*, and then independently considers the various *linear algebra* procedures available to solve it. Direct gaussian elimination is a favorite solution methodology for problems with a reasonable number of nodal unknowns. For larger problems, numerical replacement candidates include the familiar stationary iteration procedures (Picard, Gauss-Seidel, successive overrelaxation), and implicit directional methods [alternating-direction-implicit (ADI), line relaxation, strongly implicit, etc.].

Viewing these last two paragraphs, it becomes quite evident that the principal feature of the finite element method, its *full generality*, is also its principal detraction! Specifically, a certain dexterity with the calculus, including integral vector operations, is required to master the formulational details leading to the Fortran-amenable calculational statement. However, every engineering undergraduate curriculum contains this material in thoroughness, and rather than be intimidated by the formality, the reader should view this as the *opportunity* to fully utilize this academic material. Factually, the great mathematicians following the eighteenth century derived the "limit," and hence converted difference algebra to the calculus. The computer came along 150 years later, and the approximate solution of *differential equations* reverted back to difference algebra. Now, the finite element method has emerged to return the formality of calculus to approximate solution methodology for differential equations!

We have written this book to serve the need to *clearly and concisely* convey the fundamental notions supporting the finite element method without belaboring theoretical issues. We start in one dimension, and hence use only ordinary differential and integral calculus to introduce the three fundamental ideas:

1. We seek an *approximate solution*, and hence must formally state its existence.
2. Once stated, the dominating issue for the approximation is that its *error is minimum.*
3. A *discretization* is then introduced, such that the calculus statements reduce to a computable form.

These simple, practical concepts constitute the heart of finite element methodology. They are introduced in thoroughness, leaving no intermediate formulational steps for the "casual observer" to fill in, first in the one-dimensional framework. Thus, every fundamental aspect is completely covered before proceeding to multiple dimensions and attacking the numerous geometry issues. The included *LEARN.FE* computer code supports the practical exploration of every basic step, through a sequence of discretization exercises, development completion steps, and practical study problem *computational experiments* focused on accuracy and error measuring issues. The text leads you through these topics in a logical order and at a pace verified effective via the many guinea pigs that have preceded you.

In closing this preface, we quote from comments offered by a recent graduate class that helped us debug the text and code exercises.

> Finite elements is a very powerful tool that allows solving of diverse problems that can be written in terms of differential equations. The *LEARN.FE* code can solve one- and two-dimensional and axisymmetric problems. The power of this code is the fact that it is not written to solve a specific problem for the novice user. It is a basic tool which, once understood, can be modified for almost any type of problem. The basics are all there; the code is like a *wrench in the hands of a mechanic*, and how far you can go with it is limited only by the user.

We hope you find the text and the code of similar use in accomplishing your goal to gain a firm understanding and hence dexterity with the finite element method.

McGraw-Hill and the authors would like to thank the following reviewers for their many helpful comments and suggestions: James Pitaressi, State University of New York at Binghamton; M. S. Rahman, North Carolina State University; and Robert Sennett, California State University at San Luis Obispo.

Many colleagues and students have contributed greatly to the development of this book. We wish to specifically acknowledge the fertile environment provided by our longstanding association with Dr. Osama Soliman and Messrs. Paul Manhardt, Joe Orzechowski, and Roger Cooper. Messrs. Joe Iannelli, Subrata Roy, David Chaffin, Wilbert Noronha, and Terry Dishongh, all graduate students at the University of Tennessee, have helped immeasurably in bringing the *LEARN.FE* code to fruition. Finally, we are completely indebted to Ms. Jenny Wren, our word processor par excellence, who cheerfully endured the numerous revisions and editing sequences required to produce the manuscript. This has been a great team effort!

*A. J. Baker*
*D. W. Pepper*

# FINITE ELEMENTS 1-2-3

# CHAPTER 1

# INTRODUCTION

## 1.1 ABOUT THIS BOOK

This book is written specifically for the bachelor's level engineer, scientist, and/or upper-division undergraduate student with a curiosity about, but little or no experience with, the *finite element method.* It should be of value as well to the knowledgeable practitioner with the desire to examine personal computer implementation of state-of-the-art finite element algorithms.

The essential ingredients of the finite element method for solving diffusion and transport equations are developed herein, and the resultant algorithm logic is implemented within the *LEARN.FE* computer program on the included diskette. This program is operable on most IBM-compatible PCs with disk operating system (DOS) level 3.0 or higher. Of practical importance, the *LEARN.FE* code is also operable on minicomputer or mainframe computer systems. Although knowledge of a PC system is helpful, it is not needed to develop an understanding of the operational details. The *LEARN.FE* program will execute on a PC having a math coprocessor, a minimum of 640K RAM, and either a single, dual, or hard-disk drive.

Our intent is that this book be complete and self-contained, so that the novice can both understand the concepts and effectively use the *LEARN.FE* program to solve practical problems. A working knowledge of Fortran will certainly help in understanding the transition from mathematical conception to the code. However, it is not necessary to fully understand programmatic language to learn from the examples and code exercises. The reader is expected to remember undergraduate calculus, and a first course in computer science would be helpful as well. The *LEARN.FE* program, which is "command name"-driven, is not overly sophisticated. Hence, it should not be used strictly as a "blackbox" code. It is designed to exemplify the logical structure of a versatile finite element code with general applicability to problems in mechanics and engineering.

## 1.2 ORGANIZATION

This book is designed as a self-teaching guide for understanding the finite element method and for solving diffusion and transport problems using a PC. It begins at the elementary level of analysis of steady-state one-dimensional diffusion of heat. We then include boundary thermal convection, and proceed logically and directly through multidimensional steady diffusion to unsteady fluid convection–diffusion transport. By the time the reader completes the book, and runs the example and study problem code exercises, a firm understanding of finite element methodology should be well grasped. The *LEARN.FE* program, which is modular and menu-driven, is designed to exemplify finite element implementation in the PC environment. Hence, it is written for clarity and not maximum operating efficiency. The singular intention is instruction; when this goal is accomplished, the reader may, as desired, modify, optimize and/or extend *LEARN.FE* for any computer environment.

Each of the development test cases takes a problem statement, converts it into a finite element algorithm statement, and then generates a solution family using the code. By studying these examples, the manner for defining and organizing problem statement *data*, including boundary (and initial) conditions and material properties, for a specific statement becomes apparent. For the first few examples, the solutions are readily visualized; the computational experiments are designed to emphasize available control over *approximation error.* The subsequent problem statements become progressively more involved and hence require more input data specification and solution interpretation. The *LEARN.FE* code graphics displays are programmed for color. If hard copy is desired, the code asks the user to issue a command (usually yes/no) for a screen dump. A PC graphics printer will provide for both output file printing and graphical output for any generated solution.

Chapter 2 introduces the "expert system environment" aspects of the *LEARN.FE* code, as implemented in the PC venue, including instructions for demonstration of solution creation and interpretation aspects. The technical development introduction in Chap. 3 highlights the essential ingredients of the *weak statement theory* supporting finite element analysis, utilizing a simple but highly pertinent one-dimensional problem definition. The essence of *discretization* and finite element *basis* functions leads naturally to an example and hence introduction to the issues of *accuracy* and approximation *error.*

Chapters 4 to 7 expand on this introduction, as a function of problem dimension, and hence finite element basis choices. Each chapter begins with a presentation of the pertinent "methodology" which includes automated formation of discretizations using *LEARN.FE* and introduction to the basis-oriented master matrix library. These procedural matters are exemplified by code exercises to ensure that the reader grasps each concept of the development. Thereafter, practical problem statements are developed, and finite element algorithm aspects are expanded to include nonhomogeneous fixed and flux-type boundary conditions. Each consequential development is supported by a "study problem" oriented to code exercises, leading to the organization of "computational experiments" to generate data needed to verify that an adequately accurate solution has been attained.

The sequence of these chapters moves from one-dimensional problems on to two-dimensional problems using both triangular- and quadrilateral-shaped elements, and hence on to an introduction to three-dimensional problems. Chapter 8 presents the unsteady initial-value problem of diffusion with imposed fluid convection, and concludes with an accurate and efficient time-split solution procedure based on one-dimensional element methodology. Finally, the Appendix provides code documentation and key program listings. The diskette contains the *LEARN.FE* code, in compiled form and selected source, and the range of input files for various code exercises.

## 1.3 THE ESSENCE OF THE FINITE ELEMENT METHOD

The *finite element method* is a collection of theory-rich techniques which can produce near-optimal approximate solutions to the initial-boundary value partial differential equations common to engineering and mathematical physics. The implementation ultimately employs subdivision of the problem domain (geometry) into a finite number of small regions called *finite elements.* Many convenient shapes are available, such as triangles and quadrilaterals in two dimensions, and tetrahedra, pentahedra, and hexahedra for three dimensions. Using a *weak statement*, or a "weighted-residual," finite element theory simply requires that the *approximation error be a minimum*! All combinations of appropriate boundary conditions become naturally included in the process. This basic constraint produces a set of algebraic equations, written on each (i.e., the *generic*) finite element, which are then collected together using a procedure called "assembly" to form the global matrix statement. The approximation-dependent variable distribution is then determined via solution of this matrix statement using any linear algebra technique.

The finite element method was originally conceived by engineers in the 1950s to analyze aircraft structural systems using the emerging scientific digitial computer. Turner et al. (1956) published the first paper, followed by Clough (1960) and Argyris (1963), among others. As the finite element structural method matured, the concept of a "force balance" was replaced by theory founded in the variational calculus and classical Rayleigh-Ritz methods (Rayleigh, 1877; Ritz, 1909). Application of the finite element method to nonstructural problems was first reported by Zienkiewicz and Cheung (1965), and then to a wider class of problems in nonlinear mechanics by Oden (1972). There have been many contributors to the development of the mathematical theory of finite elements, including Babuska and Aziz (1972), Ciarlet and Raviart (1972), Aubin (1972), Strang and Fix (1973), Oden and Reddy (1976), and Lions and Magenes (1972).

The direct extension of these classical theoretical concepts to unsteady field transport, including nonlinear fluid mechanics problems, is not possible. The fluid convection term couples velocity to field gradients, which does not admit creation of a variational principle as required by the Rayleigh-Ritz theory. For this reason at least, the "historical" approximation procedures for nonlinear transport have utilized direct replacement of derivative terms by algebraic divided difference expressions, or *finite differences* (Richtmyer and Morton, 1967; Roache, 1972). A related procedure employs integration over cells whereby the gradient terms (cell fluxes) are replaced with *finite volume* quotients (Patankar, 1980).

Essentially *every* finite difference and/or finite volume recursion relation replacement for specific derivative terms can be reconstructed within the weighted-residual or "weak statement" framework intrinsic to the finite element method. Within the general family of weighted-residual methods (Finlayson, 1972), "collocation" reproduces the classical finite difference form, and finite volume statements are retrieved by using constant weights. Generalizing these weights to be functions, and then requiring that they be identical to the approximation functions, yields the *Galerkin weak statement criterion*, named after the original (nondiscrete) procedure derived by B. G. Galerkin (1915). For a linear elliptic partial differential equation, the Galerkin constraint exactly reproduces the algorithm statement produced by the classical variational principle. Several of these algorithm variations are illustrated in Baker (1983, Chaps. 4 and 6) using the weak statement formulation as the theoretical foundation.

In this book, the finite element algorithm implementation in one, two, and three dimensions employs the *Galerkin criterion* for the weak statement for *optimal accuracy*. The principles and mathematical rigor associated with finite element algorithm construction employs classical calculus operations, in direct opposition to finite differencing algebraic methodology. We hope you enjoy finding direct use for your calculus background; the additional "theoretical" effort more than pays for itself as issues of solution accuracy, boundary condition enforcement, and geometric flexibility unfold. There is no painless cookbook way to gain understanding of the finite element method other than to read, and perhaps reread, the development as presented herein. Every effort has been made to make presentation of the basic principles as clear and concise as possible. We sincerely hope we have been successful in this regard for your benefit.

## 1.4 REFERENCES


Argyris, J. H. (1963): *Recent Advances in Matrix Methods of Structural Analysis*, Pergamon, Elmsford, New York.

Aubin, J. P. (1972): *Approximation of Elliptic Boundary Value Problems*, Wiley-Interscience, New York.

Babuska, I. and A. K. Aziz, (1972): "Lectures on the Mathematical Foundations of the Finite Element Method," in A. K. Aziz (ed.), *Mathematical Foundations of the Finite Element Method with Applications to Partial Differential Equations*, Academic Press, New York, pp. 1–345.

Baker, A. J. (1983): *Finite Element Computational Fluid Mechanics*, Hemisphere, Washington, D.C.

Ciarlet, P. G. and P. A. Raviart, (1972): "General Lagrange and Hermite Interpolation in $R^n$ with Applications to the Finite Element Method," *Arch. Rat. Mech. Anal.*, Vol. 46, pp. 177–199.

Clough, R. W. (1960): "The Finite Element Method in Plane Stress Analysis," *Proceedings of the Second Conference on Electronic Computation*, American Society of Civil Engineers, Pittsburgh, Pa., pp. 345–378.

Finlayson, B. A. (1972): *The Method of Weighted Residuals and Variational Principles*, Academic Press, New York.

Galerkin, B. G. (1915): "Series Occurring in Some Problems of Elastic Stability of Rods and Plates," *Engrg. Bull.*, Vol. 19, pp. 897–908.

Lions, J. L. and E. Magenes, (1972): *Non-Homogeneous Boundary-Value Problems and Applications*, Vol. I (trans. from 1963 French ed. by P. Kenneth), Springer-Verlag, New York.

Oden, J. T. (1972): *Finite Elements of Non-Linear Continua*, McGraw-Hill, New York.

Oden, J. T. and J. N. Reddy, (1976): *Introduction to the Mathematical Theory of Finite Elements*, Wiley, New York.

Patankar, S. V. (1980): *Numerical Heat Transfer and Fluid Flow*, Hemisphere, Washington, D.C.

Rayleigh, J. W. S. (1877): *Theory of Sound*, 1st ed., rev. Dover, N.Y., (1945).

Richtmyer, R. D. and K. W. Morton, (1967): *Difference Methods for Initial-Value Problems*, 2d ed., Interscience, New York.

Ritz, W. (1909): "Uber Eine Neue Methode *zer* Losung Gewisser Variations-Probleme der Mathematischen Physik," *J. Reine Angew. Math.*, Vol. 135, p. 1.

Roache, P. J. (1972): *Computational Fluid Mechanics*, Hermosa, Albuquerque, N.M.

Strang, G. and G. J. Fix, (1973): *An Analysis of the Finite Element Method*, Prentice-Hall, Englewood Cliffs, N.J.

Turner, M., R. Clough, H. Martin, and L. Topp, (1956): "Stiffness and Deflection of Complex Structures," *J. Aero. Sci.*, Vol. 23, No. 9, pp. 805–823.

Zienkiewicz, O. C. and Y. K. Cheung, (1965): "Finite Elements in the Solution of Field Problems," *The Engineer*, pp. 507–510.

# CHAPTER 2

# THE *LEARN.FE* COMPUTER PROGRAM

## 2.1 OVERVIEW

The electronic digital computer, first developed in the 1940s, has exerted a most profound impact on the engineering and scientific communities. The rapid and continuing improvement of hardware and software sophistication enables increasingly more complex theoretical and engineering design problems to be computationally examined. The hand-held electronic calculator, at its time of introduction in the late 1960s, initialized a revolution for the student as well as the practicing engineer and scientist. The evolution of small inexpensive calculators, coupled with the reduction in size of microelectronic circuitry, resulted in emergence of the mini computer and subsequently the *personal computer* "PC" in the mid 1970s. Annual sales of PCs in 1990 was about 4 million at a unit cost ranging from $500 to $10,000, and they operated immediately on being plugged in. For comparison, a mainframe computer currently costs $2 to $5 million, and takes several weeks for installation and set up. At the high end, a supercomputer costs \$10 to \$25 million, usually requires a mainframe as the front end for I/O operations, and installation quite often necessitates the construction of a new building.

The first digital computers were large devices consisting of many electronic tubes and associated circuitry. Their reliability was poor, they were difficult to repair, and downtime was disproportionately large. Nevertheless, they performed their calculational tasks better than alternative methods, and made certain extensive calculations possible for the first time. The invention of the transistor and integrated circuits reduced costs and downtime significantly. Today, for example, a board containing thousands of circuits can be replaced as a modular unit in a matter of minutes. The development of the microprocessor, or large-scale integrated (LSI) chip, by the INTEL Corporation in the early 1970s, served as the catalyst for development of the personal computer. LSI technology allowed hundreds of transistors to be placed on a single silicon chip. As the evolutionary consequence, microcomputer systems are now capable of outperforming mainframe computers in use less than a decade ago.

The finite element method is ideally suited for execution on a digital computer. The aircraft industry exploited this in the mid-1960s while developing finite element structural programs for their (relatively) large computers. Quite sophisticated programs and solution techniques resulted, some of which remain in use today, such as NASTRAN. Since the procedure ultimately involves solving large matrix statements, of an order equal to the number of unknowns in the model, mainframe computers were the only venue for executing finite element analyses in the period 1960–1980. The first generation of small memory 8-bit microcomputers was not amenable to effective finite element implementation, since simple problems could require many hours to complete a calculation. The 16-bit PC of 2 to 3 years ago was capable of solving fairly large matrix statements and handling the associated data transfers. The 32-bit PC available today is significantly more powerful, rivaling engineering workstations in speed and performance.

Table 2.1 summarizes the essential differences between a representative mainframe computer and a nonenhanced PC. The modern PC is capable of running thousands of software packages available on the market today. Personal computers are most efficient in handling relatively simple single-user problems requiring only modest memory, speed, and accuracy. Accuracy is related to word length; the more bits to the computer word, the larger the number of significant digits. For example, the IBM PC-AT is a 16- to 24-bit computer, while the newer PCs on the market

**TABLE 2.1**
**Computer comparisons**

| Criteria | Mainframe computer | Personal computer |
|---|---|---|
| Size | | |
| Physical | Small–medium room | Typewriter-television |
| Precision | 32–64 bits | 8–16, 32 bits |
| Memory | <100 Mbytes | 640K–4Mbytes |
| Speed | | |
| CPU cycle | 70 ns | 200 ns |
| Memory cycle | 250 ns | 400 ns |
| Cost | \$2–\$5 million | \$500–\$10,000+ |
| Utility | | |
| Operating | Multiprogramming | Single task |
| Languages | Nearly all | BASIC, Fortran, Cobol, Pascal, C |
| Applied software: | | |
| Availability | Excellent | Voluminous |
| Cost | Expensive | \$10–\$1000+ |

today are either 80386 (INTEL), 80486, or 68020 (Motorola) chip-based and are 32-bit machines. Further, the 68030-chip and i860 machines are now entering the marketplace.

The CPU clock cycle for a PC is about an order of magnitude slower than that for a mainframe computer. However, many elementary engineering and scientific calculations require more time for I/O operations than the CPU. Therefore, overall estimates of calculation time for a problem must also consider processing time of the peripheral equipment. Efforts have been made in *LEARN.FE* to process I/O and calculations concurrently, and hard copy is left as a user option since most PC peripheral printers are slow.

## 2.2 THE *LEARN.FE* CODE

As stated previously, this book is designed as a self-teaching guide for the engineer-scientist to become acquainted with the finite element method on a *personal basis*. What better venue could therefore exist to accomplish this objective than a well-designed *computational laboratory* operating on a PC? The *LEARN.FE* finite element computer program, contained on the diskette included in the rear pocket of this book, is designed specifically to meet this objective. Actually, rather than *designed*, the *LEARN.FE* code system has *evolved* over a decade of teaching finite element methods in graduate-level engineering courses and at specialty short courses. Finite element theory is precise and the mathematics are easily expressed. Converting this elegance into a user-friendly number-generating capability is not so easy!

In cooperation with numerous colleagues and hundreds of students, the *LEARN.FE* code system has evolved as a PC-based laboratory environment to apply theoretical developments. The code uses a command name management structure that facilitates "blackbox" operation, as the student transcends the *novice* period, and then directly permits extension of capability as user comfort increases. The technical developments in each chapter are keyed to code exercise sequences. At first, these are simple executions of contained data files. Thereafter, we ask the reader to augment the files and proceed to more complicated problem definitions, and hence finite element solutions. On completion of each "hand-holding" stage, a study problem is defined for which the reader is asked to set up the finite element analysis. Thereupon, we suggest a *computational experiment*, to generate the data required to ensure that the answers are "good enough to believe;" thus the numerical approximation error is verifiably small.

The *LEARN.FE* system is thereby specifically organized to assist the reader in the transition from novice to confident user of finite element methodology. The command name input structure greatly facilitates this via data file organization keyed to basic steps in problem definition and solution. The READIT subroutine is host to the command input system. The presently implemented four-character command names and the operational sequences thereby keyed are:

*TITL:* Read a title for the problem.

*TYPE:* Read integer data defining key algorithm definitions, e.g., problem dimension, finite element basis degree, grid-generator definition keys, axisymmetric switch.

*PRIN:* Keys debug print, long- or short-form output, hard copy, etc.

*GRID:* Accesses the automated mesh generation procedure; keyed to problem dimensionality, grid nonuniformity, triangles versus quadrilaterals, isoparametric transformations, etc.

*MATL:* Define the distribution of diffusive material properties.

*SORC:* Definition of internal source term distribution.

*DIRI:* Definition of solution domain surfaces with fixed (Dirichlet) boundary conditions and these data.

*NEUM:* Definition of solution domain surfaces with fixed flux (Neumann) boundary conditions and these data.

*ROBN:* Definition of solution domain surfaces with homogeneous and nonhomogeneous gradient (Robin) boundary conditions and these data.

*FORM:* Form and assemble the finite element algorithm global matrix statement for a steady problem definition.

*SOLV:* Call the code solver, for the "formed" matrix statement, and return the nodal solution.

*GAUS:* Replace *SOLV* sequence with a Gauss-Seidel iterative procedure for larger problems.

*GRAF:* For PCs with a graphics card, output discretizations and solution arrays in graphics form.

*INIT:* Specify the initial condition required to start an unsteady or evolutionary problem statement.

*INTE:* Combine the *FORM* and *SOLV* operations for a time-split initial-value solution sequence and specify the various control data.

*UVEL:* Define the fluid convection velocity field for an unsteady transport problem.

*NORM:* Compute the energy seminorm of the computed solution.

*STOP:* Return to the top of READIT; hence define another problem statement.

*EXIT:* Leave READIT and terminate execution of the code.

Subroutine READIT is organized so that these command names can be input in any order and may be repeated to alter an input file. Similarly, new command names are easily added when the reader decides to extend the capabilities of *LEARN.FE.* The Appendix contains nomenclature and key listings for the code to provide a ready reference as needed.

## 2.3 THE *LEARN.FE* DEMONSTRATOR

The *LEARN.FE* diskette contains both input and solution files for most of the study problems and code exercises presented and discussed in the next chapters. A preview of associated grid generation and color graphics solution presentations can help orient you to the subject material of this text. Therefore, we suggest you load the code onto your PC and respond to the prompt for the "DEMO." The code will then proceed into established solution demonstrations according to your responses to the monitor prompts.

# CHAPTER 3

# WEIGHTED RESIDUALS AND THE WEAK STATEMENT

## 3.1 OVERVIEW

As briefly introduced in Chap. 1, the finite element method is a general and systematic computational procedure for approximately solving problems in physics and engineering. Application of the theory directly produces a matrix statement amenable to computation. This chapter introduces the *weak statement* as the mathematical foundation for development of the finite element method, which ultimately leads to the computable form.

The term *finite element* was coined in the late 1950s to define the method developed by aeronautical engineers for computer-based structural analyses. Even though energy methods have long been a part of strength of materials analysis, the finite element method was first developed as an approximate expression for the balance of forces on a triangulation in two dimensions. The considerable success enjoyed by this newly devised procedure soon prompted detailed theoretical analyses. In a few years, the engineer's method was determined to be strongly founded in the calculus of variational boundary value problems, of which energy methods are a special case. The variational calculus provided a rational mathematical theory explaining *why* the engineer's *finite element method* worked so well, and the applicable problem range was extended tremendously.

The emergence and use of this rich theory has suggested to some that the engineer must learn variational calculus to use the finite element method. That this is not necessary is the basic premise of this book! This chapter introduces, and illustrates with complete detail all formulational steps of the finite element algorithm for an elementary but practical study problem. The governing differential equation balances diffusion effects with a source, coupled with an applied flux and a fixed boundary condition. The problem statement is so simple that we can solve it directly, hence establish the analytical solution for comparison. Thereupon, we develop the finite element approximate solution procedure step by step through its basic ingredients as follows.

**Step 1.** State the solution *approximation* as a series of assumed-known spatial functions multiplied by an unknown expansion coefficient set.

**Step 2.** From step 1, recognize that *two decisions* face us, i.e., what are suitable spatial functions to use, and how do we determine the unknown coefficients?

**Step 3.** Answer the latter question by defining a *weak statement* to determine the expansion coefficients, to ensure that the error in the approximation is an absolute minimum for *any choice* of spatial functions.

**Step 4.** Examine candidate spatial functions and insist that they have long-term versatility, hence select piecewise continuous polynomials forming the basis of *Lagrange interpolation.*

**Step 5.** Define a *discretization* of the solution spatial domain into finite elements; hence rearrange the piecewise polynomials into basis functions that yield easy evaluation of the weak statement integrals.

**Step 6.** Complete these integrals for the specified problem data; hence construct the *matrix statement* for determining the expansion coefficients defined in step 1.

**Step 7.** Apply the appropriate boundary data and *solve* the matrix statement, hence insert the previously unknown expansion coefficients into the stated approximation, step 1.

**Step 8.** Determine the accuracy of the approximation, quantize the error, and hence validate solution quality.

We now proceed to develop and illustrate the details of each formulational step.

## 3.2 AN EXAMPLE PROBLEM

We wish to determine the temperature distribution due to conduction of heat through a thick slab with thermal conductivity $k(x)$ as shown in Fig. 3.1. The slab is thermally loaded by a prescribed heat flux $q$, applied at the surface $x = a$, while the other surface at $x = b$ is held at the constant temperature $T = T_b$. Further, assume that $s = s(x)$ defines the distribution of internal heat generation.

The differential equation, denoted as $L(\cdot)$ (see List of Symbols), governing the steady-state temperature distribution in the slab is

$$L(T) = -\frac{d}{dx}\left(k\frac{dT}{dx}\right) - s = 0 \qquad a < x < b \tag{3.1}$$

The boundary conditions communicating the problem definition to (3.1) include a first-order differential equation $l(\cdot)$ and fixed data as

$$l(T) = -k\frac{dT}{dx} - q = 0 \qquad \text{at } x = a \tag{3.2}$$

$$T = T_b \qquad \text{at } x = b \tag{3.3}$$

The positive direction of heat flow is parallel to $x$; hence, the heat goes into the slab at $x = a$, which accounts for the sign of $q$ in (3.2).

The solution to (3.1) to (3.3) can be found analytically provided $k(x)$ and $s(x)$ are simple functions. For constant $k$ and $s$, one can directly integrate (3.1) twice and determine the constants of integration by imposing the boundary conditions (3.2) and (3.3). This solution is independent of an $x$-axis origin shift; hence, for $L = b - a$ the domain span, and setting $a = 0$

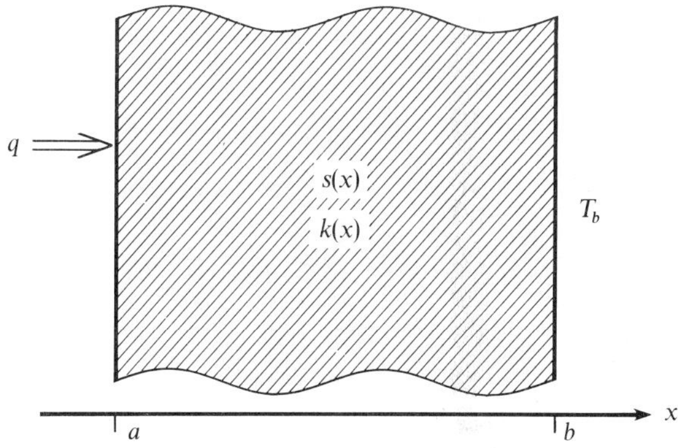


**FIGURE 3.1**
Conduction of heat in a slab of thickness $L$.

in (3.1) and (3.2), a suggested exercise to verify this solution is

$$T(x) = \frac{sL^2}{2k}\left[1 - \left(\frac{x}{L}\right)^2\right] + \frac{qL}{k}\left(1 - \frac{x}{L}\right) + T_b \tag{3.4}$$

This example problem is simple—we were able to determine the analytical solution. We can thus use (3.4) as a benchmark to compare approximate solutions obtained from the finite element procedure. Practical engineering problems almost never lend themselves to development of the analytical solution for comparison. We must then delve into enough theory to examine accuracy and learn how to estimate numerical error.

## 3.3 THE GALERKIN WEAK STATEMENT

The two fundamental theories leading to construction of approximate solutions to (3.1) to (3.3) are the Rayleigh-Ritz variational procedure and the Galerkin weak statement. Other lesser utilized weighted-residual methods include collocation, finite volume, and least squares, each of which constitutes a subset within the weak statement formulation. Regardless of which procedure one chooses, the fundamental step is recognition that one seeks an *approximate* solution for the distribution of temperature $T(x)$. The classical calculus procedure is to define a Taylor (or power) series of known functions of $x$, each multiplied by an unknown constant to be determined according to certain criteria. Hence

$$T^N(x) = a_1\phi_1(x) + a_2\phi_2(x) + \cdots + a_N\phi_N(x) \tag{3.5}$$

is an appropriate expression for an approximation. In (3.5), the $a_i$ for $1 \leq i \leq N$ are the unknown constants, the $\phi_i(x)$ are the known functions of $x$, and the superscript on $T^N(x)$ denotes there are $N$ terms in the approximation.

Note that (3.5) can contain the analytical solution if it is known. For example, for the definitions

$$a_1 = \frac{sL^2}{2k} \qquad \phi_1(x) \equiv 1 - \left(\frac{x}{L}\right)^2$$

$$a_2 = \frac{qL}{k} \qquad \phi_2(x) \equiv 1 - \frac{x}{L}$$

$$a_3 = T_b \qquad \phi_3(x) = 1$$

it is easy to verify that (3.5) is identical with (3.4) for $N = 3$.

Unfortunately, the exact solution is never known for the real-world problems of ultimate interest. However, as stated in (3.5), *any approximation* to a (one-dimensional) solution can certainly be expressed in the form

$$T^N(x) = \sum_{i=1}^{N} a_i \phi_i(x) \tag{3.6}$$

The major challenge is obvious, i.e., find or define a suitably general set of functions $\phi_i(x)$, $1 \leq i \leq N$, such that (3.6) generates a good approximation. As we develop finite element theory, the computationally convenient and mathematically robust form for the *trial function set* $\phi_i(x)$ in (3.6) will emerge.

Approximating the solution to (3.1) to (3.3) with the series expression (3.6) generally will not coincide with the exact solution. Consequently, an *error* is created which is simply the difference between $T(x)$ and $T^N(x)$. Denoting this as $e^N(x)$, then

$$T(x) = T^N(x) + e^N(x) \tag{3.7}$$

clearly states the connection between the approximation and its error. The analyst would certainly seek to make $e^N(x)$ as small as possible, and hence minimize the difference between $T(x)$ and $T^N(x)$. However, $e^N(x)$ is a function (of $x$) and its minimization requires a general approach. One "measure" of the size of $e^N(x)$ is to substitute (3.7) into (3.1), which yields a differential equation $L(e^N)$ for the error

$$L(e^N) = L(T) - L(T^N) = -L(T^N) \tag{3.8}$$

since (3.1) states that $L(T) = 0$. Equation (3.8) shows one connection between the error and the approximation (3.6). Hence

$$L(T^N) = -\frac{d}{dx}\left(k\frac{dT^N}{dx}\right) - s \neq 0 \tag{3.9}$$

as a precise measure of the distribution of the error in $T^N(x)$ over the domain of (3.1).

It makes sense to require that (3.9) be as close to zero as possible. However, it is not reasonable to try to force the error to be zero everywhere, since this amounts to determining the exact solution $T$! Instead, the pragmatic approach is to require the measure of the approximation error (3.9) to vanish in an overall integrated sense. The corresponding mathematical statement is that the weighted residual of (3.9) must vanish:

$$\int_\Omega w_i(x) L(T^N) dx \equiv 0 \qquad \text{for } 1 \leq i \leq N \tag{3.10}$$

In (3.10), $\Omega$ denotes the problem statement domain [the interval $0 < x < L$ of (3.1)], and the weight function set $w_i(x)$ is absolutely arbitrary at this point. The decision on specific $w_i(x)$ determines the type of weighted-residual technique chosen. For the *Galerkin criterion*, each weight function $w_i$ is made identical to the corresponding trial function $\phi_i$ of the approximation (3.6), hence

$$w_i(x) \equiv \phi_i(x) \qquad \text{for } 1 \le i \le N \tag{3.11}$$

Selection of the Galerkin definition (3.11) for (3.10) exactly recovers the classical Rayleigh-Ritz approximate variational formulation for a linear field problem [e.g., (3.1)]. Further, the Galerkin choice is particularly advantageous when solving any (linear or nonlinear) problem, since (3.10) guarantees that the associated approximation *error is minimum* since it is *orthogonal* to the trial function set $\phi_i(x)$, $1 \le i \le N$. The word "orthogonal" means "perpendicular" in the mathematical sense of a distribution. Both $T(x)$ and $T^N(x)$ are curves spanned by the $x$ axis, and orthogonal infers that the distribution of the "distance" between these two curves is the absolute smallest it can be for any $N$.

The *Galerkin weak statement* [(3.10) to (3.11)] is thus elegantly simple. We need only evaluate the integrals in (3.10) using the selected approximation form (3.6) and the identical weight functions (3.11). For (3.1), the resultant calculus operation is

$$\int_0^L \phi_i(x)\left[-\frac{d}{dx}\left(k\frac{dT^N}{dx}\right) - s\right]dx \equiv 0 \qquad \text{for } 1 \le i \le N \tag{3.12}$$

The second derivative term in (3.12) can be rearranged using integration by parts to reduce the order of the derivative by one. Equation (3.12) then becomes the Galerkin *symmetric weak statement* (denoted $WS$) form

$$WS = \int_0^L \frac{d\phi_i}{dx} k \frac{dT^N}{dx}\, dx - \int_0^L \phi_i s\, dx - \phi_i k \frac{dT^N}{dx}\bigg|_0^L = 0 \qquad \text{for } 1 \le i \le N \tag{3.13}$$

The symmetric expression (3.13) is the form of a weighted residual (3.10) that is termed a "weak statement." We suggest you take the time to verify (3.13) as obtained from (3.12). This terminology is selected, not because of any inadequacy (weakness) in (3.13), but because the differentiability that the approximation $T^N$ must support has been weakened. Specifically, in (3.13) only the first derivative of $T^N(x)$ is required to exist, while in (3.12) the second derivative must exist. This makes (3.13) the desired computational form, since the same order of derivative occurs for both the approximation $T^N(x)$ and each member of the weight (trial) function set $\phi_i(x)$.

Substituting (3.6), with its summation index changed to $j$ to avoid confusion, the explicit computational form for (3.13) is

$$WS = \sum_{j=1}^{N} \left( a_j \int_0^L \frac{d\phi_i}{dx} k \frac{d\phi_j}{dx} dx \right) - \int_0^L s\phi_i \, dx - q\phi_i \bigg|_{x=0} = 0 \quad \text{for } 1 \leq i \leq N \tag{3.14}$$

Note that $-k(dT^N/dx) \equiv q$ at $x = 0$, as required by the boundary condition (3.2), is incorporated *exactly* into (3.14), for *any* approximation $T^N$, confirming that the finite element method handles all flux boundary conditions in a natural way.

## 3.4 THE FINITE ELEMENT BASIS

The final step in evaluating the Galerkin weak statement (3.14) requires one to specifically define the trial function set $\phi_i(x)$, $1 \leq i \leq N$. Recalling (3.4) to (3.5), we can be very precise about the $\phi_i(x)$ if knowledge exists about the correct answer. However, practical problems are characterized by a lack of such knowledge. Therefore, we need to select a set of general-purpose functions $\phi_i(x)$ that will admit versatility.

The mathematical properties that any potential member of $\phi_i(x)$ must possess are quite elementary. Viewing (3.14), any candidate must have a nonvanishing first $x$ derivative, and the product of this derivative with all other trial space member first derivatives, i.e., $(d\phi_i/dx)(d\phi_j/dx)$, must be integrable on $(0, L)$. The first requirement simply excludes all isolated constants from the set $\phi_i$. Other than that, a vast reservoir exists from which one may choose a $\phi_i(x)$.

For example, you may recall using Fourier series in calculus to represent a function. The typical term is $\sin(n\pi x/L)$, and then (3.6) would become

$$T^N(x) = \sum_{n=1}^{N} a_n \phi_n(x) = \sum_{n=1}^{N} a_n \sin(n\pi x/L) \tag{3.6a}$$

where by convention the summation index is now $n$ rather than $i$. Every term in (3.6*a*) certainly has a nonzero first derivative, and all products of first derivatives are square integrable on $(0, L)$. In fact, the Fourier series functions $\sin n\pi x/L$ form an orthogonal set on $(0, L)$, which is used in classical analysis to great advantage. However, (3.6*a*) does not possess the desired boundary condition versatility, nor does this form for $T^N$ extend easily to multidimensional problems with nonregular boundary geometries.

An alternative approach for selection of $\phi_i(x)$ draws on interpolation theory. The first thought might be to use a simple power series in $x$; hence (3.6) becomes

$$T^N(x) = \sum_{i=0}^{N} a_i x^i = a_0 + \sum_{i=1}^{N} a_i x^i \tag{3.6b}$$

and we have $\phi_i(x) \equiv x^i$. The constant $a_0$ provides for a nonzero left-end boundary condition, all $\phi_i(x)$ possess a nonzero first derivative and all products for reasonable $N$ are square integrable on $(0, L)$. Therefore, (3.6*b*) possesses the basic mathematical requirements, but it also exhibits a fatal flaw. Specifically, for large $N$, the power series expansion (3.6*b*) becomes highly oscillatory between evaluation points.

In fact, the simple high-order power series (3.6*b*) is not generally useful for interpolation. The preference is to restrict the value of $N$ to a small integer, and to break up the function to be interpolated into small segments with as many local endpoints on the interior of $(0, L)$ as needed to achieve the desired accuracy. This is called *Lagrange interpolation*, and Fig. 3.2 illustrates the concept.

For a representative function $T(x)$, shown as the solid line, that might be a solution to (3.1) to (3.3), a candidate piecewise continuous linear interpolation $T^N(x)$ is the dashed line. It exactly matches $T(x)$ at a finite number of locations along the $x$ axis called "nodes" which are denoted as $X1$, $X2$,

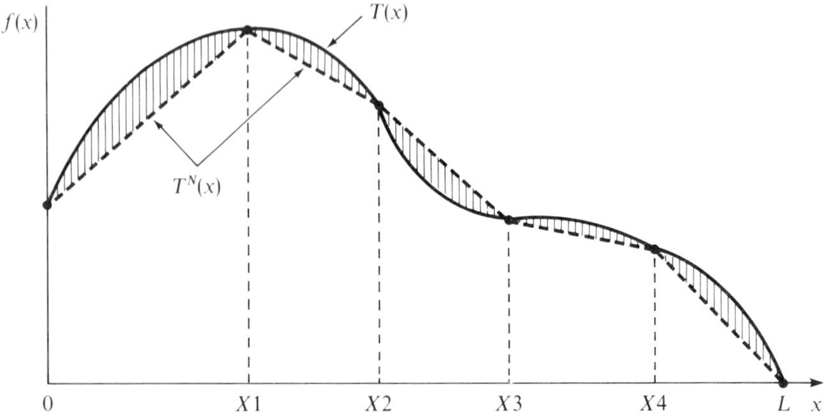


**FIGURE 3.2**
Interpolation $T^N(x)$ of a function $T(x)$.

and so on. The difference between the piecewise interpolant $T^N(x)$ and the function $T(x)$, i.e., the error $e^N(x)$, lies in the shaded areas. The error is indeed a function of $x$, and it can certainly be reduced by adjusting the number or location of the node points $XI$ at which the equality between $T^N(x)$ and $T(x)$ is enforced.

The fundamental distinction between finite element approximation theory and Lagrange interpolation is that $T(x)$ is unknown! Hence, the analyst must construct the approximation in the absence of a priori knowledge of the exact solution. However, the concept of piecewise continuous, relatively low-order polynomials is highly appropriate for establishing the trial function set $\phi_i(x)$. The endpoint is establishment of the *finite element basis*, a set of polynomials defined uniformly on *every* subdivision (finite element) of the solution domain $\Omega$, as created by placing nodes as desired (for accuracy), and hence constructing the domain *discretization* denoted as $\Omega^h$.

The introduction of the discretization $\Omega^h$ of $\Omega$ is a central concept of finite element analysis. It greatly simplifies construction of a wide range of highly suitable trial function sets $\phi_i(x)$ as well as evaluation of the integrals in the weak statement (3.14). The one-dimensional development for piecewise linear polynomials fully illustrates the concept. Figure 3.3*a* shows the domain of (3.1) discretized into two finite element domains $\Omega_e$ for $e = 1$ and $e = 2$. The nodes of this discretization $\Omega^h$ are thus $X1$, $X2$, and $X3$ as noted. The linear polynomial that is piecewise continuous on $\Omega^h$, and which is centered and of unit value at $X2$ is

$$\phi_2 = \begin{cases} \dfrac{x - X1}{X2 - X1} & \text{for } X1 \le x \le X2 \\[2ex] \dfrac{X3 - x}{X3 - X2} & \text{for } X2 \le x \le X3 \\[2ex] 0 & \text{for } x < X1 \quad \text{or} \quad x > X3 \end{cases} \tag{3.15}$$

Thus, (3.15) is the precise definition of the (linear) trial function $\phi_i(x)$ for $i = 2$, i.e., that member centered at node $X2$.

Equation (3.15) is plotted in Fig. 3.3*b* as the middle entry; it is indeed a piecewise continuous linear polynomial on $X1 \le x \le X3$. This function is then appropriate for any other subregion of $\Omega$ by a simple displacement to the left or right of $X2$. The one node displacement each way thus yields the other two trial functions $\phi_1$ and $\phi_3$, which are also shown in Fig. 3.3*b*. Note that the pieces of $\phi_1$ and $\phi_3$ that fall outside $\Omega$ are discarded since $T^N(x)$ exists only on $(0, L)$.

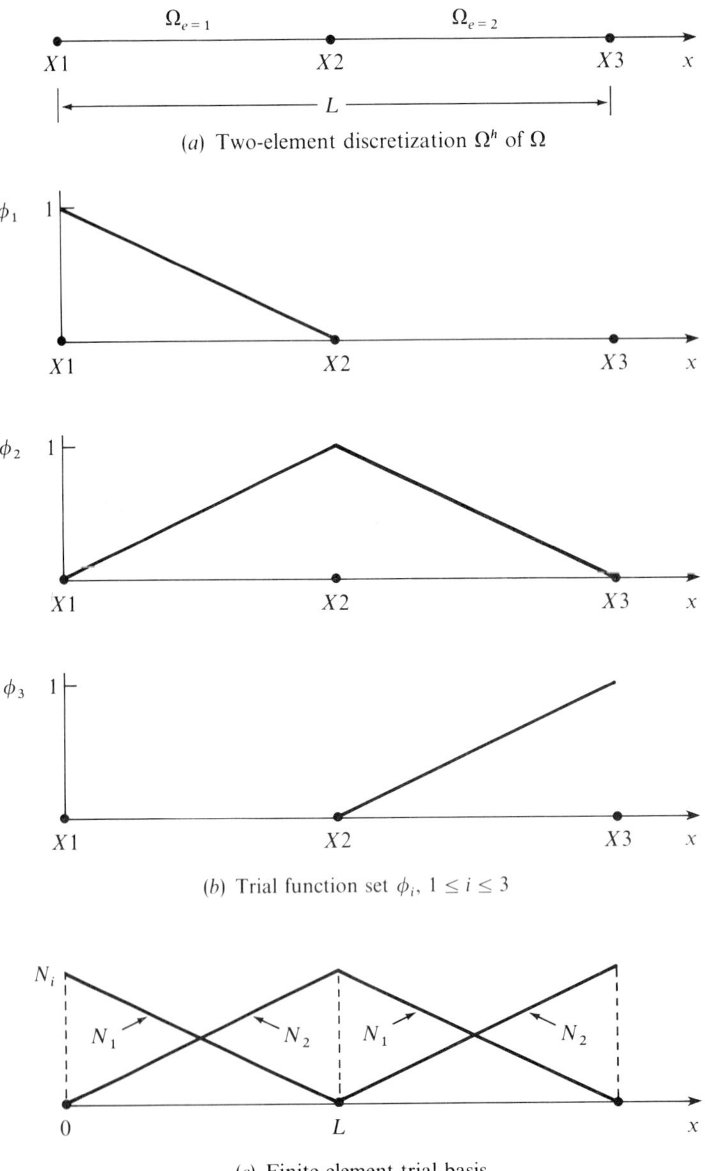


(a) Two-element discretization $\Omega^h$ of $\Omega$

(b) Trial function set $\phi_i$, $1 \leq i \leq 3$

(c) Finite element trial basis

**FIGURE 3.3**
Illustration of finite element discretization, approximation function set $\phi_i(x)$, and finite element base $N_i$.

Hence, for the selected two finite element discretization $\Omega^h$ of $\Omega$, the specific form of $T^N(x)$ for $N = 3$ is

$$T^N(x) \equiv \sum_{i=1}^{N=3} a_i \phi_i(x) = a_1 \phi_1(x) + a_2 \phi_2(x) + a_3 \phi_3(x) \tag{3.16}$$

Further, (3.15) generalized to be centered at *any* node $XJ$ of $\Omega^h$ is

$$\phi_j(x) = \begin{cases} \dfrac{x - XJM1}{XJ - XJM1} & \text{for } XJM1 \leq x \leq XJ \\[2ex] \dfrac{XJP1 - x}{XJP1 - XJ} & \text{for } XJ \leq x \leq XJP1 \\[2ex] 0 & \text{for } x < XJM1 \quad \text{or} \quad x > XJP1 \end{cases} \tag{3.17}$$

In (3.17), $XJM1$ is read "$XJ$ minus 1" and correspondingly $XJP1$ means "$XJ$ plus 1."

The central idea underlying the finite element computational implementation of (3.14) for (3.6), e.g., (3.16) to (3.17) for $N = 3$, is to focus attention on the single *generic* element domain $\Omega_e$, and hence to consider only the pieces of any $\phi_j(x)$ that are nonzero on $\Omega_e$. Figure 3.3*c* presents the overlay of the three $\phi_j$ of the $M = 2$ discretization of $\Omega$. It is quite evident that *every* finite element domain $\Omega_e$ is spanned by each of two straight line segments constituting the generic (global) trial function $\phi_j$. If we denote the endpoint coordinates of each element $\Omega_e$ as $XR$ and $XL$, where $XR$ denotes "$x$-right" while $XL$ denotes "$x$-left," then from (3.17) these two line segments have the form

$$\begin{aligned} N_1 &= \frac{XR - x}{XR - XL} \\ N_2 &= \frac{x - XL}{XR - XL} \end{aligned} \tag{3.18}$$

The two functions $N_i(x)$, $1 < i \leq 2$ given in (3.18) are called the (linear) *finite element basis* of the associated *linear trial function* $\phi_j$, (3.17). The feature of (3.18) is that these expressions hold for *every element* $\Omega_e$ of any discretization. In distinction, the global trial function set $\phi_i$ depends on specific node sets [see (3.17)], as well as how many elements are in the discretization, i.e., the parameter $N$ in (3.6) and (3.16), recall Fig. 3.3*b*.

With (3.18), evaluation of the integrals in the weak statement (3.14) is particularly easy. The first step is to use (3.18) to express the temperature approximation on the typical finite element $\Omega_e$ as,

$$T_e = N_1 Q_1^e + N_2 Q_2^e = \sum_{j=1}^{2} N_j Q_j^e \tag{3.19}$$

where $Q_1^e$ and $Q_2^e$ are the values that the temperature takes at the left $(XL)$ and the right $(XR)$ node of any $\Omega_e$. The global approximation $T^N$ given in (3.5) can then be replaced as the sum of the elemental approximations $T_e$ using (3.19) as

$$\begin{aligned} T^N(x) &= \sum_{i=1}^{N} a_i \phi_i(x) \\ &= \sum_{e=1}^{M} \sum_{j=1}^{2} N_j(x) Q_j^e \end{aligned} \tag{3.20}$$

In (3.20), $M$ denotes the total number of finite elements in the discretization. Then, the finite element replacement for (3.13) as the Galerkin weak statement can be written as the sum over $M$ finite elements of any discretization $\Omega^h$ of $\Omega$ as

$$\begin{aligned} WS^h = \sum_{e=1}^{M} \Bigg[ \sum_{j=1}^{2} \int_{\Omega_e} \frac{dN_i}{dx} k \frac{dN_j}{dx} \, dx \, Q_j^e - \int_{\Omega_e} N_i s \, dx \\ - k \frac{dT}{dx} N_i \delta_{eM} + k \frac{dT}{dx} N_i \delta_{e1} = 0 \Bigg] \quad \text{for } 1 \le i \le 2 \end{aligned} \tag{3.21}$$

We have added a superscript $h$ to $WS$ to denote the introduction of the discretization $\Omega^h$. The first two terms in (3.21) involve integrals only over the *generic* finite element domain (which are very easy to evaluate as will follow). The other two terms are the boundary fluxes at each end of $\Omega$, which become replaced directly by the boundary conditions and do not involve any integral over $\Omega_e$. Finally, the notation $\delta_{ea}$ in (3.21) is the "Kronecker delta" for boundary condition specification. Thus, $\delta_{e1} \equiv 1$ for $\Omega_e = \Omega_1$ and is zero elsewhere, while $\delta_{eM} \equiv 1$ for $\Omega_e = \Omega_M$ only, the last finite element of the mesh.

The finite element basis form of the discretized Galerkin weak statement (3.21) can be directly evaluated for the generic domain $\Omega_e$. Letting $l_e \equiv (XR - XL)_e$ denote the span (length) of the element, the spatial derivatives required for the first term in (3.21) are formed using (3.18) and yield

$$\frac{dN_i}{dx} = \begin{cases} -1/l_e & i = 1 \\ 1/l_e & i = 2 \end{cases} \tag{3.22}$$

Assuming the conductivity $k$ is a constant, the first integral in (3.21) is then evaluated and compactly written in matrix notation as

$$\begin{aligned}\int_{\Omega_e} \frac{dN_i}{dx} k \frac{dN_j}{dx} dx\, Q_j^e &= k \int_0^{l_e} \frac{1}{l_e} \begin{Bmatrix} -1 \\ 1 \end{Bmatrix} \frac{1}{l_e} \{-1, 1\} dx \{Q\}_e \\ &= \frac{k}{l_e^2} \begin{bmatrix} 1 & -1 \\ -1 & 1 \end{bmatrix} \int_0^{l_e} dx \{Q\}_e \\ &= \frac{k}{l_e} \begin{bmatrix} 1 & -1 \\ -1 & 1 \end{bmatrix} \{Q\}_e \end{aligned} \tag{3.23}$$

Comparing the index notation of (3.21) to the matrix notation of (3.23), the index $i$ denotes a matrix row while $j$ corresponds to a matrix column. The net result is the $2 \times 2$ square matrix $[\cdot]$ with entries $\pm 1$.

In the same manner, the (assumed constant) source term in (3.21) is directly evaluated, yielding

$$\int_{\Omega_e} N_i s\, dx = s \int_0^{l_e} \begin{Bmatrix} N_1 \\ N_2 \end{Bmatrix} dx = \frac{s l_e}{2} \begin{Bmatrix} 1 \\ 1 \end{Bmatrix} \tag{3.24}$$

In (3.24), the notation $\{\cdot\}$ indicates a column matrix, and we suggest you verify the result using (3.18).

Thus, using (3.23) to (3.24), the finite element matrix form for the Galerkin weak statement (3.21) is

$$\begin{aligned} WS^h &= \sum_{e=1}^{M} WS_e \\ &= \sum_{e=1}^{M} \left( \frac{k}{l_e} \begin{bmatrix} 1 & -1 \\ -1 & 1 \end{bmatrix} \{Q\}_e - \frac{s l_e}{2} \begin{Bmatrix} 1 \\ 1 \end{Bmatrix} - k \frac{dT}{dx} \begin{Bmatrix} -\delta_{e1} \\ \delta_{eM} \end{Bmatrix} \right) = \{0\} \end{aligned} \tag{3.25}$$

For the model problem (3.1), wherein the data $k$ and $s$ are indeed constants, (3.25) constitutes the linear basis finite element algorithm statement for *any discretization* and for *any* set of *boundary conditions*! For the two-element uniform discretization of $\Omega$ shown in Fig. 3.3*a*, the two terms forming the sum $1 \le e \le M = 2$ in (3.25) are:

$$\begin{aligned} \text{for } e = 1\text{: } WS_1 &= \frac{k}{l_1} \begin{bmatrix} 1 & -1 \\ -1 & 1 \end{bmatrix} \{Q\}_{e=1} - \frac{s l_1}{2} \begin{Bmatrix} 1 \\ 1 \end{Bmatrix} - k \frac{dT}{dx} \begin{Bmatrix} -\delta_{11} \\ 0 \end{Bmatrix} \\ &= \frac{k}{L/2} \begin{bmatrix} 1 & -1 \\ -1 & 1 \end{bmatrix} \{Q\}_1 - \frac{sL/2}{2} \begin{Bmatrix} 1 \\ 1 \end{Bmatrix} - q \begin{Bmatrix} 1 \\ 0 \end{Bmatrix} \end{aligned} \tag{3.26a}$$

$$\text{for } e = 2\text{: } WS_2 = \frac{k}{l_2}\begin{bmatrix} 1 & -1 \\ -1 & 1 \end{bmatrix}\{Q\}_2 - \frac{sl_1}{2}\begin{Bmatrix} 1 \\ 1 \end{Bmatrix} - k\frac{dT}{dx}\begin{Bmatrix} 0 \\ \delta_{22} \end{Bmatrix}$$

$$= \frac{2k}{L}\begin{bmatrix} 1 & -1 \\ -1 & 1 \end{bmatrix}\{Q\}_2 - \frac{sL}{4}\begin{Bmatrix} 1 \\ 1 \end{Bmatrix} - k\frac{dT}{dx}\begin{Bmatrix} 0 \\ 1 \end{Bmatrix} \tag{3.26b}$$

The element summation defined in (3.25), which is called *assembly*, is performed by matrix rows. There are three geometric nodes $XJ$ defined for the $M = 2$ discretization (Fig. 3.3*a*). Denoting the corresponding unknown nodal temperatures as $Q1$, $Q2$, and $Q3$, the global matrix statement represented by (3.25) must be of order three, and hence is of the form

$$[K]\begin{Bmatrix} Q1 \\ Q2 \\ Q3 \end{Bmatrix} = \{b\} \tag{3.27}$$

In (3.27), the column matrix $\{b\}$, often called the *global load matrix*, contains all the known data, while $[K]$ is the *global diffusion matrix* of the finite element algorithm. Since $\{Q\}_{e-1}$ in (3.23) and (3.26) contains the entries $\{Q1, Q2\}^T$, while $\{Q\}_{e=2}$ contains $\{Q2, Q3\}^T$, we can expand the order of each element thermal conduction matrix $[K]_e$ in (3.26) to three and form the global matrix $[K]$ defined in (3.27) as

$$[K] = \sum_{e=1}^{M} [K]_e$$

$$= \frac{2k}{L}\begin{bmatrix} 1 & -1 & 0 \\ -1 & 1 & 0 \\ 0 & 0 & 0 \end{bmatrix} + \frac{2k}{L}\begin{bmatrix} 0 & 0 & 0 \\ 0 & 1 & -1 \\ 0 & -1 & 1 \end{bmatrix}$$

$$= \frac{2k}{L}\begin{bmatrix} 1 & -1 & 0 \\ -1 & 2 & -1 \\ 0 & -1 & 1 \end{bmatrix} \tag{3.28}$$

Similarly, the global load matrix $\{b\}$ defined in (3.27) is formed via element *assembly* using (3.26) as

$$\{b\} = \sum_{e=1}^{2} \{b\}_e$$

$$= \frac{sL}{4}\begin{Bmatrix} 1 \\ 1 \\ 0 \end{Bmatrix} + \begin{Bmatrix} q \\ 0 \\ 0 \end{Bmatrix} + \frac{sL}{4}\begin{Bmatrix} 0 \\ 1 \\ 1 \end{Bmatrix} + \begin{Bmatrix} 0 \\ 0 \\ k\dfrac{dT}{dx} \end{Bmatrix} \tag{3.29}$$

We strongly suggest you verify the operations leading to (3.28) and (3.29), to assimilate the concept of *assembly*.

We are now in the position to determine the nodal temperature distribution $\{Q\}$ of the finite element linear basis solution. Dividing (3.28) to (3.29) through by $2k/L$ and inserting into (3.27), the global matrix statement (3.27) for the $M = 2$ uniform discretization of $0 \le x \le L$ for the model problem (3.1) to (3.2) is

$$\begin{bmatrix} 1 & -1 & 0 \\ -1 & 2 & -1 \\ 0 & -1 & 1 \end{bmatrix} \begin{Bmatrix} Q1 \\ Q2 \\ Q3 \end{Bmatrix} = \frac{sL^2}{8k} \begin{Bmatrix} 1 \\ 2 \\ 1 \end{Bmatrix} + \frac{L}{2k} \begin{Bmatrix} q \\ 0 \\ k\dfrac{dT}{dx} \end{Bmatrix} \tag{3.30}$$

Matrix solving (3.30) for $\{Q\}$ is direct, except a modification is needed since the boundary condition (3.3) requires that $Q3 \equiv T_b$, the fixed boundary temperature. Hence, as the last step before algebraic solution, the global matrix statement (3.30) must be modified in the appropriate row to account for any Dirichlet boundary conditions. Thus, deleting the third row in (3.30) and replacing it by $Q3 \equiv T_b$ yields the global matrix statement in final form

$$\begin{bmatrix} 1 & -1 & 0 \\ -1 & 2 & -1 \\ 0 & 0 & 1 \end{bmatrix} \begin{Bmatrix} Q1 \\ Q2 \\ Q3 \end{Bmatrix} = \begin{Bmatrix} \dfrac{L}{2k}\left(\dfrac{sL}{4} + q\right) \\ sL^2/4k \\ T_b \end{Bmatrix} \tag{3.31}$$

This operation for Dirichlet data, i.e., replacing the appropriate matrix row with the known constraint, is highly useful in multidimensional problems. For the one-dimensional example problem, the last row and column in (3.31) is easily eliminated to produce the rank 2 system,

$$\begin{bmatrix} 1 & -1 \\ -1 & 2 \end{bmatrix} \begin{Bmatrix} Q1 \\ Q2 \end{Bmatrix} = \begin{Bmatrix} \dfrac{L}{2k}\left(\dfrac{sL}{4} + q\right) \\ \dfrac{sL^2}{4k} + T_b \end{Bmatrix} \tag{3.32}$$

which involves only the unknown nodal temperatures. We suggest that you verify that the solution of (3.32) is

$$\begin{Bmatrix} Q1 \\ Q2 \end{Bmatrix} = \begin{bmatrix} 1 & -1 \\ -1 & 2 \end{bmatrix}^{-1} \begin{Bmatrix} \dfrac{L}{2k}\left(\dfrac{sL}{4} + q\right) \\ \dfrac{sL^2}{4k} + T_b \end{Bmatrix}$$

$$= \frac{1}{1}\begin{bmatrix} 2 & 1 \\ 1 & 1 \end{bmatrix} \begin{Bmatrix} \dfrac{L}{2k}\left(\dfrac{sL}{4} + q\right) \\ \dfrac{sL^2}{4k} + T_b \end{Bmatrix}$$

$$= \begin{Bmatrix} \dfrac{sL^2}{2k} + \dfrac{qL}{k} + T_b \\ \dfrac{3sL^2}{8k} + \dfrac{qL}{2k} + T_b \end{Bmatrix} \tag{3.33}$$

Thus, the finite element uniform $M = 2$ discrete approximate solution for temperature at the nodes of $\Omega^h$ is

$$Q1 = \frac{sL^2}{2k} + \frac{qL}{k} + T_b$$

$$Q2 = \frac{3sL^2}{8k} + \frac{qL}{2k} + T_b \tag{3.34}$$

$$Q3 = T_b$$

Equation (3.34) can now be evaluated for any specific data $s$, $L$, $k$, $q$, and $T_b$, and hence produce the resultant nodal temperatures.

## 3.5 SOLUTION ACCURACY AND APPROXIMATION ERROR

Equation (3.34) gives the finite element nodal solution for the model problem (3.1) to (3.3), as obtained using linear basis functions on a uniform $M = 2$ discretization of the domain $0 \le x \le L$. Conversely, (3.4) gives the exact solution $T(x)$ as a continuous function of $x$. On evaluation of (3.4) at $x = 0$ and $x = L/2$, the direct accuracy comparison reveals that the nodal approximate solution (3.30) is in *exact* agreement for all $k$, $s$, $q$, $L$, and $T_b$! Obtaining such exact nodal agreement occurs only when a problem is very simple, as is this example. Conversely, we note that the use of piecewise linear trial functions has accurately interpolated a quadratic function, i.e., (3.4), portending the good performance expected in the general case.

We must emphasize, however, that *any* numerical solution is *an approximation*, and hence will not agree everywhere with the exact solution. This is readily verified by using the nodal solution (3.34) to construct the continuous form of the approximate solution $T^N(x)$. Equation (3.20) states the form in terms of the basis functions $N_i(x)$ and the element nodal variables $Q_i$. Figure 3.4 plots the continuous form $T^N(x)$ for both positive and negative sources $s$, which are simply straight line segments connecting the nodal solution (3.34). Plotted for comparison are the respective exact solutions $T(x)$. You immediately observe that $T^N(x)$ is the piecewise linear Lagrange interpolant of $T(x)$ with knots at the nodes of $\Omega^h$.

One can "sense" from Fig. 3.4 that $T^N(x)$ is as close as possible to $T(x)$ for the constraint of a uniform $M = 2$ linear basis approximation. Hence, the associated approximation error $e^N(x)$ must in some sense be a minimum. Furthermore, that $e^N(x)$ is indeed a function (of $x$) is fully evident; i.e., it ranges between the solid and dashed lines. While the approximate nodal temperatures are exact, true interest in heat transfer analysis lies in heat fluxes, that is, the level of $-k(dT/dx)$, especially across boundaries. Viewing Fig. 3.4, the heat-flux distribution for $T^N(x)$ is a distinct constant between each node pair. Thus, it agrees with the analytical solution only at the locations where $dT/dx$ is parallel to the appropriate segment chord $dT^N/dx$. Specifically, for the analytical solution (3.4)

$$\begin{aligned} -k\frac{dT}{dx} &= -k\frac{d}{dx}\left\{\frac{sL^2}{2k}\left[1-\left(\frac{x}{L}\right)^2\right]+\frac{qL}{k}\left(1-\frac{x}{L}\right)+T_b\right\} \\ &= sx+q \end{aligned} \tag{3.35}$$

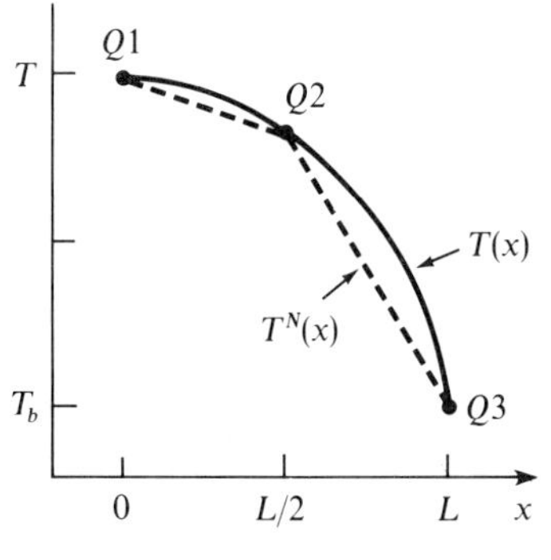


(*a*) Positive source term $s$

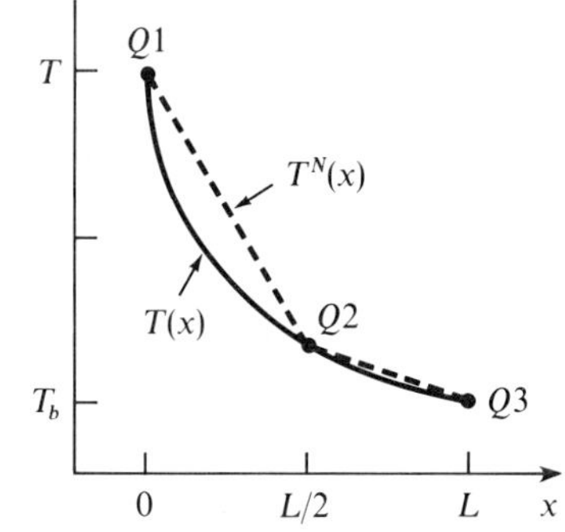


(*b*) Negative source term $s$

**FIGURE 3.4**
Comparison of exact and finite element solutions.

Conversely, in any finite element $\Omega_e$, from (3.19) and (3.18)

$$
\begin{aligned}
-k\frac{dT_e}{dx} &= -k\frac{d}{dx}(N_1Q_1^e + N_2Q_2^e) \\
&= -k\frac{d}{dx}\left(\frac{XR-x}{l_e}\right)Q_1^e - k\frac{d}{dx}\left(\frac{x-XL}{l_e}\right)Q_2^e \\
&= -\frac{k}{l_e}(Q_2^e - Q_1^e) \qquad (3.36)
\end{aligned}
$$

Recalling that $l_e = L/2$, and using (3.34) in (3.36) leads to

$$
\begin{aligned}
-k\frac{dT_{e=1}}{dx} &= -\frac{k}{L/2}\left[\frac{3sL^2}{8k} + \frac{qL}{2k} + T_b - \left(\frac{sL^2}{2k} + \frac{qL}{k} + T_b\right)\right] \\
&= -\frac{2k}{L}\left(-\frac{sL^2}{8k} - \frac{qL}{2k}\right) \\
&= \frac{sL}{4} + q \qquad \text{for } \Omega_{e=1} \qquad (3.37a)
\end{aligned}
$$

$$
\begin{aligned}
-k\frac{dT_{e=2}}{dx} &= -\frac{k}{L/2}\left[T_b - \left(\frac{3sL^2}{8k} + \frac{qL}{2k} + T_b\right)\right] \\
&= \frac{3sL}{4} + q \qquad \text{for } \Omega_{e=2} \qquad (3.37b)
\end{aligned}
$$

Comparing (3.35) and (3.37) verifies that the element heat flux for the approximation $T^N$ is in error everywhere except at the element centroid (since the discretization is uniform). Hence, it is clear that accuracy of the approximate solution $T^N(x)$ depends on how and where we measure it. It might be exact at specific select locations, but this occurrence is highly problem dependent.

Another issue is addressed in closing the introductory aspects of this subject. Returning to (3.30), recall that the third equation was eliminated on imposition of $Q3 \equiv T_b$, e.g., (3.31). In truth, the unknown eliminated in this equation was the heat flux $-k(dT/dx)$ located on the right-hand side of (3.30). This term, generated during integration by parts, was used to exactly impose the flux boundary condition on the left end of $\Omega$. The right-end flux should therefore be determined from the third equation in

(3.30). Hence

$$
\begin{aligned}
-k\frac{dT^N}{dx}\bigg|_{x=L} &= \frac{-2k}{L}\left(Q3 - Q2 - \frac{sL^2}{8k}\right) \\
&= \frac{-2k}{L}\left[T_b - \left(T_b + \frac{qL}{2k} + \frac{3sL^2}{8k}\right) - \frac{sL^2}{8k}\right] \\
&= q + sL
\end{aligned}
\tag{3.38}
$$

using (3.34). Comparing (3.38) to the analytical solution (3.35) evaluated at $x = L$ confirms that the approximate solution flux is exact (!) when computed from the global matrix statement.

We are immediately led to the left-end comparison. Replacing $q$ by $-k(dT^N/dx)$ on the right side of the first equation in (3.31) and using (3.34) leads to

$$
\begin{aligned}
-k\frac{dT^N}{dx}\bigg|_{x=0} &= \frac{2k}{L}\left(Q1 - Q2 - \frac{sL^2}{8k}\right) \\
&= \frac{2k}{L}\left[\frac{sL^2}{2k} + \frac{qL}{k} + T_b - \left(\frac{3sL^2}{8k} + \frac{qL}{2k} + T_b\right) - \frac{sL^2}{8k}\right] \\
&= q
\end{aligned}
\tag{3.39}
$$

Thus, the global matrix statement indeed imposes the applied flux exactly!

These exercises have shown that the finite element solution $T^N(x)$ yields a good approximation; however, error is present. How error is measured can have direct bearing on the validity of the solution in meeting specific requirements. The issue of accuracy, and the methods available to control approximation error, is of paramount importance once the computer code "works." We shall develop these topics in thorough detail in succeeding chapters as the finite element method becomes generalized.

## 3.6 SUMMARY

The mathematical principle underlying the finite element method is an approximate evaluation of the symmetric Galerkin weak statement constructed from the governing differential equation with boundary conditions. The finite element formulation uses basis functions to recast the global weak statement into a sum of integral expressions formed on a finite element discretization of the problem domain. These integrals are easy to evaluate for any (the generic) finite element domain $\Omega_e$. The global matrix statement is constructed by a row-wise matrix summation procedure called assembly.

The example one-dimensional heat-conduction problem has illustrated each basic building block of the finite element method one step at a time. If you are still uncomfortable with the concepts, we recommend that you reread this chapter, making sure to complete the suggested verifications. You should also work the following study problems to successful completion. Bear in mind the ancient Chinese proverb: "He who hears, forgets; he who sees, remembers; he who does, learns."

## 3.7 PROBLEMS

1. For the model problem (3.1) to (3.3), establish the finite element $WS^h$ approximate solution equivalent to (3.34) for an $M = 3$ (four-node) uniform discretization of $0 \leq x \leq L = 1$ (ft). Compare the effort required to solve this statement to that for (3.30) to (3.34). Then determine the actual nodal temperatures for the following data: $q(x = 0) = 1500$ Btu/h·ft$^2$, $s = -1400$ Btu/ft$^3$·h, $k = 4$ Btu/h·°F·ft, and $T_b(x = L) = 0$°F. Verify the improvement in $M = 3$ solution accuracy compared to the $M = 2$ solution, as shown in Fig. 3.5.

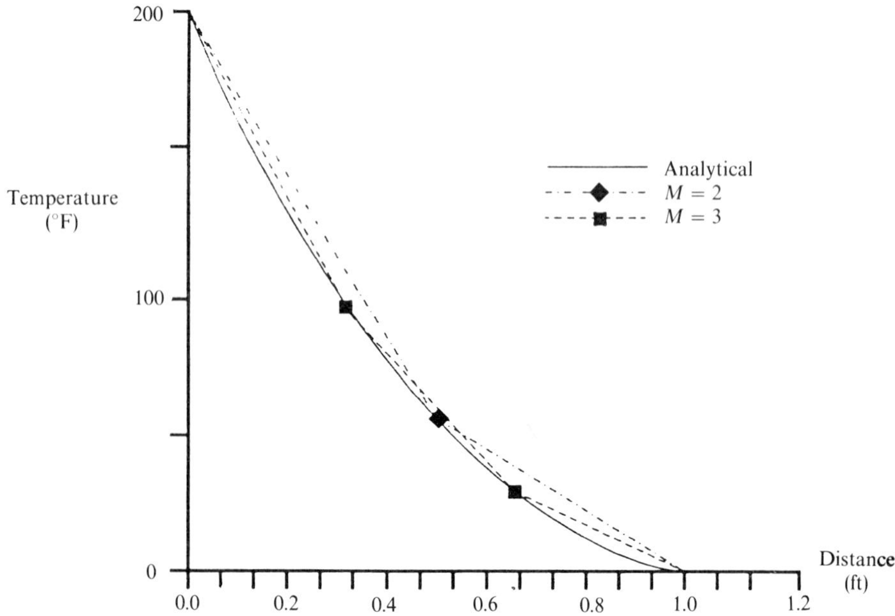


**FIGURE 3.5**
Temperature distributions for model problem discrete solutions.

**2.** For the model problem (3.1), determine the $WS^h$ approximate temperature distribution within the slab $0 \leq x \leq 1$ (ft) on a uniform $M = 3$ discretization for temperatures fixed at both endpoints; i.e., $T_b(x = 0) = T_L$ and $T_b(x = L) = T_R$. Then determine the actual nodal temperature solution for $s = -1400$ Btu/ft$^3$·h, $k = 4$ Btu/hr·°F·ft, $T_L = 200$°F and $T_R = 0$°F. Compare these results for temperature at $x = 0.33$ ft, and heat flux at $x = 0$, to the solution data obtained from Problem 1.

# CHAPTER 4

# ONE-DIMENSIONAL FINITE ELEMENTS

## 4.1 OVERVIEW

In Chap. 3, the finite element discrete approximation procedure was introduced for the symmetric Galerkin weak statement using an elementary example problem. The linear finite element basis permitted easy evaluation of the integrals in the discretized weak statement $WS^h$. Comparison with the true solution confirmed existence of discrete approximation error. The discretization $\Omega^h$ of the problem domain $\Omega$, into a union (i.e., the non-overlapping sum and denoted by the symbol "$\bigcup$") of finite element domains $\Omega_e$, was the key to setting up the computational statement. Finally, from the Chap. 3 problems, the *discrete approximation error* $e^h$ diminishes as the discretization $\Omega^h$ is refined. Throughout the superscript $h$ notation is used to firmly note the use of a discretization.

At this juncture the reader should have a basic understanding of the key formulational steps of the finite element algorithm. Our goal in this chapter is to reinforce and expand on this foundation with many additional examples detailed to illustrate the generalization of the trial basis and formalization of the underlying concepts. The problem statement domain is restricted to one dimension to prevent geometric complications from obscuring the formal elegance and versatility of finite element methodology. The principal study problem framework remains steady-state heat conduction, now with addition of the all-important thermal convection boundary condition.

## A. METHODOLOGY

### 4.2 FINITE ELEMENT MESH

Very simply stated, the acceptability of the approximate solution $\{Q\}$ depends critically on the appropriateness of the finite element discretization $\Omega^h$, called the "mesh," for the problem. A fundamental axiom of computational mechanics (and life!) is "there is no free lunch," which translates into the fact that the computational price of an adequate discretization is inescapable. An inappropriate mesh can permit approximation error to corrupt a solution. In actual fact, the mesh should adapt to the solution, i.e., move (or add) regions of refinement to location(s) where the solution exhibits steep gradients. The mathematical aspects of adaptive mesh procedures have received much attention in recent years (Babuska, et al., 1987; Demkowicz, et al., 1984, 1985). Thus, we first focus attention on a simple but highly effective method to automatically generate smoothly varying nonuniform meshes, for one-dimensional problem domains, that can be readily adjusted by the analyst to adapt to a solution.

Figure 4.1*a* illustrates a geometry of practical significance in heat transfer, specifically, a finned-tube heat exchanger. The fins conduct heat radially outward, to extend the area for (boundary) convective heat transfer to the cooling external airflow. If the finned-tube exchanger is of cast construction, the cross-sectional thickness of the fins varies radially (Fig. 4.1b). Conversely, a soldered construction would have uniform thickness fins (Fig. 4.1*c*). In either case, the fin length/thickness ratio is so large that conduction effects across the fin thickness must be negligible in comparison to radial conduction. Thus, a one-dimensional conduction problem can be assumed which generates the companion need for a (nonuniform) one-dimensional discretization (Fig. 4.1*d*).

The basic meshing requirement is to automatically, i.e., using the computer, generate smoothly varying discretizations on a line (or curve). Assuming the $x$ coordinate spans the range, Fig. 4.2 illustrates a mesh formed by $M$ elements $\Omega_e$ of individual length (measure) $l_e$. The desire

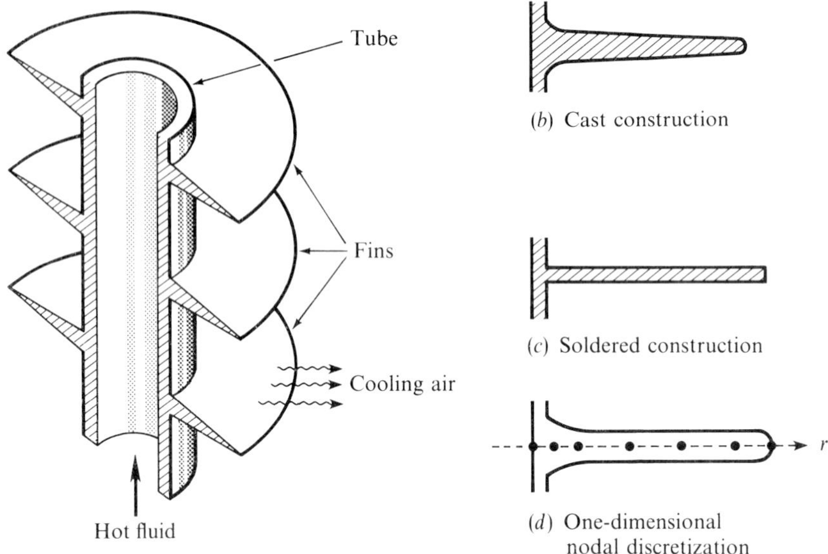


**FIGURE 4.1**
Illustration of a practical one-dimensional heat transfer problem.

is that the length progression of $l_e$, $1 \leq e \leq M$, be smoothly varying. A geometric progression is highly suitable; hence, let

$$l_{e+1} \equiv p l_e \qquad 1 \leq e \leq M \tag{4.1}$$

where $p$ is the progression rate. Note that $p = 1$ produces a uniform mesh, while for any $p > 1$, the element length increases with $e$ as illustrated in Fig. 4.2. Alternately, the mesh becomes progressively refined with increasing $e$ for any $p < 1$.

Note that for $e = 1$, $l_{e=2}$ is determined in (4.1), assuming the measure of the first element $l_{e=1}$ has been defined. The other parameter in (4.1) is $M$, the total number of elements in the discretization $\Omega^h$. Denote the total span (length) of the one-dimensional solution domain as $L$. The deterministic equation for the nodal coordinate set $\{X\}$, computed as the sequence

**FIGURE 4.2**
A nonuniform one-dimensional mesh.

of $XR_e$ on $\Omega_e$, recall (3.18), of the finite element mesh $\Omega^h$ satisfying (4.1) is (Baker, 1983, Chap. 6.6)

$$XR_{e+1} = XR_e + \frac{Lp^{e-1}}{\sum_{j=1}^{M} p^{j-1}} \qquad 1 \leq e \leq M \tag{4.2}$$

As a minor qualification, $XR_{e+1=M}$ as computed from (4.2) may differ slightly from $x_R$, the defined right $x$ coordinate of $L$, since this statement of the mesh is mathematically overspecified. The grid generator automatically corrects this procedural flaw.

Grid generation for any problem definition in this book is automated within the *LEARN.FE* program. The following computer code exercises will familiarize the reader with this aspect of code use.

**Code Exercise 4.1.** The command name *TITL* allows you to specify an appropriate title for a problem. The command name *TYPE* contains all integer data (plus a reference length scale REFL) required for a problem definition including the problem dimension and the number of grid progression ratios $p$. Command *PRIN* controls output data presentation, and *GRID* accesses the grid generator for the inputs $x_L$ and $x_R$ of $L$ and $p$. Bring up Code Exercise 4.1 to your screen, which will then display the following file.

```
TITL
***** CODE EXERCISE 4.1,1D GRID TEST  *****
TYPE   [K        N      NNODEL      REFL    NPR    NTRAN   NAXI]
        1        1        2          1.0     1
PRINT [NBUG (*)]
        1          0          0          0
GRID   [FOR N (XL     XR(*)   PR(*)), NEM (PR*)]
                 1.0       2.0         1.0
        5
STOP
```

The command name procedure recognizes only the first four left-justified character entries; hence, descriptive "Hollerith" information can be appended as shown above. Specifically, under *TYPE*, N = 1 indicates a one-dimensional problem domain, the reference length scale (REFL) is unity, and NPR = 1 indicates a single grid progression ratio ($p$) is specified. The various NBUG(*) flags under *PRIN* control output and will be discussed later as needed. The data under *GRID* should be quite readable, i.e., XL is "$x$-left," XR(∗) means "$x$-right" for PR(∗), and NEM(PR*) is the number of elements $M$. The specific data entries are 1.0, 2.0, 1.0, and 5, respectively.

All data are entered in free format, hence everything does not necessarily line up. Finally, *STOP* terminates the execution sequence with this file.

Press RETURN and the following execution output file will become created and appear:

```
TITL
***** CODE EXERCISE 4.1,1D GRID TEST  *****
TYPE    K      N     NNODEL    REFL   NPR   NTRAN  NAXI
        1      1       2        1.0    1      0      0
PRIN    NBUG(1)       (2)      (3)    (4)    (5)    (6)
               1       0        0      0      0      0
GRID    NI     NPRI   NTRAN
        1      1
        XL     XR1    PR1      XR2       PR2      XR3      PR3
        1.0    2.0    1.0
        NEM1   NEM2   NEM3
        5      0
        NODE COORDINATES(X,REAL VARIABLE)
               1.0    1.2      1.4       1.6      1.8      2.0
STOP
```

Output files also contain the command descriptive comments. The data under command name *GRID* now contain the generated nodal coordinate set $\{X\}$ for an $M = 5$ uniform discretization of $1.0 \leq x \leq 2.0$ ft. Without surprise it is $\{X\}^T = \{1.0, 1.2, 1.4, 1.6, 1.8, 2.0\}$, as shown in the last line above.

Now replace the NEM input file entry (i.e., the 5) under *GRID* with a 6 and execute. The output file you view now contains the nodal data set for an $M = 6$ uniform mesh. The last lines contain the arrays

```
GRID    .
        .
        .
       NODE COORDINATES(X,REAL VARIABLE)
       1.00   1.166   1.333  1.50   1.667   1.833   2.0
```

which is the associated $\{X\}$. Now change the PR1 entry (1.0) in the second line under *GRID* to 1.25, execute, and the *GRID* output file will contain

```
GRID    .
        .
        .
       NODE COORDINATES(X,REAL VARIABLE)
       1.00  1.0888  1.1998  1.3386  1.5121  1.7289  2.0
```

which is the $M = 6$ nonuniform discretization resulting for $p = 1.25$ in (4.1).

We suggest you experiment further with the grid generator in

*LEARN.FE.* Specifically, the generation scheme (4.2) works on any isolated segment of the $x$ axis, so it is possible to place *macro* grid blocks adjacent to each other, hence generate a computational mesh with a range of nonuniformity. For example, the $M = 6$ uniform grid test above created the mesh nodal array $\{X\}^T = \{1.0, 1.166, 1.333, 1.5, 1.667, 1.833, 2.0\}$. This mesh can also be created using two *macro* domains $\Omega^H$ of range $1 \le x \le 1.5$ and $1.5 \le x \le 2.0$. (Here, the superscript $H$ denotes a macro block to be further refined into a computational discretization.) The following modifications to the *TITL*, *TYPE*, and *GRID* command files will define the $M = 6$ uniform mesh $\Omega^h$ creation from two $\Omega^H$.

```
TITL
***** 2 MACRO GRID GENERATOR TEST*****
TYPE     K        N      NNODEL     REFL     NPR    NTRAN     NAXI
         1        1        2         1.0      2       0         0
GRID   [FOR N (XL XR(*) PR(*)),NEM(PR*)]
         1.0      1.5          1.0        2.0     1.0
         3        3
```

Make these changes, execute, and the output file will contain

```
TITL
***** 2 MACRO GRID GENERATOR TEST*****
TYPE     K        N      NNODEL   REFL      NPR     NTRAN     NAXI
         1        1        2       1.0       2        0         0
PRIN   NBUG(1)            (2)     (3)       (4)      (5)       (6)
                  1        0       0         0        0         0
GRID   NI         NPRI       NTRAN
       1          2
       XL         XR1        PR1       XR2         PR2     XR3       PR3
       1.0        1.5        1.0       2.0         1.0
       NEM1       NEM2       NEM3
       3          3
       NODE COORDINATES(X,REAL VARIABLE)
       1.00    1.1667   1.3333   1.50     1.6667    1.8333     2.0
STOP
```

Thus, we have creation of the $M = 6$ uniform mesh $\Omega^h$ using two macro domains $\Omega^H$, each of which was uniformly subdivided into three finite elements $\Omega_e$. This results from setting NPR = 2 under *TYPE* and then entering NEM = (3, 3) under *GRID*, which signifies Number of Elements per Macro for NPR macros.

Now generate a mesh with refined elements near both ends of $\Omega$, i.e., close to $x \simeq 1$ and $x \simeq 2$, by changing the progression ratio input data (1.0, 1.0) in the previous file to read

```
GRID

         1.0       1.5       1.25       2.0       0.8
         3         3
```

Execute this modification and the output file will contain

```
TITLE
***** 2 MACRO GRID GENERATOR TEST *****
TYPE     K        N    NNODEL   REFL     NPR    NTRAN   NAXI
         1        1      2        1.0     2       0       0
PRINT    NBUG(1)     (2)      (3)      (4)      (5)    (6)
              1       0        0        0        0      0
GRID     NI       NPRI     NTRAN
         1        2
         XL       XR1      PR1      XR2      PR2      XR3      PR3
         1.0      1.5      1.25     2.0      0.8
         NEM1     NEM2     NEM3
         3        3
         NODE COORDINATES(X,REAL VARIABLE)
            1.0    1.1311 1.2951   1.5 1.7049  1.8689   2.0
STOP
```

Note now that the smallest element domains $\Omega_e$ occur at each end of $\Omega$, and that they are of exactly corresponding measure by pairs, since $p_1 = 1.25$ is the reciprocal of $p_2 = 0.8$ in (4.1).

The *LEARN.FE* grid generator can handle three $\Omega^H$ definitions. Conduct some more mesh generation experiments on your PC to gain familiarity with creating uniform and nonuniform one-dimensional meshes $\Omega^h$.

## 4.3 ONE-DIMENSIONAL FINITE ELEMENT BASES

The choice of the specific trial function set $\phi_i(x)$, $1 \le i \le N$, for

$$T^N(x) = \sum_{i=1}^{N} a_i \phi_i(x) \tag{3.6}$$

is *fundamental* to creation of a discrete approximation. The finite element method provides the analyst with an unequivocal translation of trial function complexity (and completeness) into an equivalent set of universal *basis functions* $\{N\}$ appropriate to *any shape* finite element domain $\Omega_e$ for *any* problem. Herein lies the power and universality of the method. In this section the linear basis function of Chap. 3, recall (3.18) and (3.19), and the expression of the associated approximate solution (3.20), are restated as members of the generalization. The concept of basis will then be completed by including the more complex, i.e., higher-polynomial-degree, quadratic, and cubic finite element basis functions in one dimension.

## Linear Basis

Figure 4.2 illustrates a nonuniform discretization for a one-dimensional problem definition. As developed in Chap. 3, we need only focus on a single generic domain $\Omega_e$ for all mathematical development. Figure 4.3*a* defines the notation for the linear basis element; as in Chap. 3, *XL* and *XR* denote the $x$ coordinates of the left and right nodes of $\Omega_e$ of length $l_e$. As a notational refinement, let *QL* and *QR* denote the corresponding nodal values taken by the approximate solution. The linear basis was established, via (3.18), using the global coordinate $x$. In seeking the generalization, it is preferable to deal with a local coordinate $\bar{x}$, with origin at *XL*, and to normalize by $l_e$. Hence, on $\Omega_e$, the replacement of (3.18) to (3.19) as the linear finite element approximation for temperature becomes

$$T_e(\bar{x}) = \left(1 - \frac{\bar{x}}{l_e}\right)QL + \left(\frac{\bar{x}}{l_e}\right)QR \tag{4.3}$$

Thus when $\bar{x} = 0$, $T_e = QL$, and when $\bar{x} = l_e$, $T_e = QR$.

The two monomials in (4.3) play a central role in development of finite element basis families and are assigned the variable names

$$\zeta_1 \equiv \left(1 - \frac{\bar{x}}{l_e}\right)$$
$$\zeta_2 \equiv \left(\frac{\bar{x}}{l_e}\right) \tag{4.4}$$

where $\zeta$ is Greek zeta. Figure 4.3*b* graphs (4.4); note that the $\zeta_i$ are

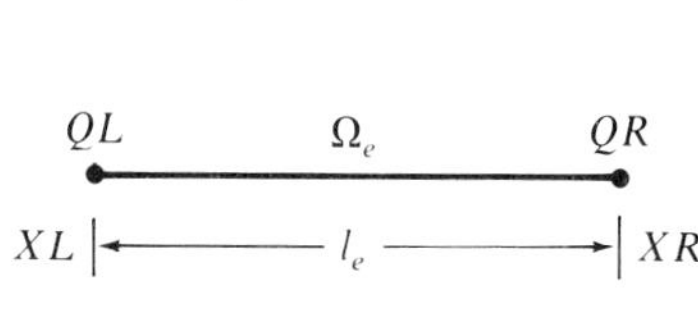


(*a*) Generic element

$1$ $\zeta_1$ $\zeta_2$ $1$

$XL$ $\Omega_e$ $XR$

(*b*) Linear basis functions $\zeta_i$

**FIGURE 4.3**
Linear basis finite element details.

normalized and hence invariant with $\Omega_e$. Combining (4.3) and (4.4) yields,

$$\begin{aligned} T_e(\bar{x}) &= \zeta_1 QL + \zeta_2 QR \\ &= \{\zeta_1, \zeta_2\}\begin{Bmatrix} QL \\ QR \end{Bmatrix}_e \\ &= \{\zeta_1, \zeta_2\}\{Q\}_e \end{aligned} \tag{4.5}$$

which identifies the column matrix $\{Q\}_e$ as the element nodal solution. As the final step, the $\zeta_i$ in (4.4) are identical to the elements of the linear basis given in (3.18) to within the origin shift for $\bar{x}$. The notation $\{N_k\}$ is in common use today for defining a finite element basis of degree (complexity) $k$. Hence, the desired generalization replaces the notation of (3.18) with,

$$\{N_{k=1}\} \equiv \begin{Bmatrix} \zeta_1 \\ \zeta_2 \end{Bmatrix} \tag{4.6}$$

and the linear basis approximation (4.5) is

$$\begin{aligned} T_e(x) &= \{\zeta_1, \zeta_2\}\{Q\}_e \\ &= \{N_1(\zeta_i)\}^T\{Q\}_e \end{aligned} \tag{4.7}$$

where $\{\cdot\}^T$ denotes a row matrix, which is the transpose of the column matrix $\{\cdot\}$.

Equations (4.4) and (4.6) to (4.7) define the linear basis approximation on $\Omega_e$. Recalling Chap. 3, the weak statement formulation will require integrals over $\Omega_e$ of products of $\{N_1(\zeta_i)\}$ and its first derivative. The latter is easy to evaluate using the chain rule,

$$\begin{aligned} \frac{d}{dx}\{N_1(\zeta_i)\} &= \frac{d}{d\zeta_i}\{N_1(\zeta_i)\}\frac{d\zeta_i}{d\bar{x}}\frac{d\bar{x}}{dx} \qquad 1 \le i \le 2 \\ &= \begin{cases} \begin{Bmatrix} -1 \\ 0 \end{Bmatrix} l_e^{-1}(1) & \text{for } i = 1 \\ \begin{Bmatrix} 0 \\ 1 \end{Bmatrix} l_e^{-1}(1) & \text{for } i = 2 \end{cases} \\ &= \frac{1}{l_e}\begin{Bmatrix} -1 \\ 1 \end{Bmatrix} \end{aligned} \tag{4.8}$$

which is in agreement with (3.22). Equations (4.6) to (4.8) complete the refined notational form for the linear basis for a one-dimensional finite element domain $\Omega_e$.

## Quadratic Basis

Viewing (4.7), which states the linear basis approximation, the quadratic basis form must be

$$T_e(x) \equiv \{N_{k=2}(\zeta_i)\}^T\{Q\}_e \tag{4.9}$$

Our only requirement is to establish the functional form of $\{N_2(\zeta_i)\}$, which will involve polynomials in the $\zeta_i$ defined in (4.4). Since (4.9) must contain quadratic terms in $x$, the form must be

$$T_e(x) = a + b\left(\frac{\bar{x}}{l_e}\right) + c\left(\frac{\bar{x}}{l_e}\right)^2 \tag{4.10}$$

Since three undetermined coefficients occur in (4.10), there must be three nodal temperatures defined for $\Omega_e$. The logical choice is to select the midpoint of $\Omega_e$ as the extra nodal location, as illustrated in Fig. 4.4. If the associated new temperature degree of freedom is denoted $QM$, for $Q$ midpoint, then (4.10) evaluated at each temperature node yields the three equations

$$\begin{aligned} T_e(\bar{x} = 0) &\equiv QL = a \\ T_e(\bar{x} = l_e/2) &\equiv QM = a + \frac{b}{2} + \frac{c}{4} \\ T_e(\bar{x} = l_e) &\equiv QR = a + b + c \end{aligned} \tag{4.11}$$

Solving (4.11) for $a$, $b$, and $c$ is straightforward; substituting this solution into (4.10) then yields

$$T_e(x) = QL + (-3QL + 4QM - QR)\frac{\bar{x}}{l_e} + (2QL - 4QM + 2QR)\left(\frac{\bar{x}}{l_e}\right)^2 \tag{4.12}$$

Equation (4.9) contains $\{Q\}_e \equiv \{QL, QM, QR\}_e^T$ as the common factor.

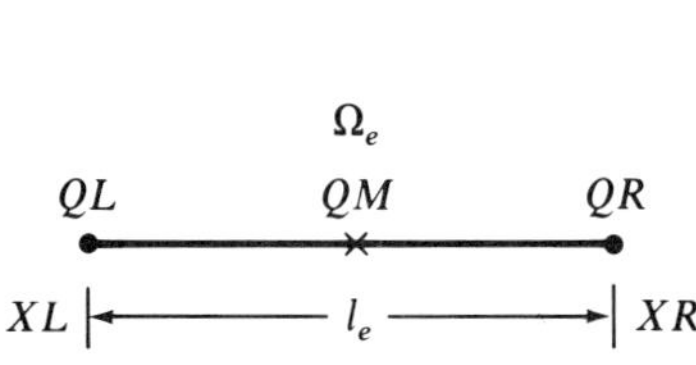


(*a*) Generic element

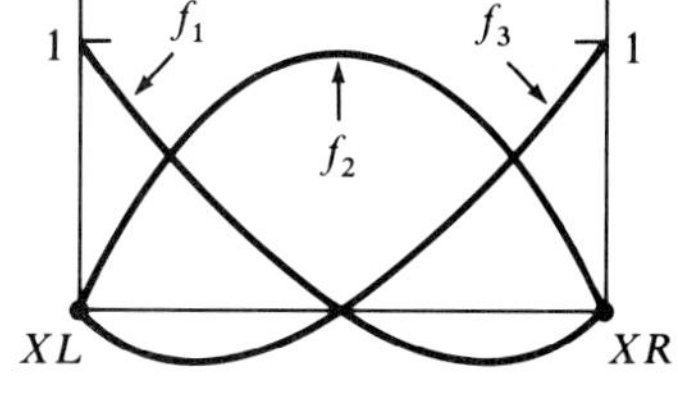


(*b*) Quadratic basis functions

**FIGURE 4.4**
Quadratic basis finite element details.

Rearranging (4.12) accordingly, and then substituting $\zeta_1 \equiv 1 - \bar{x}/l_e$ and $\zeta_2 \equiv \bar{x}/l_e$ in a symmetric fashion yields

$$T_e(x) = [\zeta_1(2\zeta_1 - 1)]QL + (4\zeta_1\zeta_2)QM + [\zeta_2(2\zeta_2 - 1)]QR$$
$$\equiv \{N_2(\zeta_i)\}^T\{Q\}_e \tag{4.13}$$

using the definition (4.9). A suggested exercise for the reader is to fill in each detail producing (4.13) from (4.11) to (4.12). The three quadratic polynomials (in $\zeta_i$) in (4.13) are labeled $f_1$, $f_2$, and $f_3$ in Fig. 4.4.

The derivation endpoint is at hand in (4.13); the elements of the quadratic basis are

$$\{N_2(\zeta_i)\} \equiv \begin{Bmatrix} \zeta_1(2\zeta_1 - 1) \\ 4\zeta_1\zeta_2 \\ \zeta_2(2\zeta_2 - 1) \end{Bmatrix} \tag{4.14}$$

The three polynomials constituting $\{N_2\}$ are graphed in Fig. 4.4*b*. Note that each is a quadratic function that is unity at the corresponding temperature node, and zero at both of the other temperature nodes. Note also that the quadratic basis element possesses only two geometric (vertex) nodes defining the element geometry. Hence, the mesh generation function remains absolutely unchanged from the linear basis procedure.

This simplification results from selection of the midpoint of $\Omega_e$ for the additional degree of freedom. The word *subparametric* is used in the literature to characterize definition and use of a higher-degree approximation for the dependent variable than for the geometry. An *isoparametric* formulation would require embedding of the mesh progression ratio $p$, defined in (4.1) to (4.2), into the basis definition, whereupon the geometric location of $QM$ becomes element-dependent. This is viewed as an unnecessary complication for the one-dimensional formulation.

The quadratic finite element formulation of the weak statement requires evaluation of integrals over $\Omega_e$ of products of $\{N_2\}$ and its first derivative. The derivative evaluation is again direct using the chain rule. The reader should complete the details to verify that

$$\frac{d}{dx}\{N_2(\zeta_i)\} = \frac{d}{d\zeta_i}\{N_2(\zeta_i)\}\frac{d\zeta_i}{d\bar{x}}\frac{d\bar{x}}{dx} \qquad 1 \le i \le 2$$
$$= \frac{1}{l_e}\begin{Bmatrix} \zeta_2 - 3\zeta_1 \\ 4(\zeta_1 - \zeta_2) \\ 3\zeta_2 - \zeta_1 \end{Bmatrix} \tag{4.15}$$

Thus, (4.9), (4.14), and (4.15) complete the quadratic basis formulation.

## Cubic Basis

The reader may now sense that creation of any one-dimensional finite element basis is an algebraic exercise. The computer experiments on approximation error, discussed later in this chapter, will confirm that error reduction is dramatically accelerated using a more complete (i.e., higher-degree) basis. The "no free lunch" axiom still holds, however, since development of the element matrices becomes more detailed. The increased bandwidth of the global matrix statement also requires greater computer resources to solve for $\{Q\}$. However, the formulation is done only once, to create the element *master matrix library*. Hence, development of the more complicated (i.e., complete) formulation is exactly the direction to move in the quest for improved solution reliability! One must thus be willing to pay for the increase in computational time and storage, a cost factor that decreases in importance every "computer generation."

In many instances, switching to the quadratic basis is adequate for a dramatic improvement in accuracy. In a few instances, even additional accuracy may be required. For completeness, then, the cubic basis function should be developed to cement the formulational structure in the reader's mind. The element approximation definition is

$$T_e(x) \equiv \{N_3(\zeta_i)\}^T\{Q\}_e \tag{4.16}$$

Recalling (4.3) and (4.10), the direct algebraic expression is

$$T_e(x) \equiv a + b\left(\frac{\bar{x}}{l_e}\right) + c\left(\frac{\bar{x}}{l_e}\right)^2 + d\left(\frac{\bar{x}}{l_e}\right)^3 \tag{4.17}$$

In this instance, (4.17) requires four nodal temperature degrees of freedom for $\Omega_e$ as shown in Fig. 4.5*a*. The resultant algebraic relations from (4.17) are

$$\begin{aligned}
T_e(\bar{x} = 0) &\equiv QL = a \\
T_e\left(\bar{x} \equiv \frac{l_e}{3}\right) &\equiv Q1 = a + \frac{b}{3} + \frac{c}{9} + \frac{d}{27} \\
T_e\left(\bar{x} = \frac{2l_e}{3}\right) &\equiv Q2 = a + \frac{2b}{3} + \frac{4c}{9} + \frac{8d}{27} \\
T_e(\bar{x} = l_e) &\equiv QR = a + b + c + d
\end{aligned} \tag{4.18}$$

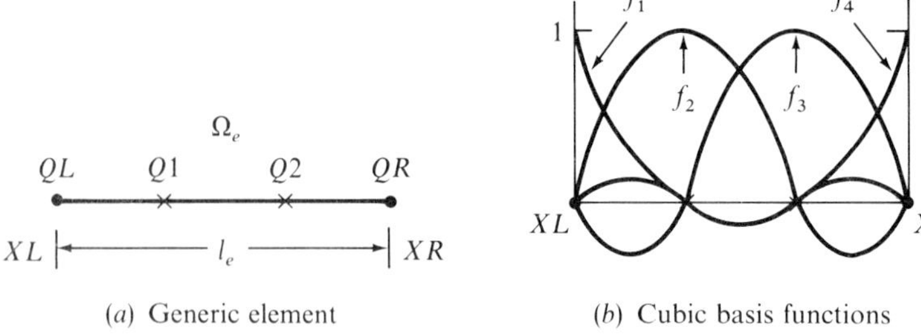


(*a*) Generic element

(*b*) Cubic basis functions

**FIGURE 4.5**
Cubic basis finite element details.

The solution of (4.18) for $a$, $b$, $c$, and $d$ is readily achieved, whereupon (4.17) can be rewritten in terms of $\{Q\}_e = \{QL, Q1, Q2, QR\}^T$. Rearrangement of this expression to expose $\{Q\}_e$ as the common multiplier [recall (4.13)] then yields the desired cubic basis definition:

$$\{N_3(\zeta_i)\} = \frac{9}{2}\begin{Bmatrix} \zeta_1(\zeta_2^2 - \zeta_2 + \frac{2}{9}) \\ \zeta_1\zeta_2(2 - 3\zeta_2) \\ \zeta_1\zeta_2(3\zeta_2 - 1) \\ \zeta_2(\zeta_2^2 - \zeta_2 + \frac{2}{9}) \end{Bmatrix} \tag{4.19}$$

The four cubic polynomials in (4.19) are labeled $f_1$ to $f_4$ in Fig. 4.5.

Formation of the element thermal conduction matrix requires the first derivative of (4.19). A suggested reader exercise is to verify that

$$\frac{d}{dx}\{N_3(\zeta_i)\} = \frac{9}{2l_e}\begin{Bmatrix} -3\zeta_2^2 + 4\zeta_2 - \frac{11}{9} \\ 9\zeta_2^2 - 10\zeta_2 + 2 \\ -9\zeta_2^2 + 8\zeta_2 - 1 \\ 3\zeta_2^2 - 2\zeta_2 + \frac{2}{9} \end{Bmatrix} \tag{4.20}$$

Equations (4.16), (4.19), and (4.20) complete the cubic basis formulation. The reason for refining the notational structure of the linear basis form into (4.6) and (4.7) should now be apparent. Hopefully, also, the reader now "sees" how the finite element procedure transforms the *fundamental decision of trial function* into the hierarchy of *bases* $\{N_k(\zeta_i)\}$, $1 \le k \le 3$, written on the *natural coordinate* set $\zeta_i$, $1 \le i \le 2$, for the generic one-dimensional finite element domain $\Omega_e$.

## 4.4 STEADY-STATE DIFFUSION EQUATION

Recall the example problem discussed in Chap. 3; specifically, determine the temperature distribution on the interval $0 \leq x \leq L$ that satisfies

$$L(T) = -\frac{d}{dx}\left(k\frac{dT}{dx}\right) - s = 0 \qquad 0 < x < L \tag{3.1}$$

$$l(T) = -k\frac{dT}{dx} - q = 0 \qquad x = 0 \tag{3.2}$$

$$T(x_b) = T_b \qquad x = L \tag{3.3}$$

The Galerkin weak statement recasts the problem into the following form. For the approximation to $T(x)$ defined as

$$T^N(x) = \sum_{i=1}^{N} a_i \phi_i(x) \tag{3.6}$$

where $\phi_i(x)$ is the *trial function set*, require the approximate symmetric weak form of (3.1) and (3.2) to vanish, i.e.,

$$WS = \sum_{j=1}^{N}\left(a_j \int_0^L \frac{d\phi_i}{dx} k \frac{d\phi_j}{dx}\,dx\right) - \int_0^L s\phi_i\,dx + q\phi_i\Big|_0^L = 0 \qquad 1 \leq i \leq N \tag{3.12}$$

The finite element implementation of (3.6) and (3.12) replaces the (global) trial function set $\phi_i(x)$ with the finite element basis $\{N_k(\zeta_i)\}$ on a discretization $\Omega^h$. The resultant expression for temperature approximation on $\Omega_e$ is

$$T_e(x) = \{N_k(\zeta_i)\}^T\{Q\}_e \tag{4.21}$$

which is the general form of (4.7), (4.9), and (4.16) for arbitrary degree $k$. When a discretization $\Omega^h$ is used to construct the approximation, then $T^N(x)$ in (3.6) is replaced by $T^h(x)$ to emphasize the existence of $\Omega^h$. Then $T^h$ is expressed in every element $\Omega_e$ by $T_e$, but one cannot simply sum the $T_e$ to obtain $T^h$ since a double contribution would occur on every finite element boundary. The mathematically precise expression of summation without overlap is given the name *union* and denoted by the symbol

$\bigcup$. Thus

$$T^h = \bigcup_e T_e(x) = \sum_e T_e(x) \text{ without boundary overlap} \tag{4.22}$$

precisely expresses the finite element approximation.

With (4.21) and (4.22) as the formal replacement of (3.6), formation of the Galerkin symmetric weak statement (3.12) is accomplished by the *assembly* of the integrals evaluated on each $\Omega_e$. Like *union*, *assembly* is a special form of summation which is accomplished on rows in the global matrix statement. To distinguish this procedure from scalar summation it is given the symbol $S_e$. Hence, for $T^h$ defined in (4.22), the weak statement (3.12) is replaced by the expression

$$WS^h = S_e\left(\int_{\Omega_e} \frac{d\{N_k\}}{dx} k \frac{d\{N_k\}^T}{dx} dx\{Q\}_e - \int_{\Omega_e} \{N_k\}s_e\, dx - k\frac{dT}{dx}\begin{Bmatrix} -\delta_{e1} \\ \delta_{eM} \end{Bmatrix}\right)$$
$$\equiv \{0\} \tag{4.23}$$

Thus, (4.21) to (4.23) completes restatement of the finite element algorithm for the model problem for arbitrary basis $\{N_k\}$ of polynomial degree $k$. The first term in (4.23) is the element thermal conduction matrix, which is conventionally denoted as $[K]_e$. Realizing that the source term and the boundary convection terms constitute user-supplied data, and denoting their evaluated sum as the element column (load) matrix $\{b\}_e$, the matrix statement produced from (4.23) is

$$S_e([K]_e\{Q\}_e - \{b\}_e) = \{0\} \tag{4.24}$$

The solution of (4.24) yields the global nodal temperature distribution $\{Q\}$ and completes the finite element solution (4.21) and (4.22) for example problem (3.1) to (3.3).

We now fully illustrate the details of formation of (4.24) for the model problem using the various trial bases $\{N_k\}$. This permits focusing on the integral operations forming $[K]_e$ and $\{b\}_e$ while promoting establishment of a firm understanding of *union* and *assembly*.

### Linear Basis

In Chap. 3, (3.23) and (3.24) presented formation of $[K]_e$ and $\{b\}_e$ for the linear basis $\{N_1\}$, while (3.28) and (3.29) illustrated *assembly* of the global matrix statement (3.27). At the risk of some repetition, we evaluate the ingredients of (4.23) assuming no prior knowledge. The conductivity matrix using linear basis elements is defined as

$$[K]_e \equiv \int_{\Omega_e} \frac{d\{N_1\}}{dx} k \frac{d\{N_1\}^T}{dx} dx \tag{4.25}$$

The derivative of $\{N_1\}$ is given in (4.8); hence, for assumed constant thermal conductivity $k$, (4.25) becomes

$$\begin{aligned}[K]_e &= k \int_0^{l_e} \frac{1}{l_e} \begin{Bmatrix} -1 \\ 1 \end{Bmatrix} \frac{1}{l_e} \{-1, 1\}\, dx \\ &= \frac{k}{l_e} \begin{bmatrix} 1 & -1 \\ -1 & 1 \end{bmatrix}\end{aligned} \tag{4.26}$$

which agrees with the appropriate term in (3.25). The distributed source term yields

$$\{b\}_e \equiv \int_{\Omega_e} \{N_1\} s_e \, dx \tag{4.27}$$

Assuming $s_e$ is constant (a severe restriction), then using (4.4) and (4.6) and noting that $d\bar{x} = dx$ one obtains directly

$$\begin{aligned}\{b\}_e &= s \int_{\Omega_e} \begin{Bmatrix} \zeta_1 \\ \zeta_2 \end{Bmatrix} dx = s \int_0^{l_e} \begin{Bmatrix} 1 - \bar{x}/l_e \\ \bar{x}/l_e \end{Bmatrix} d\bar{x} \\ &= s l_e \begin{Bmatrix} 1/2 \\ 1/2 \end{Bmatrix}\end{aligned} \tag{4.28}$$

Equation (4.28) also agrees with the appropriate term in (3.25).

Equations (4.26) and (4.28) constitute the element master matrix library for the linear basis finite element algorithm for (3.1) to (3.3) with a constant source. The boundary heat flux, the last term in (4.23), involves no integration for one-dimensional domains. Thus, the linear basis algorithm (4.24) becomes, for *any discretization* $\Omega^h$ of the domain $0 \le x \le L$

$$\begin{aligned}&S_e([K]_e\{Q\}_e - \{b\}_e) = \{0\} \\ &\quad = S_e\left(\frac{k}{l_e}\begin{bmatrix} 1 & -1 \\ -1 & 1 \end{bmatrix}\{Q\}_e - \frac{s l_e}{2}\begin{Bmatrix} 1 \\ 1 \end{Bmatrix} - k\frac{dT}{dx}\begin{Bmatrix} -\delta_{e1} \\ \delta_{eM} \end{Bmatrix}\right)\end{aligned} \tag{4.29}$$

**Example 4.1.** Determine the nodal temperature distribution for an $M = 1$ discretization of the model problem domain for arbitrary (constant) $k$ and $s$. Since $l_e$ is thus unity, (4.29) becomes

$$S_e([K]_e\{Q\}_e - \{b\}_e) = S_e\left(\frac{k}{l_e}\begin{bmatrix} 1 & -1 \\ -1 & 1 \end{bmatrix}\{Q\}_e - \frac{sl_e}{2}\begin{Bmatrix} 1 \\ 1 \end{Bmatrix} + \begin{Bmatrix} -q \\ -kT_n \end{Bmatrix}\right) \quad \text{for } e = 1$$

$$= \frac{k}{L}\begin{bmatrix} 1 & -1 \\ 0 & 1 \end{bmatrix}\begin{Bmatrix} Q1 \\ Q2 \end{Bmatrix} - \frac{sL}{2}\begin{Bmatrix} 1 \\ 0 \end{Bmatrix} + \begin{Bmatrix} -q \\ -T_b k/L \end{Bmatrix} = \{0\} \tag{4.30}$$

where the last equation in (4.30) has been modified to impose the Dirichlet condition $Q2 = T_b$, as defined by (3.3). The solution of (4.30) is

$$Q1 = \frac{sL^2}{2k} + \frac{qL}{k} + T_b \tag{4.31}$$

which again is in exact agreement with the analytical solution (3.4), since the model problem is so elementary. However, a plot of the associated $M = 1$ continuous approximation (recall Fig. 3.3) would indeed confirm that the associated error $e^h$ is larger than that for the $M = 2$ solution. We suggest you verify this comment by plotting the $M = 1$ solution $T^h$ on Fig. 3.5 using (4.31) and $T_b$.

**Example 4.2.** Determine the nodal solution for the model problem on the $M = 2$ nonuniform discretization $\Omega^h$ with nodes $\{X\} = \{0, L/3, L\}$. Then (4.29) becomes

$$S_e([K]_e\{Q\}_e - \{b\}_e)$$

$$= S_e\left(\frac{k}{l_e}\begin{bmatrix} 1 & -1 \\ -1 & 1 \end{bmatrix}\{Q\}_e - \frac{sl_e}{2}\begin{Bmatrix} 1 \\ 1 \end{Bmatrix} + \begin{Bmatrix} -q \\ -kT_n \end{Bmatrix}\right) \quad \text{for } 1 \le e \le 2 \tag{4.32}$$

and by terms

$$S_e([K]_e) = S_e\left(\frac{k}{L/3}\begin{bmatrix} 1 & -1 \\ -1 & 1 \end{bmatrix}, \frac{k}{2L/3}\begin{bmatrix} 1 & -1 \\ -1 & 1 \end{bmatrix}\right)$$

$$= S_e\left(\frac{k}{L}\begin{bmatrix} 3 & -3 \\ -3 & 3 \end{bmatrix}, \frac{k}{L}\begin{bmatrix} 3/2 & -3/2 \\ -3/2 & 3/2 \end{bmatrix}\right)$$

$$= \frac{k}{L}\begin{bmatrix} 3 & -3 & 0 \\ -3 & 3 + 3/2 & -3/2 \\ 0 & -3/2 & 3/2 \end{bmatrix} \tag{4.33}$$

$$S_e(\{b\}_e) = S_e\left(\frac{sL/3}{2}\begin{Bmatrix}1\\1\end{Bmatrix}, \quad \frac{s2L/3}{2}\begin{Bmatrix}1\\1\end{Bmatrix}\right)$$

$$= \frac{sL}{6}\begin{Bmatrix}1\\1+2\\2\end{Bmatrix} \tag{4.34}$$

Using (4.33) and (4.34), and enforcing $Q3 = T_b$, the $M = 2$ global matrix statement is

$$S_e(\cdot) = \frac{k}{L}\begin{bmatrix}3 & -3 & 0\\ -3 & 4.5 & -1.5\\ 0 & 0 & 1\end{bmatrix}\begin{Bmatrix}Q1\\Q2\\Q3\end{Bmatrix} - \frac{sL}{6}\begin{Bmatrix}1\\3\\0\end{Bmatrix} + \begin{Bmatrix}-q\\0\\-kT_b/L\end{Bmatrix} \tag{4.35}$$

Substituting the last equation into the second in (4.35) and clearing constants and fractions yields the final hand-calculation statement

$$\begin{bmatrix}6 & -6\\ -6 & 9\end{bmatrix}\begin{Bmatrix}Q1\\Q2\end{Bmatrix} = \frac{sL^2}{3k}\begin{Bmatrix}1\\3\end{Bmatrix} + \begin{Bmatrix}2qL/k\\3T_b\end{Bmatrix} \tag{4.36}$$

The solution of (4.36) is

$$\begin{Bmatrix}Q1\\Q2\end{Bmatrix} = \frac{1}{18}\begin{bmatrix}9 & 6\\ 6 & 6\end{bmatrix}\left(\frac{sL^2}{3k}\begin{Bmatrix}1\\3\end{Bmatrix} + \begin{Bmatrix}2qL/k\\3T_b\end{Bmatrix}\right)$$

$$= \frac{sL^2}{3k}\begin{Bmatrix}1/2+1\\1/3+1\end{Bmatrix} + \begin{Bmatrix}qL/k + T_b\\2qL/3k + T_b\end{Bmatrix}$$

$$= \frac{sL^2}{k}\begin{Bmatrix}1/2\\4/9\end{Bmatrix} + \frac{qL}{k}\begin{Bmatrix}1\\2/3\end{Bmatrix} + T_b\begin{Bmatrix}1\\1\end{Bmatrix} \tag{4.37}$$

We strongly suggest you take the time to fill in all the details involved in (4.32) to (4.37). It should come as no surprise that $Q1$ is again determined exactly. Evaluating the analytical solution (3.4) at $x = L/3$ yields

$$T(x = L/3) = \frac{sL^2}{k}\left(\frac{4}{9}\right) + \frac{qL}{k}\left(\frac{2}{3}\right) + T_b \tag{4.38}$$

hence the $M = 2$ solution on the nonuniform mesh also produces the exact nodal temperature $Q2$ at $x = L/3$.

## Quadratic Basis

The requirement is to generate the element master matrix library for the $\{N_2\}$ basis. The element conduction matrix is

$$[K]_e \equiv \int_{\Omega_e} \frac{d\{N_2\}}{dx} k \frac{d\{N_2\}^T}{dx} dx$$

$$= \int_{\Omega_e} \frac{k}{l_e^2} \begin{Bmatrix} \zeta_2 - 3\zeta_1 \\ 4(\zeta_1 - \zeta_2) \\ 3\zeta_2 - \zeta_1 \end{Bmatrix} \{\zeta_2 - 3\zeta_1, 4(\zeta_1 - \zeta_2), 3\zeta_2 - \zeta_1\} \, dx$$

$$= \frac{k}{l_2^2} \int_0^{l_e} \begin{bmatrix} (\zeta_2 - 3\zeta_1)^2 & (\zeta_2 - 3\zeta_1)4(\zeta_1 - \zeta_2) & (\zeta_2 - 3\zeta_1)(3\zeta_2 - \zeta_1) \\ 4(\zeta_1 - \zeta_2)(\zeta_2 - 3\zeta_1) & 16(\zeta_1 - \zeta_2)^2 & 4(\zeta_1 - \zeta_2)(3\zeta_2 - \zeta_1) \\ (3\zeta_2 - \zeta_1)(\zeta_2 - 3\zeta_1) & (3\zeta_2 - \zeta_1)4(\zeta_1 - \zeta_2) & (3\zeta_2 - \zeta_1)^2 \end{bmatrix} d\bar{x} \tag{4.39}$$

for $k$ a constant and using (4.15). Note that $[K]_e$ is a symmetric matrix and of order three, to match the three entries in the element nodal temperature matrix $\{Q\}_e = \{QL, QM, QR\}_e^T$.

Integrals of six polynomials in the $\zeta_i$ are required to evaluate (4.39), which could be accomplished by transformation to $\bar{x}$ in the manner of (4.28). However, an easier method exists, since integrals over $\Omega_e$ of *all* polynomials in the $\zeta_i$ can be analytically evaluated as

$$\int_{\Omega_e} \zeta_1^p \zeta_2^q \, dx = l_e \frac{p!q!}{(1 + p + q)!} \tag{4.40}$$

where $(p, q)$ are integers and $l_e$ is the length (measure) of $\Omega_e$. [*Note*: The expression (4.40) generalizes directly to $n$-dimensional element domains $\Omega_e$ for triangles and tetrahedra. It is truly a useful formula, as we shall see later.] As a check, recall (4.28) as the distributed source term contribution to $\{b\}_e$ for $\{N_1\}$. The computation using (4.40) yields

$$\int_{\Omega_e} \begin{Bmatrix} \zeta_1 \\ \zeta_2 \end{Bmatrix} dx = \begin{Bmatrix} 1! \; 0! \\ 0! \; 1! \end{Bmatrix} \frac{l_e}{(1 + 1 + 0)!} = \frac{l_e}{2} \begin{Bmatrix} 1 \\ 1 \end{Bmatrix} \tag{4.41}$$

With (4.40), evaluation of the integrals in (4.39) is reduced to an arithmetic detail, albeit a tedious one. For example, the (1, 1) term in (4.39) is

$$\frac{k}{l_e^2}\int_{\Omega_e}(\zeta_2 - 3\zeta_1)^2\,dx = \frac{k}{l^2}\int_{\Omega_e}(\zeta_2^2 - 6\zeta_1\zeta_2 + 9\zeta_1^2)\,dx$$

$$= \frac{k}{l_e}\left(\frac{2!0!}{3!} - \frac{6\times 1!1!}{3!} + \frac{9\times 0!2!}{3!}\right) \tag{4.42}$$

$$= \frac{k}{6l_e}(2 - 6 + 18) = \frac{7k}{3l_e}$$

It is recommended you proceed through several other terms in (4.39), hence verify that the $\{N_2\}$ basis element matrix for constant conductivity is

$$[K]_e = \frac{k}{3l_e}\begin{bmatrix} 7 & -8 & 1 \\ -8 & 16 & -8 \\ 1 & -8 & 7 \end{bmatrix} \tag{4.43}$$

One should also verify that for a uniform source, the element master matrix is

$$\{b\}_e \equiv \int_{\Omega_e}\{N_2\}s_e\,dx = \frac{sl_e}{6}\begin{Bmatrix} 1 \\ 4 \\ 1 \end{Bmatrix} \tag{4.44}$$

Equations (4.43) and (4.44) constitute the master matrix library for the $\{N_2\}$ basis for *any* discretization $\Omega^h$ of the model problem.

**Example 4.3.** Determine the $\{N_2\}$ basis nodal temperature distribution for an $M = 1$ discretization of the model problem. Thus, $l_e = L$ and the finite element algorithm statement (4.24) is

$$S_e([K]_e\{Q\}_e - \{b\}_e) = \{0\}$$

$$= S_e\left(\frac{k}{3l_e}\begin{bmatrix} 7 & -8 & 1 \\ -8 & 16 & -8 \\ 1 & -8 & 7 \end{bmatrix}\{Q\}_e - \frac{sl_e}{6}\begin{Bmatrix} 1 \\ 4 \\ 1 \end{Bmatrix} - k\frac{dT}{dx}\begin{Bmatrix} -\delta_{e1} \\ 0 \\ \delta_{eM} \end{Bmatrix}\right) \quad \text{for } e = 1 \tag{4.45}$$

Enforcing the Dirichlet boundary condition and applying the imposed heat flux $q$ yields the global matrix statement

$$\frac{k}{3L}\begin{bmatrix} 7 & -8 & 1 \\ -8 & 16 & -8 \\ 0 & 0 & 1 \end{bmatrix}\begin{Bmatrix} Q1 \\ Q2 \\ Q3 \end{Bmatrix} = \frac{sL}{6}\begin{Bmatrix} 1 \\ 4 \\ 0 \end{Bmatrix} + \begin{Bmatrix} q \\ 0 \\ kT_b/3L \end{Bmatrix} \tag{4.46}$$

Clearing fractions and directly substituting $T_b$ for $Q3$ reduces (4.46) to the hand-computable rank 2 expression

$$\begin{bmatrix} 7 & -8 \\ -8 & 16 \end{bmatrix}\begin{Bmatrix} Q1 \\ Q2 \end{Bmatrix} = \frac{sL^2}{2k}\begin{Bmatrix} 1 \\ 4 \end{Bmatrix} + \begin{Bmatrix} 3qL/k - T_b \\ 8T_b \end{Bmatrix} \tag{4.47}$$

the solution of which is

$$\begin{Bmatrix} Q1 \\ Q2 \end{Bmatrix} = \frac{1}{48}\begin{bmatrix} 16 & 8 \\ 8 & 7 \end{bmatrix}\left(\frac{sL^2}{2k}\begin{Bmatrix} 1 \\ 4 \end{Bmatrix} + \begin{Bmatrix} 3qL/k - T_b \\ 8T_b \end{Bmatrix}\right)$$

$$= \frac{sL^2}{k}\begin{Bmatrix} 1/2 \\ 3/8 \end{Bmatrix} + \frac{qL}{k}\begin{Bmatrix} 1 \\ 1/2 \end{Bmatrix} + T_b\begin{Bmatrix} 1 \\ 1 \end{Bmatrix} \tag{4.48}$$

It is no surprise that the approximate solution nodal temperatures are again exact. On the surface, one might therefore conclude that the extra work in forming the $\{N_2\}$ basis solution statement is an unrewarded effort. But recall that the approximate solution is not $\{Q\}$ but the continuous form $T^h$, (4.21) and (4.22). Figure 3.5 clearly illustrates the error in the $M = 2$ linear basis solution. The necessary next step therefore is to evaluate (4.22) using the $M = 1$ quadratic basis functions. Substituting (4.48) into (4.22) yields

$$\begin{aligned} T^h &= \bigcup_e T_{e=1}(x) \\ &= \{N_2(\zeta_i)\}^T\{Q\}_{e=1} \\ &= \{\zeta_1(2\zeta_1 - 1), 4\zeta_1\zeta_2, \zeta_2(2\zeta_2 - 1)\}\begin{Bmatrix} Q1 \\ Q2 \\ T_b \end{Bmatrix} \\ &= \zeta_1(2\zeta_1 - 1)\left(\frac{sL^2}{2k} + \frac{qL}{k} + T_b\right) + 4\zeta_1\zeta_2\left(\frac{3sL^2}{8k} + \frac{qL}{2k} + T_b\right) \\ &\quad + \zeta_2(2\zeta_2 - 1)T_b \end{aligned} \tag{4.49}$$

Combining terms in (4.49) with like multipliers yields

$$\begin{aligned} T^h(x) = \frac{sL^2}{2k}&(2\zeta_1^2 - \zeta_1 + 3\zeta_1\zeta_2) \\ &+ \frac{qL}{k}(2\zeta_1^2 - \zeta_1 + 3\zeta_1\zeta_2) \\ &+ T_b(2\zeta_1^2 - \zeta_1 + 4\zeta_1\zeta_2 + 2\zeta_2^2 - \zeta_2) \end{aligned} \tag{4.50}$$

Now substituting the definitions for $\zeta_i$ in terms of $\bar{x}/L$ [recall (4.4)], the $M = 1$ quadratic basis finite element approximate solution (4.50) becomes

$$T^h(x) = \frac{sL^2}{2k}\left(1 - \left(\frac{\bar{x}}{L}\right)^2\right) + \frac{qL}{k}\left(1 - \frac{\bar{x}}{L}\right) + T_b \qquad (4.51)$$

Equation (4.51) looks very familiar! In fact, it is identical with the *exact solution* $T(x)$, (3.4), since the origin of element coordinate $\bar{x}$ is at $x = 0$. In retrospect, this occurrence should come as no surprise, since both $T(x)$ and $T^h(x)$ formed with $\{N_2\}$ are quadratic in $x$. The role played by the choice of *basis* $\{N_k\}$ is indeed *fundamental*! We must contain our overt enthusiasm, however, because this has occurred for an extremely elementary problem. In a practical situation, one will *never* obtain the analytically exact solution. We must thus learn to quantify approximation error, and then verify theoretical predictions by designing and executing carefully designed computational experiments using *LEARN.FE.*

## Cubic Basis

One detail remains, which is implementation of the cubic basis $\{N_3\}$ into the finite element algorithm statement (4.24). Using (4.20), the thermal conduction matrix is

$$[K]_e \equiv \int_{\Omega_e} \frac{d\{N_3\}}{dx}\, k\, \frac{d\{N_3\}^T}{dx}\, dx$$

$$= k\left(\frac{9}{2l_e}\right)^2 \int_0^{l_e} \begin{Bmatrix} -3\zeta_2^2 + 4\zeta_2 - 11/9 \\ 9\zeta_2^2 - 10\zeta_2 + 2 \\ -9\zeta_2^2 + 8\zeta_2 - 1 \\ 3\zeta_2^2 - 2\zeta_2 + 2/9 \end{Bmatrix} \{\,\cdot\,\}^T\, d\bar{x} \qquad (4.52)$$

where the notation $\{\cdot\}^T$ denotes the column matrix, detailed in (4.52), written as a row matrix. Equation (4.52) requires integration of polynomials in $\zeta_i$ of degree 4. It takes care to avoid arithmetic mistakes, and the utility of the integration formula (4.40) certainly becomes appreciated! The reader should verify at least one of the entries below.

$$[K]_e = \frac{k}{40l_e}\begin{bmatrix} 148 & -189 & 54 & -13 \\ & 432 & -297 & 54 \\ & & 432 & -189 \\ (\text{sym}) & & & 148 \end{bmatrix} \qquad (4.53)$$

The uniform source term contribution to the cubic master matrix library is

$$\{b\}_e \equiv \int_{\Omega_e} \{N_3\} s_e \, dx = \frac{9s}{2} \int_0^{l_e} \begin{Bmatrix} \zeta_1(\zeta_2^2 - \zeta_2 + 2/9) \\ \zeta_1\zeta_2(2 - 3\zeta_2) \\ \zeta_1\zeta_2(3\zeta_2 - 1) \\ \zeta_2(\zeta_2^2 - \zeta_2 + 2/9) \end{Bmatrix} d\bar{x}$$

$$= \frac{sl_e}{8} \begin{Bmatrix} 1 \\ 3 \\ 3 \\ 1 \end{Bmatrix} \tag{4.54}$$

Equations (4.53) and (4.54) complete the cubic basis library. Note that each matrix is of order four, in agreement with the element nodal temperature array $\{Q\}_e = \{QL, Q1, Q2, QR\}_e^T$. We need not proceed through the $M = 1$ example problem, since the exact solution will again result. You may want to verify this.

## B. PRACTICAL PROBLEMS

### 4.5 STEADY HEAT CONDUCTION WITH BOUNDARY CONVECTION

The steady-state diffusion equation examined to this point has been appropriate, but the restriction to uniform properties and the absence of the thermal convection boundary condition is not. A more practical problem statement is to find the temperature distribution $T(x)$ that satisfies

$$L(T) = -\frac{d}{dx}\left[k(x)\frac{dT}{dx}\right] - s(x) = 0 \qquad \text{on } \Omega = (0 < x < L) \tag{4.55}$$

$$l(T) = k\frac{dT}{dn} + h(T - T_r) = 0 \qquad \text{on } \partial\Omega_1 \tag{4.56}$$

and

$$T(x_b) = T_b \qquad \text{on } \partial\Omega_2 \tag{4.57}$$

In (4.55), $k(x)$ and $s(x)$ are assumed variable, while in (4.56), $h$ is the boundary thermal convection coefficient for energy exchange with a medium at reference temperature $T_r$. Further, in (4.56), $dT/dn$ is the normal derivative, i.e., $\pm dT/dx$, on $\partial\Omega$. As before (4.57) admits that a fixed temperature may exist. One might also elect to nondimensionalize (4.55) to (4.57), which would introduce the Nusselt number, $\mathrm{Nu} = hL/k$. Finally, $\Omega$ is the problem domain with boundary $\partial\Omega$ made up of the (non-overlapping) sum of $\partial\Omega_1$ and $\partial\Omega_2$, i.e., their "union ($\bigcup$)."

The procedure to develop the finite element algorithm for (4.55) to (4.57) is as presented. The approximation definition is (3.6)

$$T^N(x) = \sum_{i=1}^{N} a_i \phi_i(x) \tag{3.6}$$

The symmetric Galerkin weak statement [recall (3.12) to (3.13)] is the straightforward development

$$\begin{aligned}
WS &= \int_\Omega \phi_i L(T^N)\, dx \qquad \text{for } 1 \le i \le N \\
&= \int_\Omega \phi_i \left[ -\frac{d}{dx}\left( k(x) \frac{dT^N}{dx} \right) - s(x) \right] dx \\
&= \int_\Omega \left[ \frac{d\phi_i}{dx} k(x) \frac{dT^N}{dx} - \phi_i s(x) \right] dx - \int_{\partial\Omega} \phi_i \left( k \frac{dT}{dn} \right) d\sigma \\
&= \int_\Omega \left[ \frac{d\phi_i}{dx} k(x) \frac{dT^N}{dx} - \phi_i s(x) \right] dx + \int_{\partial\Omega_1} \phi_i h(T^N - T_r)\, d\sigma \\
&\quad - \int_{\partial\Omega_2} \phi_i k \frac{dT^N}{dn}\, d\sigma = 0
\end{aligned} \tag{4.58}$$

The divergence theorem form of (4.58) is a generalization on (3.13), as the integration-by-parts endpoint evaluations have been replaced by a surface integral on $\partial\Omega$ with differential element $d\sigma$. This is required in progressing to axisymmetric geometries which involve boundary convection (4.56) over various surfaces.

The finite element method defines the mesh $\Omega^h$ and replaces (3.6) with $T^h$, (4.22), which is constituted of the union of element approximations

$$T_e(x) = \{N_k(\zeta_i)\}^T \{Q\}_e \tag{4.59}$$

The corresponding *discrete* approximation to the Galerkin weak statement (4.58) is

$$WS^h = S_e\left[\int_{\Omega_e} \frac{d\{N_k\}}{dx}\, k_e(x)\, \frac{d\{N_k\}^T}{dx}\, dx\{Q\}_e - \int_{\Omega_e} \{N_k\}s_e(x)\, dx \right.$$
$$\left. + \int_{\partial\Omega_1} \{N_k\}h(\{N_k\}^T\{Q\}_e - T_r)\, d\sigma - \int_{\partial\Omega_2} \{N_k\}k\, \frac{dT^h}{dn}\, d\sigma \right]$$
$$= \{0\} \tag{4.60}$$

In (4.60), the last term again verifies that the approximate normal heat flux $-k(dT^h/dn)$ will be predicted by the finite element solution on $\partial\Omega_2$ where a Dirichlet condition is specified.

As before, implementation of the finite element algorithm (4.60) requires identification of the basis degree $k$ and evaluation of the defined integrals. The new aspect is that the variable thermal conductivity and distributed heat-source parameters (data) now reside within these integrals. The question thus arises, "Must we know the functional dependence of data for a specific problem statement before the master matrix library can be completed?" The answer is a resounding *no*, since all versatility would be lost by this restriction.

The resolution is at hand in the statement of the finite element approximation (4.59). Since the basis can support the approximate solution, it can certainly interpolate *any* data specification. Thus, we have at our disposal

$$k_e(x) = \{N_k(\zeta_i)\}^T\{K\}_e \tag{4.61}$$

$$s_e(x) = \{N_k(\zeta_i)\}^T\{S\}_e \tag{4.62}$$

The column matrix $\{K\}_e$ contains the nodal values of element thermal conductivity, while $\{S\}_e$ contains the source term nodal values on $\Omega_e$. Note carefully that the column matrix $\{K\}_e$ is totally distinct from $[K]_e$, the element square matrix expressing the complete conduction contribution. These data $\{K\}_e$ and $\{S\}_e$ are determined by interrogating the problem-specific functional forms at the nodes of the (*any*) computational mesh $\Omega^h$. Algorithm versatility is thus fully preserved!

With (4.61) and (4.62), we are prepared to complete the algorithm statement and examine its application using the *LEARN.FE* code. The following developments are indexed on the *fundamental choice* of basis.

### Linear Basis

The integrals in (4.60) are particularly straightforward for $k = 1$ in (4.59). The element thermal conduction master matrix is

$$[K]_e \equiv \int_{\Omega_e} \frac{d\{N_1\}}{dx} k_e(x) \frac{d\{N_1\}^T}{dx} dx$$
$$= \frac{1}{l_e^2}\begin{bmatrix} 1 & -1 \\ -1 & 1 \end{bmatrix} \int_{\Omega_e} k_e(x)\, dx \tag{4.63}$$

recalling (4.26). The integral remaining in (4.63) has already been evaluated as the uniform source term contribution to $\{b\}_e$; recall (4.28). Thus

$$\int_{\Omega_e} k_e(x)\, dx = \int_{\Omega_e} \{N_1\}^T dx \{K\}_e = \int_{\Omega_e} \{\zeta_1, \zeta_2\}\, dx \{K\}_e$$
$$= l_e \left\{\frac{1}{2}, \frac{1}{2}\right\} \{K\}_e = l_e \left(\frac{KL}{2} + \frac{KR}{2}\right)_e \tag{4.64}$$

where $KL$ and $KR$ are the *left* and *right* nodal values of $k_e$ on $\Omega_e$. Thus, (4.64) confirms that the average value of $k_e(x)$ is the consistent linear basis implementation. Combining (4.63) and (4.64) then yields the element master matrix expression

$$[K]_e = \frac{1}{l_e} \left\{\frac{1}{2}, \frac{1}{2}\right\} \{K\}_e \begin{bmatrix} 1 & -1 \\ -1 & 1 \end{bmatrix} \tag{4.65}$$

and the distinct roles of $[K]_e$ and $\{K\}_e$ are clearly specified.

The distributed source term contribution to $\{b\}_e$ is similarly direct to evaluate as

$$\{b\}_e = \int_{\Omega_e} \{N_1\} s_e(x)\, dx = \int_{\Omega_e} \{N_1\}\{N_1\}^T dx \{S\}_e$$
$$= \frac{l_e}{6} \begin{bmatrix} 2 & 1 \\ 1 & 2 \end{bmatrix} \{S\}_e \tag{4.66}$$

Thus, the simple average suitable for $k_e(x)$ does not carry over as the consistent evaluation of a variable source term $s_e(x)$.

The thermal convection boundary condition term is new but readily yields to evaluation. This term must be applied at either $x = 0$ or $x = L$ in a pure one-dimensional problem; hence we'll again have use for $\delta_{e1}$ and $\delta_{eM}$. Recalling Fig. 4.3, assume application is to element $\Omega_{e=1}$, and hence the convection is applied at $XL$. The third term in (4.60) is then

$$\int_{\partial\Omega_{e=1}} \{N_1\}h(\{N_1\}^T\{Q\}_e - T_r)\,d\sigma = \int_{\partial\Omega_{e=1}} \begin{Bmatrix}\zeta_1\\ \zeta_2\end{Bmatrix} h(\{\zeta_1, \zeta_2\}\{Q\}_e - T_r)\,d\sigma$$

$$= \int_{\partial\Omega_{e=1}} \begin{Bmatrix}1\\ 0\end{Bmatrix} h(\{1, 0\}\{Q\}_e - T_r)\,d\sigma$$

$$= h\left(\begin{bmatrix}1 & 0\\ 0 & 0\end{bmatrix}\{Q\}_e - T_r\begin{Bmatrix}1\\ 0\end{Bmatrix}\right)_{e=1} \tag{4.67}$$

The fact that $\zeta_2 = 0$ and $\zeta_1 = 1$ at $XL$ (recall Fig. 4.3*b*) reduces the integral to a simple endpoint evaluation. If we replace the 1s in (4.67) by $\delta_{e1}$, and then repeat the operations in (4.67) at $XR$ for $\Omega_{e=M}$ (a suggested exercise), the thermal convection master matrix library is thus

$$[H]_e \equiv h_e\begin{bmatrix}\delta_{e1} & 0\\ 0 & \delta_{eM}\end{bmatrix} \tag{4.68}$$

$$\{b\}_e = h_e T_r\begin{Bmatrix}\delta_{e1}\\ \delta_{eM}\end{Bmatrix} \tag{4.69}$$

The subscript $e$ has been appended to $h_e$ to indicate that the left- and right-end convection coefficients can certainly be distinct.

The last term in (4.60) again does not require an evaluation, since the Dirichlet data imposition will remove its appearance. Thus, the linear basis finite element algorithm for the generalized heat transfer problem is

$$S_e(([K]_e + [H]_e)\{Q\}_e - \{b(s, h, T_r)\}_e) = \{0\} \tag{4.70}$$

The element master matrix library now consists of $[K]_e$ with variable conductivity, (4.65), $[H]_e$ for coupled convection, (4.68), and $\{b\}_e$ as generalized data including distributed source and boundary convection, (4.66) and (4.69). These cited equations completely define the linear basis algorithm for *any* discretization $\Omega^h$ of the problem domain $\Omega$.

**STUDY PROBLEM 1.** Determine the steady-state temperature distribution in a slab of thickness $L = 1$ ft subject to a convective thermal load at the left end with $h = 20$ Btu/(h · ft$^2$ · °F) and $T_r = 1500$°F to an accuracy of 0.1°F. The thermal conductivity of the slab material is a linear function of $x$ such that $k(x = 0) = 10$ Btu/(h · ft · °F) while $k(x = L) = 20$ Btu(h · ft · °F). There is no distributed heat source present, and at $x = L$ the slab temperature is held fixed at $T_b = 306.85$°F.

The analytical solution to this study problem is logarithmic in $x$; hence no union of piecewise continuous polynomials of any degree $k$ can reproduce the exact solution. For the given data, the exact solution temperature at the convection surface is $T(x = 0) = 1000.0°\mathrm{F}$. We can use this to compare approximate solution accuracy, the control of which resides in *our choice* of basis degree $k$ and the discretization $\Omega^h$.

**Example 4.4** Obtain the linear basis solution to Study Problem 1 for an $M = 1$ discretization. Hence, $l_{e=1} = L = 1$, $\{K\}_{e=1} = \{10, 20\}^T$ and the algorithm statement (4.70) components are

$$[K]_e = \frac{1}{l_e}\left\{\frac{1}{2}, \frac{1}{2}\right\}\{K\}_e\begin{bmatrix} 1 & -1 \\ -1 & 1 \end{bmatrix} \qquad \text{for } e = 1$$

$$= \frac{1}{1}\left\{\frac{1}{2}, \frac{1}{2}\right\}\begin{Bmatrix} 10 \\ 20 \end{Bmatrix}\begin{bmatrix} 1 & -1 \\ -1 & 1 \end{bmatrix} = \left(\frac{10}{2} + \frac{20}{2}\right)\begin{bmatrix} 1 & -1 \\ -1 & 1 \end{bmatrix}$$

$$= (15)\begin{bmatrix} 1 & -1 \\ -1 & 1 \end{bmatrix}$$

$$= \begin{bmatrix} 15 & -15 \\ -15 & 15 \end{bmatrix}$$

$$[H]_e = h\begin{bmatrix} \delta_{e1} & 0 \\ 0 & \delta_{eM} \end{bmatrix} = \begin{bmatrix} 20 & 0 \\ 0 & 0 \end{bmatrix} \qquad \text{for } e = 1$$

$$\{b\}_e = hT_r\begin{Bmatrix} \delta_{e1} \\ \delta_{eM} \end{Bmatrix} = \begin{Bmatrix} 30{,}000 \\ 0 \end{Bmatrix} \qquad \text{for } e = 1$$

Hence, on imposition of $Q2 = T_b = 306.85$, the finite element statement (4.70) for the $M = 1$ discretization is

$$\begin{bmatrix} 15 + 20 & -15 \\ 0 & 1 \end{bmatrix}\begin{Bmatrix} Q1 \\ Q2 \end{Bmatrix} = \begin{Bmatrix} 30{,}000.0 \\ 306.85 \end{Bmatrix}$$

and the solution is

$$\{Q\} = \begin{Bmatrix} Q1 \\ Q2 \end{Bmatrix} = \begin{Bmatrix} 988.65 \\ 306.85 \end{Bmatrix} °\mathrm{F} \tag{4.71}$$

Recall that the exact temperature at node 1 is $1000.0°\mathrm{F}$. Since $Q1 = 988.65°\mathrm{F}$, the $M = 1$ nodal solution is approximately 1.2 percent underpredicted.

For the interesting information it provides, alter Study Problem 1 by replacing the $x = L$ Dirichlet boundary condition with an adiabatic wall. Then repeat the $M = 1$ formulation and determine the associated solution.

Of course, $[K]_e$ is unchanged as is $\{b\}_e$. The convection matrix becomes

$$[H]_e = h\begin{bmatrix} \delta_{e1} & 0 \\ 0 & \delta_{eM} \end{bmatrix} = 20\begin{bmatrix} 1 & 0 \\ 0 & 0 \end{bmatrix} + 0\begin{bmatrix} 0 & 0 \\ 0 & 1 \end{bmatrix}$$

Since no fixed temperature exists, the terminal matrix statement does not alter the last equation in (4.70). The solution form is

$$\begin{bmatrix} 15+20 & -15 \\ -15 & 15 \end{bmatrix}\begin{Bmatrix} Q1 \\ Q2 \end{Bmatrix} = \begin{Bmatrix} 30{,}000 \\ 0 \end{Bmatrix}$$

The second equation above is directly solvable yielding $Q1 = Q2$. Plugging this into the first equation yields $Q1 = 1500°\text{F} = T_r$. The finite element algorithm has thus predicted that the steady-state slab temperature for an adiabatic wall is uniform and equal to $T_r$, the *exact* analytical solution! The truly important points are (1) the distinction between adiabatic wall and fixed temperature is a matrix row modification, and (2) the *adiabatic* wall boundary condition comes *for free*, i.e., no extra specification is required. These points generalize directly to $n$ dimensions, hence constitute an extremely useful ingredient of the methodology.

Now returning to the study problem at hand, consider the following.

**Example 4.5.** Generate the $M = 2$ uniform mesh finite element solution to Study Problem 1. The data modifications are $l_e = L/2 = 1/2$ and $\{K\} = \{10, 15, 20\}^T$. Thus,

$$\Omega_{e=1}\colon\quad [K]_1 = \frac{1}{l_e}\left\{\frac{1}{2}, \frac{1}{2}\right\}\{K\}_e\begin{bmatrix} 1 & -1 \\ -1 & 1 \end{bmatrix} \qquad \text{for } e = 1$$

$$= \frac{1}{1/2}\left\{\frac{1}{2}, \frac{1}{2}\right\}\begin{Bmatrix} 10 \\ 15 \end{Bmatrix}\begin{bmatrix} 1 & -1 \\ -1 & 1 \end{bmatrix}$$

$$= \begin{bmatrix} 25 & -25 \\ -25 & 25 \end{bmatrix}$$

$$[H]_1 = \begin{bmatrix} 20 & 0 \\ 0 & 0 \end{bmatrix}$$

$$\{b\}_1 = \begin{Bmatrix} 30{,}000 \\ 0 \end{Bmatrix}$$

$$\Omega_{e=2}\colon\quad [K]_2 = \frac{1}{1/2}\left\{\frac{1}{2}, \frac{1}{2}\right\}\begin{Bmatrix} 15 \\ 20 \end{Bmatrix}\begin{bmatrix} 1 & -1 \\ -1 & 1 \end{bmatrix}$$

$$= \begin{bmatrix} 35 & -35 \\ -35 & 35 \end{bmatrix}$$

You should verify that the assembly of the algorithm statement (4.70) is, using the $M = 2$ data and imposing the Dirichlet constraint $Q3 = T_b$

$$\begin{bmatrix} 25+20 & -25 & 0 \\ -25 & 25+35 & -35 \\ 0 & 0 & 1 \end{bmatrix} \begin{Bmatrix} Q1 \\ Q2 \\ Q3 \end{Bmatrix} = \begin{Bmatrix} 30{,}000 \\ 0 \\ 306.85 \end{Bmatrix} \tag{4.72}$$

The solution is

$$\{Q\} = \begin{Bmatrix} Q1 \\ Q2 \\ Q3 \end{Bmatrix} = \begin{Bmatrix} 996.86 \\ 594.36 \\ 306.85 \end{Bmatrix} {}^{\circ}\mathrm{F} \tag{4.73}$$

As expected, the $M = 2$ solution has generated a nodal approximation to the midslab temperature. Of greater significance, the surface temperature prediction has increased *substantially* to 996.86°F, which is only about 0.3 percent in error. Specifically, refining $\Omega^h$ by a factor of 2 has decreased the error by about a factor of 4. As will be developed, the asymptotic convergence theory states that this factor is precisely 4 for the linear basis algorithm operating on meshes $\Omega^h$ of sufficient refinement.

The following exercise suggests that you use *LEARN.FE* to generate the numerical data base necessary to confirm that the error decreases precisely by a factor of four upon halving of the mesh measure ($l_e$).

**Code Exercise 4.2.** The goal is to conduct a numerical experiment to verify that the finite element surface temperature for Study Problem 1 converges to the exact solution in direct proportion to discretization refinement. Your requirement is to access the input file for this exercise, execute the code, and then proceed through the grid refinement study by modification of the input file. On access of the base file, the following will appear on your screen.

```
TITL
***** CODE EXERCISE 4.2,M=2,K=1 *****
TYPE    [K        N      NNODEL   REFL    NPR     NTRAN   NAXI]
         1        1        2       1.0     1
PRIN   [NBUG(*)]
         1        1        1        3
GRID    [FOR N (XL XR(*)PR(*)),NEM(PR*)]
                 1.0      2.0        1.0
         2
MATL    [FOR NPR, KL(*),KR(*)]
         1        10.0     20.0
ROBN    [NROB, NODE(*), H(*), TR(*))]
         1
         1
                 20.0    1500.0
```

```
DIRI    [NDIR, NODE(*), QB(*)]
         1
         3
         306.85282
FORM
SOLV
STOP
```

You are already familiar with the command name functions *TITL*, *TYPE*, *PRIN*, *GRID*, and *STOP*. Under *GRID*, *LEARN.FE* will establish a uniform $M = 2$ discretization of the interval $1.0 \le x \le 2.0$, which contains an origin shift from before but $L$ remains equal to unity. The next command instruction is *MATL* (*MATeriaL*), which defines that the range of thermal conductivity on $\Omega$ is $10 \le k \le 20$. Following is the command name *ROBN* (*ROBiN* boundary condition); the sequential data states that there is one specification (NROB), that it is applied at node number 1, i.e., the left end of $\Omega$ (NODE($*$)), and the values of $h$ and $T_r$ are 20.0 and 1500.0, respectively. The next instruction is *DIRI* (*DIRI*chlet boundary condition); the sequence of data states that one specification exists, it is applied at node 3, i.e., the right end of the $M = 2$ discretization, and the value is 306.85282.

This completes specification of the problem data. Thereupon, the command name *FORM* instructs the code to *form* the matrix statement (4.70), i.e., evaluate and assemble $[K]_e$, $[H]_e$, and $\{b\}_e$ for $1 \le e \le M$. The instruction *SOLV* then calls the equation *solv*er in *LEARN.FE* with the global matrix statement assembled by *FORM*. It returns with the nodal temperature matrix $\{Q\}$ and outputs solution data according to the NBUG($*$) instructions keyed under *PRIN*. *STOP* then returns the code to its initiation status.

After you have accessed this data file, execute the code by pressing RETURN, and after a few seconds the following execution file will appear.

```
TITL
***** CODE EXERCISE 4.2,M=2,K=1*****
TYPE    K       N      NNODEL   REFL    NPR   NTRAN   NAXI
        1       1        2       1.0     1      0      0
PRIN    NBUG(1)       (2)       (3)    (4)    (5)    (6)
                1       1        1      3      0      0
GRID    NI      NPRI       NTRAN
        1       1
        XL      XR1        PR1       XR2       PR2       XR3      PR3
        1.0     2.0        1.0
        NEM1    NEM2     NEM3
        2
```

```
             ELEMENT NODE CONNECTION ARRAY.(MEL)
             ELEMENT NO     NODE1    NODE2  NODE3    NODE4
                  1           1        2
                  2           2        3
             NODE COORDINATES(X,REAL VARIABLE)
                    1.0      1.5      2.0
             NODE COORDINATES(X,INTEGER MAP)EXPONENT E:-2
                    100      150      200

MATL         MACRO NODAL CONDUCTIVITIES
              1      10.0    20.0
             AK:
                    10.00   15.00   20.00
ROBN         NROB      1
             JROB      1
                       H      TREF
                    20.00 1500.0

DIRI         NFIX      1
             JFIX      3
                 306.85282

FORM         SYSTEM MATRIX[K+H]COLUMN FORMAT
                 0.000        45.000  -25.000
               -25.000        60.000  -35.000
                 0.000         1.000    0.000

             DATA MATRIX (b) ROW FORMAT
              30000.00           0.00    306.85

SOLV         NODAL SOLUTION(REAL VARIABLE)
                996.87         594.36    306.85
             NODAL SOLUTION(INTEGER MAP)EXPONENT E:0
                  996     594      306
STOP
```

This execution file presents all intermediate data since most debug print flags are on. The output under *TITL* and *TYPE* needs no further comment, and the NBUG(*) entries under *PRIN* key output as follows. Under *GRID*, the element/node connection array (MEL) has been printed [NBUG(3) = 1], confirming that $\Omega_{e=1}$ has nodes 1 and 2 while $\Omega_{e=2}$ contains nodes 2 and 3. (Later on in Chaps. 5 and 6, the triangle and quadrilateral two-dimensional elements will have entries under the remaining column headings.) The next two data strings contain the mesh nodal coordinates in real variable form (F format) and in a truncated integer print format that suppresses the decimals (for later use). The output under *MATL*, *ROBN*, and *DIRI* is reflection of input data with descriptive names appended.

After *FORM*, the *PRIN* command NBUG(2) = 1 requests output of the assembled system matrix $[K + H]$ and the right-side data matrix $\{b\}$, i.e., the $M = 2$ global statement (4.70). These data arrays are directly comparable, entry by entry, to the hand assembly (4.72) of Example 4.5. A presentation distinction exists, however, for the system matrix $S_e([K]_e + [H]_e)$. To enhance visualization, *LEARN.FE* assembles this matrix using a column format rather than the conventional diagonal matrix format of (4.72). Hence, the entries in the central vertical column of the output $[K + H]$ array correspond to the matrix diagonal entries in (4.72). For the $M = 2$ case only, both matrices are $3 \times 3$. As $M$ increases, the *LEARN.FE* matrix presentation allows screen viewing of quite large assembled matrix statements.

Finally, the solution nodal temperature matrix $\{Q\}$ is output at the end of *SOLV*. *LEARN.FE* provides the data in both full significance real ($E$ and $F$) formats and in a truncated (not rounded) integer format. The latter output form is particularly useful for two-dimensional problems where it appears geometrically similar to the discretization. Note that the code solution is identical to (4.73) to the presented significance.

Hopefully, you have a basic understanding of these input/output processes for *LEARN.FE*. Study Problem 1 asks you to execute a grid sensitivity study to verify that the error in $Q1$ can be reduced to 0.1 degree. In the process, you will also verify that the approximation error decreases by about a factor of 2 times the (uniform) grid refinement. The input file changes are restricted to *TITL*, $M$ (i.e., NEM1) under *GRID*, and the node temperature to be set to $T_b$ under *DIRI*. Specifically, to set up the $M = 4$ solution, the modifications are

```
TITL
***** CODE EXERCISE 4.2,M=4,K=1*****
GRID
                    1.0         2.0    1.0
          4
DIRI
         1
         5
         306.8582
```

Make these changes to the input file for Code Exercise 4.2, execute *LEARN.FE*, and record the solution value for $Q1$. Note also the expanded data array for $\{K\}$, i.e., AK, and the larger system matrix $[K + H]$. Then set up and run the $M = 8$ and $M = 16$ solutions and observe the output data. The nodal data for $Q1$ will produce about a factor of 4 improvement in solution accuracy for each refinement, and the $M = 16$ solution indeed achieves the requested accuracy.

At this point you may well benefit from additional experimentation with *LEARN.FE*. Think up some variations on the problem statement [such as $L$, $k(x)$, $h$, $T_r$, $T_b$], construct the command files and execute a convergence study. You will not know the exact solution, but any well-posed problem approximate solution will exhibit convergence. More details are presented later on this issue.

We now return to the fundamental choice of basis degree and its impact on algorithm performance, especially accuracy.

## Quadratic Basis

Recalling (4.60) as the generic matrix statement, the requirement is to establish the master matrix library for the $k = 2$ implementation. Formation and evaluation of the integrals requires close attention to detail, but the operations are straightforward. The thermal conduction master matrix is

$$
\begin{aligned}
[K]_e &\equiv \int_{\Omega_e} \frac{d\{N_2\}}{dx} k_e(x) \frac{d\{N_2\}^T}{dx}\, dx \\
&= \frac{1}{l_e^2} \int_{\Omega_e} k_e(x) \begin{Bmatrix} \zeta_2 - 3\zeta_1 \\ 4(\zeta_1 - \zeta_2) \\ 3\zeta_2 - \zeta_1 \end{Bmatrix} \{\zeta_2 - 3\zeta_1, 4(\zeta_1 - \zeta_2), 3\zeta_2 - \zeta_1\}\, dx \\
&= \frac{1}{l_e^2} \int_{\Omega_e} k_e(x) \\
&\quad \times \begin{bmatrix} (\zeta_2 - 3\zeta_1)^2 & (\zeta_2 - 3\zeta_1)4(\zeta_1 - \zeta_2) & (\zeta_2 - 3\zeta_1)(3\zeta_2 - \zeta_1) \\ 4(\zeta_1 - \zeta_2)(\zeta_2 - 3\zeta_1) & 16(\zeta_1 - \zeta_2)^2 & 4(\zeta_1 - \zeta_2)(3\zeta_2 - \zeta_1) \\ (3\zeta_2 - \zeta_1)(\zeta_2 - 3\zeta_1) & (3\zeta_2 - \zeta_1)4(\zeta_1 - \zeta_2) & (3\zeta_2 - \zeta_1)^2 \end{bmatrix} dx
\end{aligned}
\tag{4.74}
$$

In distinction to the linear basis expression (4.63), observe that its evaluation involves products of $k_e(x)$ with the $3 \times 3$ matrix of quadratic polynomials in $\zeta_i$. Hence, variable thermal conductivity will not be implemented as the simple element average that occurred in the $k = 1$ formulation. However, all we need do is specify the form for (4.61) and proceed through the calculus details.

To illustrate use of mixed-degree approximations, and since $k(x)$ is usually of rather mild variation, let's assume the interpolation for $k_e(x)$ can utilize the linear basis for (4.61). Therefore, (4.74) becomes

$$[K]_e = \frac{1}{l_e^2}\int_{\Omega_e} k_e[\cdot]\,dx \equiv \frac{1}{l_e^2}\int_{\Omega_e} \{N_1\}^T\{K\}_e[\cdot]\,dx$$

$$= \frac{\{K\}_e^T}{l_e^2}\int_{\Omega_e} \{N_1\}[\cdot]\,dx \tag{4.75}$$

where $[\cdot]$ denotes the $3 \times 3$ square matrix detailed in (4.74). The rearrangements in the second line of (4.75) recognize that the matrix row–column product $\{\cdot\}^T\{\cdot\}$ is a scalar, and hence equals its transpose, and that the elements of $\{K\}_e$ are independent of $\zeta_i$, and hence can be extracted from under the integral.

The matrix expression (product) constituting the integrand in (4.75) is not of standard form. However, all it states is that every element of the $3 \times 3$ matrix $[\cdot]$ is multiplied by the array $\{\cdot\}$. If this array was a scalar, say, $\alpha$, then $\alpha[\cdot]$ is a matrix every element of which is $\alpha$ times the entries in $[\cdot]$. Hence, $\{\cdot\}[\cdot]$ is also a matrix, every element of which is an array of length equal to that of $\{\cdot\}$. Therefore, for example, expanding the (1, 1) term in the integral in (4.75) yields

$$\int_0^{l_e}\{N_1\}\begin{bmatrix}(\zeta_2 - 3\zeta_1)^2 & \cdots \\ \cdots\cdots & \cdots\cdots \\ (\text{sym}) & \cdots\end{bmatrix}dx = \int_0^{l_e}\begin{Bmatrix}\zeta_1 \\ \zeta_2\end{Bmatrix}\begin{bmatrix}(\zeta_2 - 3\zeta_1)^2 & \cdots \\ \cdots\cdots & \cdots\cdots \\ (\text{sym}) & \cdots\end{bmatrix}dx$$

$$= \int_0^{l_e}\begin{bmatrix}\begin{Bmatrix}\zeta_1(\zeta_2 - 3\zeta_1)^2 \\ \zeta_2(\zeta_2 - 3\zeta_1)^2\end{Bmatrix} & \cdots \\ \cdots\cdots & \cdots\cdots \\ (\text{sym}) & \cdots\end{bmatrix}dx$$

$$= \frac{l_e}{6}\begin{bmatrix}\begin{Bmatrix}11 \\ 3\end{Bmatrix} & \cdots \\ \cdots\cdots & \cdots\cdots \\ (\text{sym}) & \cdots\end{bmatrix} \tag{4.76}$$

using (4.40) to evaluate the second-line terms. It is suggested that you verify that the $k = 2$ basis master matrix (4.75) is

$$[K]_e = \frac{\{K\}_e^T}{6l_e} \begin{bmatrix} \begin{Bmatrix} 11 \\ 3 \end{Bmatrix} & \begin{Bmatrix} -12 \\ -4 \end{Bmatrix} & \begin{Bmatrix} 1 \\ 1 \end{Bmatrix} \\ & \begin{Bmatrix} 16 \\ 16 \end{Bmatrix} & \begin{Bmatrix} -4 \\ -12 \end{Bmatrix} \\ \text{(sym)} & & \begin{Bmatrix} 3 \\ 11 \end{Bmatrix} \end{bmatrix} \tag{4.77}$$

**Exercise 4.1.** Verify that (4.77) reduces to (4.43) under the restriction that $k_e(x)$ is the uniform constant $k$. In this instance, each entry in $\{K_e\}$ is $k$, hence $\{K\}_e^T = \{k, k\} = k\{1, 1\}$. Inserting this into (4.77) then yields

$$[K]_e = \frac{k\{1, 1\}}{6l_e}[\cdot] = \frac{k}{6l_e} \begin{bmatrix} 11 + 3 & -12 - 4 & 1 + 1 \\ & 16 + 16 & -4 - 12 \\ \text{(sym)} & & 3 + 11 \end{bmatrix}$$

$$= \frac{k}{3l_e} \begin{bmatrix} 7 & -8 & 1 \\ -8 & 16 & -8 \\ 1 & -8 & 7 \end{bmatrix} \quad \text{QED.}$$

The completion of $[K]_e$ in (4.77) illustrates how functional complications (including nonlinearities) are handled by the finite element method. The product of terms that are spatially variable always produces forms such as (4.75), the evaluation of which is always direct on extension of conventional matrix nomenclature. Of course, the inner product $\{\cdot\}^T\{\cdot\}$ that produces a scalar in (4.77) must always be completed prior to any operation on the matrix $[\cdot]$.

Evaluation of the distributed source term for the $\{N_2\}$ basis implementation is quite direct, since it does not involve any algebraic nonlinearity. Using (4.40), you should verify an entry or two in the following form.

$$\{b\}_e = \int_{\Omega_e} \{N_2\} s_e(x)\, dx = \int_0^{l_e} \{N_2\}\{N_2\}^T\, dx \{S\}_e$$

$$= \frac{l_e}{30} \begin{bmatrix} 4 & 2 & -1 \\ 2 & 16 & 2 \\ -1 & 2 & 4 \end{bmatrix} \{S\}_e \tag{4.78}$$

In (4.78), we assume that $s_e(x)$ is interpolated by the quadratic basis. Conversely, a linear basis interpolation could certainly be chosen, whereupon $[\cdot]$ in (4.78) becomes a $3 \times 2$ nonsquare matrix, to match the two nodal variables in $\{S\}_e$ for $k = 1$. This simplification is not recommended, however, since a source term can be sharply varying, hence solution accuracy could become compromised by interpolation error.

The final contribution to the algorithm statement (4.60) is the boundary thermal convection term. For initial development, assume it to be located at $x = x_L$; hence

$$\int_{\partial\Omega_{e=1}} \{N_2\} h(\{N_2\}^T\{Q\}_e - T_r)\, d\sigma = \int_{\partial\Omega_{e=1}} \begin{Bmatrix} 1 \\ 0 \\ 0 \end{Bmatrix} h(\{1, 0, 0\}\{Q\}_e - T_r)\, d\sigma$$

$$= h\left(\begin{bmatrix} 1 & 0 & 0 \\ 0 & 0 & 0 \\ 0 & 0 & 0 \end{bmatrix}\{Q\}_e - T_r\begin{Bmatrix} 1 \\ 0 \\ 0 \end{Bmatrix}\right)\delta_{e1} \tag{4.79}$$

since only the first entry in $\{N_2\}$ is nonvanishing at $XL_{e=1}$.

A very striking conclusion can now be made by comparing (4.79) with the linear basis form (4.67). Specifically, the finite element algorithm predicts that imposition of the general nonhomogeneous Robin boundary condition is *absolutely unaffected* by the choice of basis degree for $T^h$ on $\Omega^h$! On some thought, this is exactly what should occur, since boundary conditions are totally independent of the domain differential equation; hence its discrete approximation. (This is a truly significant distinction from finite difference methods, which use inward interpolation of $T^h$ to estimate the boundary normal derivative.) Hence, replacing the 1's by $\delta_{e1}$, and generalizing to either end of $\Omega$, the $\{N_2\}$ basis boundary convection master matrix expressions are

$$[H]_e = h\begin{bmatrix} \delta_{e1} & 0 & 0 \\ 0 & 0 & 0 \\ 0 & 0 & \delta_{eM} \end{bmatrix} \tag{4.80}$$

$$\{b\}_e = hT_r\begin{Bmatrix} \delta_{e1} \\ 0 \\ \delta_{eM} \end{Bmatrix} \tag{4.81}$$

Equations (4.77), (4.78), (4.80), and (4.81) constitute the element master matrix library for the quadratic basis implementation for the generalized heat transfer problem in (4.55) to (4.57). The only requirement now is to define the discretization $\Omega^h$, and assemble and solve the global matrix statement

$$S_e(([K]_e + [H]_e)\{Q\}_e - \{b(s, h, T_r)\}_e) = \{0\} \tag{4.70}$$

**Example 4.6.** Generate the quadratic basis solution to Study Problem 1 for an $M = 1$ discretization of $0 \le x \le L = 1$. Hence, $l_{e=1} = L = 1$, $\{K\}_{e=1} = \{10, 20\}^T$, and the algorithm statement components are

$$[K]_e = \frac{\{K\}_e^T}{6l_e}[\cdot] \qquad \text{for } e = 1$$

$$= \frac{\{10, 20\}}{(6)(1)} \begin{bmatrix} \begin{Bmatrix} 11 \\ 3 \end{Bmatrix} & \begin{Bmatrix} -12 \\ -4 \end{Bmatrix} & \begin{Bmatrix} 1 \\ 1 \end{Bmatrix} \\ & \begin{Bmatrix} 16 \\ 16 \end{Bmatrix} & \begin{Bmatrix} -4 \\ -12 \end{Bmatrix} \\ \text{(sym)} & & \begin{Bmatrix} 3 \\ 11 \end{Bmatrix} \end{bmatrix}$$

$$= \frac{10}{6} \begin{bmatrix} 17 & -20 & 3 \\ & 48 & -28 \\ \text{(sym)} & & 25 \end{bmatrix}$$

$$[H]_e = h \begin{bmatrix} \delta_{e1} & 0 & 0 \\ 0 & 0 & 0 \\ 0 & 0 & \delta_{eM} \end{bmatrix} = \begin{bmatrix} 20 & 0 & 0 \\ 0 & 0 & 0 \\ 0 & 0 & 0 \end{bmatrix} \qquad \text{for } e = 1$$

$$\{b\}_e = hT_r \begin{Bmatrix} \delta_{e1} \\ 0 \\ \delta_{eM} \end{Bmatrix} = \begin{Bmatrix} 30{,}000 \\ 0 \\ 0 \end{Bmatrix} \qquad \text{for } e = 1$$

Forming (4.70) with these data, clearing the fraction and modifying the last equation to enforce the Dirichlet condition $Q3 = T_b = 306.85$ then yields the global matrix statement

$$\begin{bmatrix} 17 + 12 & -20 & 3 \\ -20 & 48 & -28 \\ 0 & 0 & 1 \end{bmatrix} \begin{Bmatrix} Q1 \\ Q2 \\ Q3 \end{Bmatrix} = \begin{Bmatrix} 18{,}000 \\ 0 \\ 306.85 \end{Bmatrix} \tag{4.82}$$

the solution of which produces

$$\{Q\} = \begin{Bmatrix} Q1 \\ Q2 \\ Q3 \end{Bmatrix} = \begin{Bmatrix} 999.65 \\ 595.52 \\ 306.85 \end{Bmatrix} {}^\circ\text{F} \tag{4.83}$$

Recalling that the exact solution for $Q1$ is 1000.0°F, and comparing (4.83) with the $M = 2$ linear basis solution (4.73), yields the conclusion that the $M = 1$ quadratic basis solution accuracy is a truly notable improvement. The error in $Q1$ is only 0.03 percent, an order of magnitude smaller than the $k = 1$, $M = 2$ result. We had to pay the price of forming the $k = 2$

basis master matrix library, but a direct benefit was reaped for *no additional* computational effort; in other words, the matrix ranks of (4.82) and (4.72) are identical. Recall that, from the data produced by Code Exercise 4.2, an $M = 8$ discretization was required to predict $Q1$ to 0.02 percent accuracy using the $k = 1$ formulation. The $k = 2$ algorithm yields comparable accuracy for $M = 1$.

It is worthwhile to evaluate the importance of accurate evaluation of $[K]_e$, and hence the somewhat complicated form and code handling of (4.77). If we assume that the simple element-average value $\bar{k}_e$ of $k_e$ is adequate, then (4.77) takes the simplified form (denoted $[\bar{K}]_e$).

$$[\bar{K}]_e \equiv \bar{k}_e \int_{\Omega_e} \frac{d\{N_2\}}{dx}\frac{d\{N_2\}^T}{dx}\,dx$$

$$= \frac{\bar{k}_e}{3l_e}\begin{bmatrix} 7 & -8 & 1 \\ -8 & 16 & -8 \\ 1 & -8 & 7 \end{bmatrix}$$

The important assessment is the following.

**Example 4.7.** Generate the $M = 1$ quadratic basis solution for the study problem using $[\bar{K}]_e$ rather than $[K]_e$. Thus

$$[\bar{K}]_e = \frac{15}{3 \times 1}[\,\cdot\,] = 5\begin{bmatrix} 7 & -8 & 1 \\ -8 & 16 & -8 \\ 1 & -8 & 7 \end{bmatrix}$$

The terms in $[H]_e$ and $\{b\}_e$ are unchanged. The replacement for (4.82) is

$$\begin{bmatrix} 7+4 & -8 & 1 \\ -8 & 16 & -8 \\ 0 & 0 & 1 \end{bmatrix}\begin{Bmatrix} Q1 \\ Q2 \\ Q3 \end{Bmatrix} = \begin{Bmatrix} 6{,}000 \\ 0 \\ 306.85 \end{Bmatrix}$$

the solution of which is

$$\{Q\} = \begin{Bmatrix} Q1 \\ Q2 \\ Q3 \end{Bmatrix} = \begin{Bmatrix} 988.65 \\ 647.75 \\ 306.85 \end{Bmatrix} {}^{\circ}\mathrm{F} \tag{4.84}$$

Comparing (4.84) and (4.83) confirms that a very serious loss in solution accuracy for $Q1$ has resulted from using the element-average approximation for conductivity. In fact, the $Q1$ solution is now identical to the linear basis $M = 1$ result, (4.71), and it required twice as much work (global matrix rank) to obtain. The inaccuracy results directly from *data interpolation error*. This error will decrease in significance quite rapidly as the discretization is refined. However, it will always be present at some level to pollute the high-order accurate quadratic (or cubic) basis solution. Since the investment in forming $[K]_e$ accurately is a one-time event, there appears to be no need to willfully commit interpolation error for one-dimensional algorithm formulations.

The previous two examples lead naturally to a code exercise for evaluation of the rate of accuracy improvement of the quadratic basis algorithm with discretization refinement. The asymptotic theory predicts that the factor should be about 16 for refining the mesh measure $l_e$ uniformly by a factor of 2.

**Code Exercise 4.3.** The goal is to generate the numerical database for Study Problem 1 needed to confirm quadratic basis solution accuracy improvement with grid refinement. Call up the corresponding input file, whereupon your screen display will be

```
TITL
***** CODE EXERCISE 4.3,M=2,K=2*****

TYPE   [K      N     NNODEL    REFL    NPR    NTRAN    NAXI]
        2      1       3        1.0     1
PRIN   [NBUG(*)]
        1        1        1        3

GRID   [FOR N(XL,XR(*),PR(*)), NEM(PR*)]
              1.0      2.0       1.0
        2
MATL   [FOR NPR(KL(*),KR(*))]
        1      10.0     20.0
ROBN   [NROB, NODE(*),(H(*),TR(*))]
        1
        1
             20.0    1500.0
DIRI   [NDIR, NODE(*), QB(*)]
        1
        5
        306.85282

FORM
SOLV
STOP
```

Under command *TYPE*, the first integer is now a 2, defining choice of the $k = 2$ basis for the solution. Further, NNODEL = 3 signifies there are now three nodes in the element. The last entry under *GRID* specifies an $M = 2$ element discretization. Thus, the temperature node at $x_R$ of $\Omega$ is now number 5; the corresponding entry has been made under *DIRI*.

Execute this file, and the following output file will appear on your monitor.

```
TITL
***** CODE EXERCISE 4.3,M=2,K=2 *****
TYPE     K        N      NNODEL    REFL    NPR    NTRAN   NAXI
         2        1      3          1.0    1        0       0
PRIN NBUG(1)    (2)      (3)       (4)      (5)     (6)
         1        1        1         3        0       0

GRID     NI       NPRI     NTRAN
         1        1
         XL       XR1      PR1      XR2      PR2      XR3      PR3
         1.0      2.0      1.0
         NEM1     NEM2     NEM3
         2

         ELEMENT NODE CONNECTION ARRAY (MEL)
         ELEMENT NO.       NODE1     NODE2     NODE3     NODE4
              1              1         2         3
              2              3         4         5

         NODE COORDINATES(X,REAL VARIABLE)
         1.00     1.25     1.50     1.75     2.0

         NODE COORDINATES(X,INTEGER MAP)EXPONENT E:-2
         100     125     150     175     200

MATL     MACRO NODAL CONDUCTIVITIES
          1       10.00       20.00
         AK:
         10.00    12.50    15.00    17.50    20.00
ROBN     NROB       1
         JROB       1
                    H       TREF
                 20.00 1500.0
DIRI     NFIX      1
         JFIX      5
             306.85282
FORM     SYSTEM MATRIX      [K+H],  COLUMN FORMAT
         0.00      0.00      71.67     -60.00      8.33
         0.00    -60.00     133.33     -73.33      0.00
         8.33    -73.33     140.00     -86.67     11.67
         0.00    -86.67     186.67    -100.00      0.00
         0.00      0.00       1.00       0.00      0.00

         DATA MATRIX (b), ROW FORMAT
    30000.00       0.00       0.00       0.00    306.85
SOLV     NODAL SOLUTION(REAL VARIABLE)
          999.97     776.98     594.54     440.42     306.85

          NODAL SOLUTION(INTEGER MAP)EXPONENT E:0
          999     776     594     440     306
STOP
```

The debug output from *GRID*, accessed by NBUG(3) = 1, displays the node-element connection table (MEL) for this discretization as well as the mesh nodal array $\{X\}$ for $k = 2$ and $M = 2$ in both real and integer formats. Further, NBUG(2) = 1 has requested the assembled system matrix printout. The output from *FORM* shows the pentadiagonal system matrix $[K + H]$ in column format. Under *SOLV*, the nodal temperature solution displayed in integer form (INTEGER MAP) provides a quick and easy view, and the real data confirm that $Q1 = 999.97°\text{F}$, which meets the error requirement for Study Problem 1.

Recall this input file, change *PRIN* data to NBUG(4) = 1 and reexecute to produce nodal temperature output in $E$ format. Then sequentially change *GRID* data to NEM1 = 4, 8, and 16, ie., uniform $M = 4$, 8, and 16 element discretizations, and execute the code. Tabulate the output data for $Q1$, which should agree with that shown in Table 4.1. Hence, note indeed that for $2 \leq M \leq 8$, the decrease in error for $Q1$ is about a factor of sixteen when halving the measure $l_e$ of a uniform mesh. Note also that the $M = 16$ prediction fails to agree with this trend. In fact, $Q1$ is predicted to exceed the exact solution!

The new (dominating) error mechanism is lack of PC precision in the arithmetic solution operations, which will be further discussed after developing the cubic basis formulation. The second set of data in Table 4.1 is for double-precision calculations, whereupon the $M = 16$ solution indeed obeys the error reduction factor.

## Cubic Basis

The next step is to implement the algorithm statement (4.60) for the cubic basis $\{N_3(\zeta_i)\}$. The element thermal conductivity matrix, and its ensuing

**TABLE 4.1**
**Quadratic basis prediction of $Q1$ and its error**

| | Single precision | | Double precision | |
|---|---|---|---|---|
| $M$ | $Q1$ | Error | $Q1$ | Error |
| 1 | 999.65 | 0.35 | | |
| 2 | 999.9703 | 0.0297 | | |
| 4 | 999.9982 | 0.0018 | | |
| 8 | 999.9999 | 0.0001 | 999.999868 | 0.000132 |
| 16 | 1000.004 | 0.004 | 999.999992 | 0.000008 |

evaluation assuming $k_e(x) = \{N_1\}^T\{K\}_e$ is, using integration formula (4.40)

$$[K]_e \equiv \int_{\Omega_e} \frac{d\{N_3\}}{dx} k_e(x) \frac{d\{N_3\}^T}{dx} dx$$

$$= \frac{81\{K\}_e^T}{4l_e^2} \int_0^{l_e} \begin{Bmatrix} \zeta_1 \\ \zeta_2 \end{Bmatrix} \begin{Bmatrix} -3\zeta_2^2 + 4\zeta_2 - 11/9 \\ 9\zeta_2^2 - 10\zeta_2 + 2 \\ -9\zeta_2^2 + 8\zeta_2 - 1 \\ 3\zeta_2^2 - 2\zeta_2 + 2/9 \end{Bmatrix} \{\cdot\}^T d\bar{x}$$

$$= \frac{\{K\}_e^T}{80l_e} \begin{bmatrix} \begin{Bmatrix} 262 \\ 34 \end{Bmatrix} & \begin{Bmatrix} -327 \\ -51 \end{Bmatrix} & \begin{Bmatrix} 78 \\ 30 \end{Bmatrix} & \begin{Bmatrix} -13 \\ -13 \end{Bmatrix} \\ & \begin{Bmatrix} 594 \\ 270 \end{Bmatrix} & \begin{Bmatrix} -297 \\ -297 \end{Bmatrix} & \begin{Bmatrix} 30 \\ 78 \end{Bmatrix} \\ & & \begin{Bmatrix} 270 \\ 594 \end{Bmatrix} & \begin{Bmatrix} -51 \\ -327 \end{Bmatrix} \\ \text{(sym)} & & & \begin{Bmatrix} 34 \\ 262 \end{Bmatrix} \end{bmatrix} \tag{4.85}$$

The corresponding (tedious) exercise produces the distributed source term master matrix

$$\{b\}_e \equiv \int_{\Omega_e} \{N_3\} s_e \, dx = \int_{\Omega_e} \{N_3\}\{N_3\}^T dx \{S\}_e$$

$$= \frac{l_e}{1680} \begin{bmatrix} 128 & 99 & -36 & 19 \\ 99 & 648 & -81 & -36 \\ -36 & -81 & 648 & 99 \\ 19 & -36 & 99 & 128 \end{bmatrix} \{S\}_e \tag{4.86}$$

Finally, recalling the $k = 2$ basis implementation for the thermal convection boundary condition, it should come as no surprise that the cubic basis form is identical except for using a rank 4 matrix to match the order of $\{Q\}_e$. Thus

$$[H]_e \equiv h \begin{bmatrix} \delta_{e1} & 0 & 0 & 0 \\ & 0 & 0 & 0 \\ & & 0 & 0 \\ \text{(sym)} & & & \delta_{eM} \end{bmatrix} \tag{4.87}$$

$$\{b\}_e = hT\begin{Bmatrix} \delta_{e1} \\ 0 \\ 0 \\ \delta_{eM} \end{Bmatrix} \tag{4.88}$$

**Example 4.8.** Generate the $M = 1$ solution for Study Problem 1 using the cubic basis. Thus, $l_e = L = 1$, $\{K\}_e = \{10, 20\}^T$ and using (4.85)

$$[K]_e \equiv \frac{\{K\}_e^T}{80 l_e}[\cdot] = \frac{\{10, 20\}_e}{80 \times 1}[\cdot]$$

$$= \frac{1}{8}\begin{bmatrix} 330 & -429 & 138 & -39 \\ & 1134 & -891 & 186 \\ & & 1458 & -675 \\ (\text{sym}) & & & 558 \end{bmatrix}$$

The evaluation of $[H]_e$ and $\{b\}_e$, (4.87) and (4.88), is unchanged; hence, the global matrix statement is

$$\begin{bmatrix} 330+160 & -429 & 138 & -39 \\ -429 & 1134 & -891 & 186 \\ 138 & -891 & 1458 & -675 \\ 0 & 0 & 0 & 1 \end{bmatrix}\begin{Bmatrix} Q1 \\ Q2 \\ Q3 \\ Q4 \end{Bmatrix} = \begin{Bmatrix} 240{,}000 \\ 0 \\ 0 \\ 306.85282 \end{Bmatrix} \tag{4.89}$$

and its solution yields

$$\{Q\} = \begin{Bmatrix} 999.9893 \\ 712.1133 \\ 488.9065 \\ 306.8528 \end{Bmatrix} {}^\circ\text{F} \tag{4.90}$$

Viewing (4.90), and recalling that for the exact solution $T(x = 0) = 1000.0^\circ$F, the accuracy of the $M = 1$ cubic basis solution is truly remarkable, i.e., $Q1 = 999.9893^\circ$F meets the accuracy requirement for $M = 1$! The following code exercise suggests that you develop the $M = 2$ solution, whereupon $Q1 = 999.9998^\circ$F will result. However, when you proceed to the $M = 4$ solution, you will determine that $Q1$ exceeds $1000.0^\circ$F by an amount larger than the error in the $M = 2$ solution. The inconsistency is not in the theory, which states that the solution accuracy will improve by a factor of about 64 for halving the uniform mesh measure $l_e$. The root cause is that the precision of the PC computer is inadequate for solving (4.70) for the cubic basis implementation.

Recalling Chap. 2, the current-generation PC computer word length is 16 to 32 bits, which yields roughly six to nine significant digits. This is inadequate for the $k = 2$ and $k = 3$ formulation global matrix calculations for $M > 8$ and $M > 2$, respectively. A double-precision calculation would reduce numerical round-off error to a nonsignificant level, and *LEARN.FE* can be edited to operate in double precision. You may so alter the code, if you care to, but then computational speed will decrease substantially as will maximum problem size capacity.

**Code Exercise 4.4.** Use *LEARN.FE* to generate the numerical data base for Study Problem 1, to confirm improvement of the cubic basis solution accuracy for $M = 2$. Call up the corresponding input file, whereupon your screen display will be

```
TITL
***** CODE EXERCISE 4.4,M=2,K=3 *****

TYPE  [K       N       NNODEL  REFL    NPR    NTRAN   NAXI]
       3       1         4       1.0    1
PRIN  [NBUG(*)]
       1       1         1          3

GRID  [FOR N(XL,XR(*), XL(*)),  NEM(PR*)]
               1.0      2.0        1.0
       2
MATL  [FOR NPR(KL(*), KR(*))]
       1       10.0     20.0
ROBN  [NROB, NODE(*), (H(*), TR(*))]
       1
       1
               20.0  1500.0
DIRI  [NDIR, NODE(*), QB(*)]
       1
       7
       306.85282
FORM
SOLV
STOP
```

Execute this file, and hence generate the $M = 2$ cubic basis solution output file. Note in particular the $k = 3$ basis alterations to the MEL array, and to the bandwidth of the *FORM* system matrix, which is given below.

```
FORM SYSTEM MATRIX [K+H], COLUMN FORMAT

   0.00    0.00     0.00     98.25  -100.88    30.75    -8.13
   0.00    0.00  -100.88    249.75  -185.63    36.75     0.00
   0.00   30.75  -185.63    290.25  -135.38     0.00     0.00
  -8.13   36.75  -135.38    222.00  -148.13    44.25   -11.38
   0.00    0.00  -148.13    357.75  -259.88    50.25     0.00
   0.00   44.25  -259.88    398.25  -182.63     0.00     0.00
   0.00    0.00     0.00      1.00     0.00     0.00     0.00

DATA MATRIX(b), ROW FORMAT

30000.00    0.00    0.00      0.00      0.00      0.00   306.85
```

## 4.6 ACCURACY AND CONVERGENCE FORMALIZED

We now summarize the results of the computational experiments you have completed, and hence firm up measured aspects of accuracy and convergence of the finite element approximate solution as the mesh $\Omega^h$ is refined. Your generated $Q1$ temperature error data for uniform mesh refinement are plotted in Fig. 4.6. The linear basis data agree very well with the theoretical quadratic convergence rate for $1 \leq M \leq 16$; i.e., on a log-log plot these data lie parallel to a straight line of slope equal to 2. The quadratic basis data for $2 \leq M \leq 8$ also agree well with the theory slope of 4. The coarsest $M = 1$ mesh solution error lies below the line, and round-off error pollutes all solutions for $M > 8$. The cubic basis data agree with theory for $1 \leq M \leq 2$ only and for $M = 4$ when using double precision.

When not affected by round-off, the fact that these data are well interpolated by straight lines on a log-log plot indicates the functional form of the approximation error is

$$\text{error}^h \simeq C_k l_e^{2k} \tag{4.91}$$

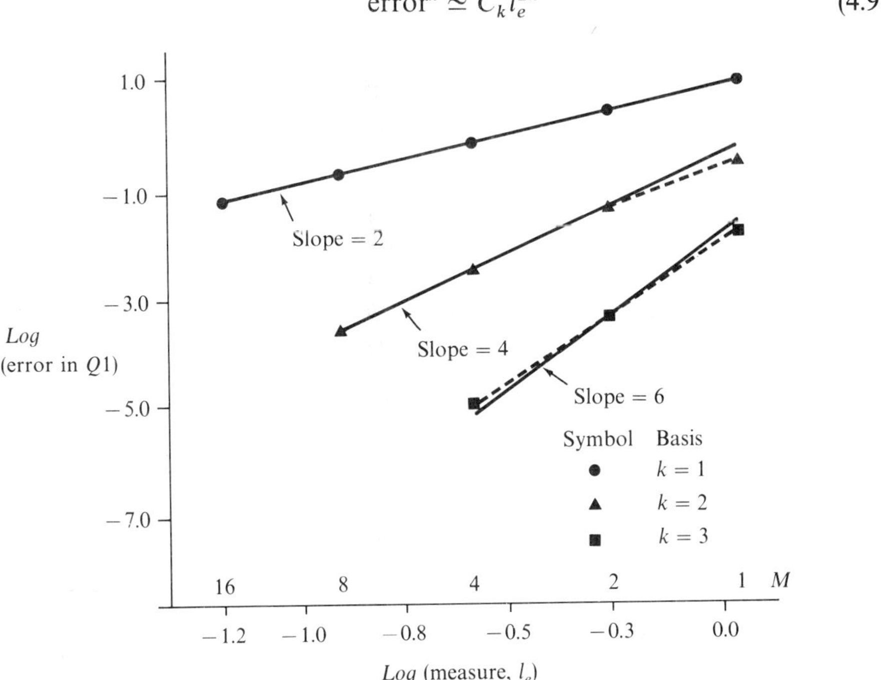


**FIGURE 4.6**
*Log-log* plot of finite element solution temperature error in $Q1$ versus uniform mesh measure $l_e$.

Here, $k$ is the finite element basis degree and $C_k$ is a constant which is independent of the element measure $l_e$ but does depend on basis degree $k$. Specifically, $C_k$ decreases with $k$ since the error is relatively smaller for any given $\Omega^h$, that is, $l_e$, for higher basis degree, see Fig. 4.6.

The existence of relation (4.91) is of singular importance, since it provides the means to *quantitatively* estimate the error in an approximate solution. To establish Fig. 4.6, hence (4.91), we utilized knowledge of the exact solution value for $Q1$. However, the mathematical asymptotic convergence theory for linear elliptic boundary value problems independently predicts (4.91) on a sufficiently refined mesh $\Omega^h$ (Oden and Reddy, 1976, Chap. 8). Therefore, an error estimation procedure can function *without* knowing the exact solution, but instead using solutions constructed on uniform mesh refinements. Specifically, for one mesh and its double refinement $\Omega^h$ and $\Omega^{h/2}$, the corresponding solutions $T^h$ and $T^{h/2}$ satisfy the definition (3.7):

$$T^h + e^h = T_{\text{exact}} = T^{h/2} + e^{h/2} \tag{4.92}$$

Substituting $l_e = h$ and $l_e = h/2$ into (4.91), you may easily verify that

$$e^h = (2^{2k})e^{h/2} \tag{4.93}$$

provided the solutions $T^h$ and $T^{h/2}$ lie on the convergence curve. The key feature of (4.93) is that it is independent of the (unknown problem-dependent) constant $C_k$ present in (4.91). Substituting (4.93) into (4.92) then yields

$$T^{h/2} - T^h = (2^{2k} - 1)e^{h/2} \tag{4.94}$$

Letting $\Delta T^{h/2} \equiv T^{h/2} - T^h$ denote the computed difference in the approximate solutions on the two meshes, the approximate error in the finer grid solution is

$$e^{h/2} = \frac{\Delta T^{h/2}}{2^{2k} - 1} \tag{4.95}$$

Equation (4.95) is a *quantitative error estimate* independent of $C_k$, and valid anywhere on $\Omega$ provided the approximate solutions $T^h$ and $T^{h/2}$ lie on the convergence curve [i.e., (4.91)]. You may easily *confirm* whether the "adequate mesh" requirement is satisfied by evaluating (4.95) with two pairs of refined mesh solutions, i.e., for $\Omega^h$, $\Omega^{h/2}$, and $\Omega^{h/4}$. Then, if the resultant two data points $e^{h/2}$ and $e^{h/4}$ lie on a straight line of slope $2k$ on a log-log plot, all solutions $T^h$, $T^{h/2}$, and $T^{h/4}$ satisfy the adequate mesh requirement. The following exercise illustrates the procedure.

**Example 4.9.** Estimate the uniform discretization $\Omega^{h/M}$ required for the $k = 1$ basis finite element approximate solution error for Study Problem 1 to be 0.1 at the left end of the domain $\Omega$. Further, estimate the exact solution temperature at this location. The needed data are available from your computational experiments and the resultant evaluations of (4.95) and (3.7) produce the data set summarized in Table 4.2.

Plot the first two entries in the right column of Table 4.2 on the *log-log* graph $e^{h/M}$ versus $l_e$. Then confirm that they *do not* lie on a line of slope equal to 2. Therefore, the $\Omega^h$ mesh is too coarse to produce an approximate solution $T^h$ that satisfies (4.91) or can be used in (3.7) to estimate $T_{\text{exact}}$. Then plot the second two data points, to confirm they indeed lie close to the required line slope. Thus, $T^{h/2}$, $T^{h/4}$, and $T^{h/8}$ each satisfies (4.91) and the associated meshes $\Omega^{h/2}$, $\Omega^{h/4}$, and $\Omega^{h/8}$ are "adequate." You may thus safely extrapolate the data to predict that the discretization $\Omega^{h/M}$ for $M \simeq 12$ meets the accuracy requirement. Of even greater utility, the relatively coarse mesh $\Omega^{h/4}$ solution and its error estimate (4.95) permit prediction of $T_{\text{exact}}$ to well within the requirement.

A slight theoretical flaw exists in this development to which the purist should take exception. Specifically, (4.91) may not always hold at every location $x$ within $\Omega$, especially where material properties change abruptly, e.g., a thermal conductor-insulator interface, or on a boundary. However, (4.91) will be consistently reliable if one selects the mathematician's expression for "error" which elegantly includes all such issues. For elliptic boundary value problems, one such *error measure* is the "energy" in the solution. The "energy seminorm" is the proper name, and for the weak statement finite element approximation (4.70), to the problem statement (4.55) to (4.57), its definition is

$$\|T^h\|_E \equiv \tfrac{1}{2} \sum_e \left(\{Q\}_e^T [K + H]_e \{Q\}_e\right) \tag{4.96}$$

For the approximation error measured in the energy seminorm, i.e.,

**TABLE 4.2**
**Linear basis solution accuracy assessments**

| Mesh | $M$ | $l_e$ | $Q1$ | $\Delta Q1$ | $e^{h/2}$, (4.95) | $T_{\text{exact}}$(est.), (3.7) |
|---|---|---|---|---|---|---|
| $\Omega^h$ | 1 | 1 | 988.65 | | | |
| $\Omega^{h/2}$ | 2 | 0.5 | 996.8657 | 8.2157 | 2.7386 | 999.6043 |
| $\Omega^{h/4}$ | 4 | 0.25 | 999.1910 | 2.3253 | 0.7751 | 999.9651 |
| $\Omega^{h/8}$ | 8 | 0.125 | 999.7962 | 0.6051 | 0.2017 | 999.9979 |
| $\Omega^{h/16}$ | 16 | 0.0625 | 999.9487 | 0.1529 | 0.0510 | 999.9997 |
| $\Omega^{h/32}$ | 32 | 0.03125 | 999.9871 | 0.0381 | 0.0127 | 999.9998 |

$\|e^h\|_E \equiv \|T\|_E - \|T^h\|_E$, the revised form of the estimate (4.95) becomes

$$\|e^{h/2}\|_E = \frac{\Delta\|T^{h/2}\|_E}{2^{2k} - 1} \tag{4.95a}$$

Equation (4.96) is easy to evaluate, since the element master matrices $[K]_e$ and $[H]_e$ already exist as well as the DO loop in FORMIT to sum the indicated matrix products with the nodal solution array $\{Q\}_e$. *LEARN.FE* handles the computation of $\|T^h\|_E$ via the command name *NORM*. The following code exercise illustrates its use to refine error estimation.

**Code Exercise 4.5.** Call up the input data file for Code Exercise 4.2 for the $k = 1$ basis analysis of Study Problem 1. Insert command *NORM* after *SOLV* and reexecute the $1 \leq M \leq 16$ uniform mesh solutions after changing NBUG(2) and NBUG(3) to zero in *PRIN* (to suppress the debug print). The additional line of output following the *SOLV* arrays contains the evaluation of (4.96). Repeat the data manipulations of Example 4.9 to confirm use of the energy norm to validate convergence, and hence to verify the adequate mesh needed to estimate the exact solution energy norm.

Table 4.3 contains the data for $\|T^h\|_E$ that *LEARN.FE* will create and illustrates the manipulations needed to estimate $\|T^h\|_E$, the exact value of which is 134.6574(5). Note that the $M = 8$ solution error estimate yields this value, to PC single precision, and that $\Delta\|T^{h/2}\|_E$ decreases by essentially a factor of 4 for each mesh refinement $M \geq 4$. Table 4.4 contains the corresponding data for the $k = 2$ basis solution sequence. The convergence trend is evident; however, the $M = 4$ solution is accurate to PC precision, and hence most data are polluted by round-off error. This will be evident as you proceed through the convergence exercises in this chapter. The coarse mesh data will not agree with the convergence estimate because of mesh inadequacy, while in proceeding to much finer meshes, and/or higher-degree bases, round-off error will totally pollute the estima-

TABLE 4.3
**Linear basis solution convergence in $\|T^h\|_E$**

| Mesh | $M$ | $l_e$ | $\|T^h\|_E$ | $\Delta\|T^{h/2}\|_E$ | $\|e^{h/2}\|$, (4.95*a*) | $\|T\|_E$ (est.) |
|---|---|---|---|---|---|---|
| $\Omega^h$ | 1 | 1 | 132.607 (5) | | | |
| | | | | 1.48406 (5) | | |
| $\Omega^{h/2}$ | 2 | 0.5 | 134.09106 (5) | | 0.49468 (5) | 134.586 (5) |
| | | | | 0.42014 (5) | | |
| $\Omega^{h/4}$ | 4 | 0.25 | 134.51120 (5) | | 0.14005 (5) | 134.651 (5) |
| | | | | 0.10935 (5) | | |
| $\Omega^{h/8}$ | 8 | 0.125 | 134.62055 (5) | | 0.03645 (5) | 134.657 (5) |
| $\Omega^{h/16}$ | 16 | 0.0625 | | | | |
| $\Omega^{h/32}$ | 32 | 0.03125 | | | | |

**TABLE 4.4**
**Quadratic basis solution convergence in $\| T^h \|_E$**

| Mesh | $M$ | $l_e$ | $\| T^h \|_E\ 10^5$ | $\| \Delta T^h \|_E\ 10^5$ | $\| e^{h/2} \|\ 10^5$, (4.95$a$) | $\| T \|_E\ 10^5$(est.) |
|---|---|---|---|---|---|---|
| $\Omega^h$ | 1 | 1 | 134.597 | | | |
| $\Omega^{h/2}$ | 2 | 0.5 | 134.65201 | 0.05501 | 0.003667 | 134.65567 |
| $\Omega^{h/4}$ | 4 | 0.25 | 134.65700 | 0.00499 | 0.0003327 | 134.65733 |
| $\Omega^{h/8}$ | 8 | 0.125 | 134.65758 | 0.00058 | 0.0000387 | 134.65761 |

tion process. Our intention is to have you gain first-hand experience now, to build your foundation in understanding the error mechanisms lurking behind all practical problem computational analyses.

With these additional considerations in hand, we now proceed to a second study problem to amplify on computational detail and accuracy.

**STUDY PROBLEM 2.** This study problem requests an accuracy/convergence assessment for a problem with distributed source. For simplicity assume the conductivity is constant ($k = 0.1$) and that Dirichlet data $T_b$ are applied at both endpoints of $\Omega$ of span $0 \leq x \leq L$. The source distribution is $s(x) = A \sin(\pi x/L)$, where $A$ is an input constant which sets the overall level. Use *LEARN.FE* to predict the exact solution maximum temperature.

The input file for this study (as well as any one-dimensional problem) is created by modification to the file for Study Problem 1. Hence, bring the $k = 1$ file to the screen, delete the *ROBN* command name and data, and replace it with *SORC*. The data sequence for *SORC* can be input directly as nodal data or can be keyed to $A$ and $x/L$. Using the direct *SORC* input stream for now, alter *TITL* and add *SORC* and *NORM* to make the file appear as follows.

```
TITL
***** STUDY PROBLEM 2,M=4,K=1 *****
TYPE   [K       N    NNODEL   REFL    NPR    NTRAN   NAXI]
        1       1      2        1.0     1
PRIN   [NBUG(*)]
        1       1       0        3
GRID   [FOR N (XL,XR(*),PR(*)),NEM(PR*)]
                1.0       2.0      1.0
        4
MATL   [FOR NPR(KL(*),KR(*))]
        1       0.1      0.1
SORC   [NSORC, NODE(*),S(*)]
        5
        1      2        3        4       5
        0.0    70.7     100      70.7    0.0
```

```
DIRI    [NDIR, NODE(*), QB(*)]
        2
        1       5
        0.0     0.0
FORM
SOLV
NORM
STOP
```

This file defines an $M = 4$, $k = 1$ solution for Study Problem 2 with source magnitude $A = 100$. Execute *LEARN.FE* and the output file will appear as

```
TITL
*** STUDY PROBLEM 2,SOURCE M=4, K=1 ***
TYPE     K        N    NNODEL     REFL   NPR    NTRAN   NAXI
         1        1      2         1.0    1       0       0

PRINT    NBUG(1)      (2)      (3)      (4)      (5)    (6)
               1        1        0        3        0      0

GRID  NI       NPRI     NTRAN
      1        1
      XL       XR1      PR1      XR2      PR2      XR3     PR3
      1.0      2.0      1.0
      NEM1     NEM2     NEM3
      4
      NODE COORDINATES (X, INTEGER MAP) EXPONENT E:-2
      100      125      150      175      200
MATL  MACRO NODAL CONDUCTIVITIES
      1      0.10000E+00     0.10000E+00
      AK:
             0.10      0.10  0.10      0.10     0.10
SORC  NSORC: 5 USER SPECIFICATION
      JSORC: 1         2     3         4        5
             0.00      70.7  100       70.7     0.00

      SOURCE DISTRIBUTION (INTEGER MAP) EXPONENT E:0
             0         70          100    70         0
DIRI  NFIX:  2
      JFIX:  1         5
             0.000     0.000
FORM  SYSTEM MATRIX [K + H], COLUMN FORMAT
             0.000   1.000    0.000
            -0.400   0.800   -0.400
            -0.400   0.800   -0.400
            -0.400   0.800   -0.400
             0.000   1.000    0.000
      DATA MATRIX (b), ROW FORMAT
             0.00    15.9518   22.5592  15.9518  0.00
SOLV  NODAL SOLUTION (REAL VARIABLES)
             0.00000 68.07847  96.27750  68.07847  0.00000

      NODAL SOLUTION (INTEGER MAP) EXPONENT E:-1
             0        680       962      680    0

NORM  F.E.SOLN. ENERGY SEMI-NORM: E(QH)=0.21719458E+04
STOP
```

This problem data set predicts a solution that is quite inaccurate, mainly because the $M = 4$ mesh poorly resolves the source term distribution. You can determine that the adequate mesh requirement is not met until the linear basis uniform discretization reaches $M \simeq 16$ elements. Manually inserting the corresponding source nodal data becomes burdensome and prone to error. Therefore, the *LEARN.FE* input stream permits definition of a source in functional form, e.g., $A \sin(\pi x/L)$. The nodal values are then internally computed once *GRID* has generated $\{X\}$, the nodes of the mesh $\Omega^h$. The currently embedded functions in subroutine SOURCE include sine and a gaussian, and others can be added by mimicking the code structure.

Source functions are accessed via subroutine READIT, on specification of amplitude ($A$), function half-width and the $x$ coordinate of the peak. Further, multiple sources can be defined in this manner. Hence, for Study Problem 2 alter the input data file under *SORC* and *DIRI* to read,

```
SORC   [(NSORC,TYPE),X(MAX),(PEAK,HALF.W)]
       1      1
       1.5
       100.0  0.5
DIRI   [(NDIR,REPEAT),NODE(*),(NRPT, QB(*))]
       2      1
       1      5
       2      0.0
```

and reexecute the file. The output will be identical in appearance except for the comments following *SORC*. Further, the *DIRI* input modification above illustrates how *LEARN.FE* accepts repetitive data strings.

Now proceed with the $k = 1$ discretization refinement study by stacking input files back-to-back for $4 \leq M \leq 64$ and changing command name *STOP* to *EXIT* except for $M = 64$. This allows *LEARN.FE* to cycle through the files without interruption, and hence yields rapid acquisition of the desired data base.

Once these data are at hand, your requirement is to predict solution accuracy, and hence to determine the range of adequate meshes $\Omega^h$ before round-off pollutes the data. You should elect to verify this in terms of both $Q_{\max}$, recall Table 4.1, and in the energy semi-norm (4.96). Note now that no contributions stem from $[H]_e$ since no convection boundary condition is applied. Tables 4.5 and 4.6 contain some of the data you will generate for these computations. Use your data to fill them out.

The data in Tables 4.5 and 4.6 provide a significant characterization of the accuracy-convergence character of the finite element algorithm solutions. The last column is added to provide quick verification of the

TABLE 4.5
**Linear basis source solution problem convergence study**
**(*a*) Solution maximum**

| Mesh | $M$ | $l_e$ | $Q_m$ | $\Delta Q_m$ | $e^{h/M}$ | $Q_{\text{exact}}$ (est.) | Slope |
|---|---|---|---|---|---|---|---|
| $\Omega^h$ | 4 | 1/4 | 96.277504 | | | | |
| | | | | 3.751663 | | | |
| $\Omega^{h/2}$ | 8 | 1/8 | 100.029167 | | 1.250554 | 101.27972 | |
| | | | | 0.967072 | | | 1.9558 |
| $\Omega^{h/4}$ | 16 | 1/16 | 100.996239 | | 0.322357 | 101.318596 | |
| | | | | 0.243434 | | | 1.98998 |
| $\Omega^{h/16}$ | 32 | 1/32 | 101.239693 | | 0.081151 | 101.320844 | |
| | | | | 0.061439 | | | 1.98640 |
| $\Omega^{h32}$ | 64 | 1/64 | 101.301132 | | 0.024801 | 101.321612 | |

**(b) Solution energy seminorm**

| Mesh | $M$ | $l_e$ | $\lVert T^h \rVert_E\, 10^3$ | $\Delta \lVert T^{h/2} \rVert_E\, 10^3$ | $\lVert e^{h/2} \rVert_E\, 10^3$ | $\lVert T \rVert_E\, 10^3$ (est.) | Slope |
|---|---|---|---|---|---|---|---|
| $\Omega^h$ | 4 | 1/4 | 2.1719458 | | | | |
| | | | | 0.2653306 | | | |
| $\Omega^{h/2}$ | 8 | 1/8 | 2.4372764 | | 0.0884435 | 2.525719 | |
| | | | | | | | 1.892 |
| $\Omega^{h/4}$ | 16 | 1/16 | 2.5087332 | | | 2.532552 | |
| | | | | | | | 1.974 |
| $\Omega^{h/8}$ | 32 | 1/32 | 2.5269268 | | | 2.532991 | |
| | | | | | | | 1.986 |
| $\Omega^{h16}$ | 64 | 1/64 | 2.5315186 | | | 2.533049 | |

adequate mesh requirement, for validity of (4.95) and (4.95*a*), by computing the piecewise value of the data slope via

$$\text{slope} = \frac{\log(e^{h/M}) - \log(e^{h/2M})}{\log(l_e) - \log(l_e/2)} = \frac{\log(e^{h/M}/e^{h/2M})}{\log(2)}$$

A quick glance in this column confirms that (1) all linear basis data obtained on uniform discretizations $M > 4$ satisfy mesh adequacy and (2) round-off error does not appear as a factor in the finer mesh data. Conversely, even the $M = 2$ quadratic basis solution is marginally mesh-adequate, and round-off has certainly polluted the $M = 32$ solution.

Therefore, the data obtained between confirmed mesh inadequacy and round-off is quality, and hence available for prediction of the accurate solution. On the basis of maximum temperature $Q_m$, these data predict that $Q_{\text{exact}} \approx 101.321$ to single-precision significance based on the $k = 2$, $M = 16$, and/or $k = 1$, $M = 32$ estimates of $e^{h/2}$. The quadratic basis actually achieves this as the nodal solution, while the verified convergence of the linear basis solution permits extrapolation of the $M = 32$ nodal solution to this value. It is of interest that the analytical solution value is $q(x/L = 0.5) = 101.3211836$; hence the estimated value is good to six significant digits, which is PC single precision. Note that the $M = 32$, $k = 2$ nodal solution is an overprediction, as suspected by the convergence slope error, as is the $M = 64$, $k = 1$ extrapolation, although the nodal value

TABLE 4.6
**Quadratic basis source problem convergence study**
**(*a*) Solution maximum**

| Mesh | $M$ | $l_e$ | $Q_m$ | $\Delta Q_m$ | $e^{h/2}$ | $Q_{\text{exact}}$ (est.) | Slope |
|---|---|---|---|---|---|---|---|
| $\Omega^h$ | 2 | 0.5 | 100.592216 | | | | |
| $\Omega^{h/2}$ | 4 | 0.25 | 101.279778 | 0.687562 | 0.045837 | 101.325615 | |
| $\Omega^{h/4}$ | 8 | | 101.318649 | | 0.002591 | 101.321240 | 4.145 |
| $\Omega^{h/8}$ | 16 | | 101.321014 | | 0.000158 | 101.321172 | 4.039 |
| $\Omega^{h16}$ | 32 | | 101.322105 | | 0.000073 | 101.322178 | 1.116 |

**(b) Solution energy seminorm**

| Mesh | $M$ | $l_e$ | $\lVert T^h \rVert_E\ 10^3$ | $\Delta \lVert T^{h/2} \rVert_E\ 10^3$ | $\lVert e^{h/2} \rVert_E\ 10^3$ | $\lVert T \rVert_E\ 10^3$ (est.) | Slope |
|---|---|---|---|---|---|---|---|
| $\Omega^h$ | 2 | 1/2 | 2.485360 | | | | |
| $\Omega^{h/2}$ | 4 | 1/4 | 2.529778 | 0.044418 | 0.002962 | 2.532740 | |
| $\Omega^{h/4}$ | 8 | 1/8 | 2.532820 | | | 2.533023 | 3.863 |
| $\Omega^{h/8}$ | 16 | 1/16 | 2.533008 | | | 2.533021 | 4.012 |
| $\Omega^{h16}$ | 32 | 1/32 | 2.533033 | | | 2.533035 | 2.920 |

remains underpredicted. Thus, round-off error is also of (marginal) significance in the linear basis solution process for $M > 32$.

As a final point, the convergence indications in the energy seminorm data predict solution performance in agreement with $Q_m$. To seven significant digits, the estimate is $\lVert T \rVert_E \simeq 2.533021 \times 10^3$ for $k = 2$, $M = 16$, and $\lVert T \rVert_E \simeq 2.532991 \times 10^3$ for $k = 1$, $M - 32$. Thc cxact sulutiön energy seminorm is $\lVert T(x) \rVert_E = 2.533029591 \times 10^3$; hence both estimates are good to PC single precision. Both basis solution evaluations approach the exact value from below, as do the asymptotic convergence estimates. The theory predicts this as required, hence norm (or nodal) computations that oscillate are undoubtedly polluted by round-off. This will typically (in fact, virtually always) occur when the finite element discrete solution is essentially identical to the Lagrange interpolation of the exact solution.

You may wonder why Study Problem 2 requires relatively finer meshes to give accurate solutions in comparison to Study Problem 1. The answer lies in the more precise expression for the asymptotic error estimate (4.91), specifically

$$\lVert \text{error}^h \rVert \simeq C_k l_e^{2k} (c_1 \lVert \text{data} \rVert_\Omega + c_2 \lVert \text{data} \rVert_{\partial\Omega}) \qquad (4.91a)$$

In (4.91$a$), the notation $\lVert \cdot \rVert$ means any suitable measure (norm), $c_1$ and $c_2$ are constants, and subscript $\Omega$ and $\partial\Omega$ signify data evaluations on the problem domain and its boundaries. As previously stated, "data" for a problem constitute what you specify in advance, e.g., conductivity $k$ and source $s$ on $\Omega$, heat transfer $h$, $T_r$ and fixed data $T_b$ on $\partial\Omega$, and domain size $L$. Up to now, all such data have been constants or linear functions not subject to interpolation error. Conversely, the sine source term is quite inaccurately interpolated on the coarser $k = 1$ basis meshes. Hence the $WS^h$ evaluation $\{b\}$ of $\|\text{data}\|_\Omega$ in (4.91$a$) changes substantially until $M$ becomes sufficiently large. The "message" is that $\Omega^h$ must also be sufficiently adequate to accurately interpolate given *data* distributions.

The use of higher-degree finite element bases to generate approximate solutions becomes quite appropriate for problems with smooth source distributions, owing to the improvement in interpolation accuracy. We suggest you verify this assertion by completing the appropriate $k = 1$ and $k = 2$ accuracy-convergence analyses for variations on Study Problem 2. Specifically, extend the problem statement to multiple sources including those that do not span all of $\Omega$, and/or use of the gaussian source specification to gain further familiarity with these accuracy issues using *LEARN.FE*.

## 4.7 AXISYMMETRIC HEAT CONDUCTION WITH CONVECTION

Many problems in steady-state heat transfer involve fluids flowing in pipes (see Fig. 4.1). The corresponding axisymmetric conservation law statement is

$$L(T) = -\frac{1}{r}\frac{d}{dr}\left[rk\frac{dT}{dr}\right] - s = 0 \qquad \text{on } \Omega = (r_1 < r < r_2,\ 0 \leq \theta \leq 2\pi) \tag{4.97}$$

$$l(T) = k\frac{dT}{dn} + h(T - T_r) = 0 \qquad \text{on } \partial\Omega_1 \tag{4.98}$$

$$T(x_b) = T_b \qquad \text{on } \partial\Omega_2 \tag{4.99}$$

As in the previous development (4.55) to (4.57), the general case requires that $k$ and $s$ be variable, hence dependent on the radial coordinate $r$, while $h$ and $T_r$ remain the coefficients of the energy-exchange medium. Again, $dT/dn$ is the temperature derivative normal to the boundary segment $\partial\Omega_1$, which can now be oriented either perpendicular or parallel to the radial coordinate, recall the fins in Fig. 4.1. For this reason, we must consider the domain $\Omega$ to lie in the complete axisymmetric plane rather than just along the radial coordinate span, as noted in (4.97).

Development of the finite element algorithm for (4.97) to (4.99) follows the developed recipe. The statement of approximation is

$$T^N(x) = \sum_{i=1}^{N} a_i \phi_i(x) \tag{3.6}$$

where $x$ will now be interpreted as $r$. The symmetric Galerkin weak statement becomes, multiplying by $2\pi$ for later convenience,

$$\begin{aligned}
(2\pi)WS &\equiv \int_\Omega \phi_i L(T^N)\, d\tau \qquad \text{for } 1 \le i \le N \\
&= \int_\Omega \phi_i \left[ -\frac{1}{r}\frac{d}{dr}\left( rk \frac{dT^N}{dr} \right) - s \right] r\, dr\, d\theta \\
&= \int_\Omega \phi_i \left[ -\frac{d}{dr}\left( rk \frac{dT^N}{dr} \right) - rs \right] dr\, d\theta \\
&= \int_\Omega \left( \frac{d\phi_i}{dr} rk \frac{dT^N}{dr} - \phi_i rs \right) dr\, d\theta - \int_{\partial\Omega} \phi_i k \left( \frac{dT^N}{dn} \right) r\, d\sigma \\
&= \int_\Omega \left( \frac{d\phi_i}{dr} k \frac{dT^N}{dr} - \phi_i s \right) r\, dr\, d\theta + \int_{\partial\Omega_1} \phi_i h(T^N - T_r) r\, d\sigma \\
&\quad - \int_{\partial\Omega_2} \phi_i k \frac{dT^N}{dn} r\, d\sigma = 0
\end{aligned} \tag{4.100}$$

The finite element approximation $T^h$ is the definition for (3.6) with (4.100) evaluated on the discretization $\Omega^h$ using the basis form

$$T_e(r) = \{N_k(\zeta_i)\}^T \{Q\}_e \tag{4.59}$$

The finite element discrete approximation to (4.100) is then

$$\begin{aligned}
WS^h = (2\pi)^{-1} S_e \Bigg( &\int_{\Omega_e} \frac{d\{N_k\}}{dr} k_e \frac{d\{N_k\}^T}{dr} r\, dr\, d\theta \{Q\}_e \\
&- \int_{\Omega_e} \{N_k\}\{N_k\}^T r\, dr\, d\theta \{S\}_e \\
&+ \int_{\partial\Omega_e \cap \partial\Omega_1} \{N_k\} h(\{N_k\}^T \{Q\}_e - T_r) r\, d\sigma \\
&- \int_{\partial\Omega_e \cap \partial\Omega_2} \{N_k\} k \frac{dT^h}{dn} r\, d\sigma \Bigg) = \{0\}
\end{aligned} \tag{4.101}$$

The notation in (4.101) coincides with that developed for (4.60). The last term verifies again that the approximate normal heat flux $-k(dT^h/dn)$ will be predicted on a domain boundary segment where temperature is constrainted by Dirichlet data.

The appearance of the axisymmetric discrete weak statement (4.101) is a minor variation on (4.60). Specifically, the differential elements for the integrals are now $r\,dr\,d\theta$ and $r\,d\sigma$ instead of $dx$ and the simple endpoint evaluation. For the case where heat is convected across only the inner and outer radial surfaces, specifically, the endpoints of $\Omega$, then $d\sigma = d\theta$ and the integral over $\theta$ appears uniformly on every term in (4.101). Following integration, then, the resultant uniform multiplier $2\pi$ can be canceled throughout. For constant conductivity $k_e$, and neglecting the source term for the moment, (4.101) and (4.60) become functionally identical by replacing $\{K\}_e$ with $\{R\}_e$, the radial coordinate set of the geometric nodes of $\Omega_e$, as the data multiplier for formation of the conduction matrix $[K]_e$.

Thus, for this elementary axisymmetric problem with uniform conductivity, the linear basis master matrix library definitions become

$$k = 1: \quad [K]_e = \frac{k_e}{l_e}\left\{\frac{1}{2}, \frac{1}{2}\right\}\{R\}_e\begin{bmatrix} 1 & -1 \\ -1 & 1 \end{bmatrix} \tag{4.65a}$$

$$[H]_e = h\begin{bmatrix} RL_1\delta_{e1} & 0 \\ 0 & RR_M\delta_{eM} \end{bmatrix} \tag{4.68a}$$

$$\{b\}_e = hT_r\begin{Bmatrix} RL_1\delta_{e1} \\ RR_M\delta_{eM} \end{Bmatrix} \tag{4.69a}$$

In (4.68$a$) and (4.69$a$), $RL_1$ denotes the radial coordinate of the left node of $\Omega_{e=1}$, for example, the inner radius of the pipe in Fig. 4.1$a$. Thus, $RR_M$ denotes the radius of the right node of $\Omega_{e=M}$ for convection applied to the outer radius of a cylinder.

The axisymmetric definition for the source term master matrix requires an extension on the cartesian form (4.66). For $k = 1$, this term in (4.101) is now evaluated as

$$\begin{aligned} \{b\}_e &= \int_{\Omega_e} \{N_k\}\{N_k\}^T r_e\, dr\{S\}_e \\ &= \{R\}_e^T \int_{\Omega_e} \{N_1\}\{N_1\}\{N_1\}^T\, dr\{S\}_e \end{aligned}$$

$$= \frac{l_e\{R\}_e^T}{12}\begin{bmatrix} \begin{Bmatrix}3\\1\end{Bmatrix} & \begin{Bmatrix}1\\1\end{Bmatrix} \\ \begin{Bmatrix}1\\1\end{Bmatrix} & \begin{Bmatrix}1\\3\end{Bmatrix}\end{bmatrix}\{S\}_e \qquad (4.66a)$$

As before, the radius distribution on $\Omega_e$ has been interpolated in terms of its nodal values $\{R\}_e$. As the consequence, (4.66$a$) involves a degree one hypermatrix, as introduced earlier in forming the quadratic and cubic basis element conduction matrices, recall (4.77) and (4.85).

For the higher-degree basis formulations, the convection boundary condition matrices remain as defined by (4.68$a$) and (4.69$a$) with the addition of one or two rows and columns of zeros to match the order of $\{Q\}_e$; recall (4.49), (4.87), and (4.88). The corresponding axisymmetric element conduction master matrix statements are

$$k=2: [K]_e = \frac{k_e}{6l_e}\{R\}_e^T\begin{bmatrix} \begin{Bmatrix}11\\3\end{Bmatrix} & \begin{Bmatrix}-12\\-4\end{Bmatrix} & \begin{Bmatrix}1\\1\end{Bmatrix} \\ & \begin{Bmatrix}16\\16\end{Bmatrix} & \begin{Bmatrix}-4\\-12\end{Bmatrix} \\ (\text{sym}) & & \begin{Bmatrix}3\\11\end{Bmatrix}\end{bmatrix} \qquad (4.77a)$$

$$k=3: [K]_e = \frac{k_e}{80l_e}\{R\}_e^T\begin{bmatrix} \begin{Bmatrix}262\\34\end{Bmatrix} & \begin{Bmatrix}-327\\-51\end{Bmatrix} & \begin{Bmatrix}78\\30\end{Bmatrix} & \begin{Bmatrix}-13\\-13\end{Bmatrix} \\ & \begin{Bmatrix}594\\270\end{Bmatrix} & \begin{Bmatrix}-297\\-297\end{Bmatrix} & \begin{Bmatrix}30\\78\end{Bmatrix} \\ & & \begin{Bmatrix}270\\594\end{Bmatrix} & \begin{Bmatrix}-51\\-327\end{Bmatrix} \\ (\text{sym}) & & & \begin{Bmatrix}34\\262\end{Bmatrix}\end{bmatrix} \qquad (4.85a)$$

Establishment of the corresponding master matrices for the source term contribution to $\{b\}_e$ is left as a suggested exercise.

**Example 4.10.** Superheated liquid sodium at 1500 K is flowing through a high-pressure pipe of inside radius $r_1 = 5$ cm and thickness $\Delta r = 5$ cm. The thermal conductivity of the pipe is $k = 100$ W/(cm · K) and the thermal convection coefficient is $h = 40$ W/(cm$^2$ · K). The pipe is immersed in a liquid that undergoes an endothermic phase change at $T = 306.85$ K. Determine the temperature distribution through the pipe thickness.

We suggest you always keep the first analysis as simple as possible. Hence, try the linear basis formulation on an $M = 2$ discretization. Thus

$$[K]_e = \frac{k_e}{l_e}\left\{\frac{1}{2}, \frac{1}{2}\right\}\{R\}_e\begin{bmatrix} 1 & -1 \\ -1 & 1 \end{bmatrix} \qquad e = 1, 2$$

$$\Omega_{e=1}: [K]_1 = \frac{100}{5/2}\left\{\frac{1}{2}, \frac{1}{2}\right\}\begin{Bmatrix} 5.0 \\ 7.5 \end{Bmatrix}\begin{bmatrix} 1 & -1 \\ -1 & 1 \end{bmatrix}$$

$$= \begin{bmatrix} 250 & -250 \\ -250 & 250 \end{bmatrix}$$

$$\Omega_{e=2}: [K]_2 = \frac{100}{5/2}\left\{\frac{1}{2}, \frac{1}{2}\right\}\begin{Bmatrix} 7.5 \\ 10.0 \end{Bmatrix}\begin{bmatrix} 1 & -1 \\ -1 & 1 \end{bmatrix}$$

$$= \begin{bmatrix} 350 & -350 \\ -350 & 350 \end{bmatrix}$$

$$[H]_e = h\begin{bmatrix} RL_1\delta_{e1} & 0 \\ 0 & RR_M\delta_{eM} \end{bmatrix} \qquad e = 1$$

$$= 40\begin{bmatrix} 5 & 0 \\ 0 & 0 \end{bmatrix}$$

$$\{b\}_e = hT_r\begin{Bmatrix} RL_1\delta_{e1} \\ RR_M\delta_{eM} \end{Bmatrix} \qquad e = 1$$

$$= (40)(1500)\begin{Bmatrix} 5 \\ 0 \end{Bmatrix}$$

Recalling that the finite element global matrix statement form is

$$S_e(([K]_e + [H]_e)\{Q\}_e - \{b\}_e) = \{0\} \tag{4.70}$$

assembling the generated element data for $M = 2$ and imposing the phase change coolant temperature as a Dirichlet constraint yields for (4.70)

$$\begin{bmatrix} 450 & -250 & 0 \\ -250 & 600 & -350 \\ 0 & 0 & 1 \end{bmatrix}\begin{Bmatrix} Q1 \\ Q2 \\ Q3 \end{Bmatrix} = \begin{Bmatrix} 300{,}000 \\ 0 \\ 306.85 \end{Bmatrix} \tag{4.102}$$

By now you may sense these data have appeared before, as indeed they have. Equation (4.102) is simply 10 times (4.72), the variable conductivity slab $M = 2$, $k = 1$ matrix statment. Hence, the solution to (4.102) is

$$\{Q\} = \begin{Bmatrix} Q1 \\ Q2 \\ Q3 \end{Bmatrix} = \begin{Bmatrix} 996.87 \\ 594.36 \\ 306.85 \end{Bmatrix} \text{K} \tag{4.103}$$

The entire computational experiment project conducted for Study Problem 1 is immediately applicable. Hence, you know that the surface temperature error is approximately 0.3 percent, and that it can be predicted to an accuracy of 0.1 by using the $M = 4$, $k = 1$ solution with error estimate.

**Code Exercise 4.6.** *LEARN.FE* is easily adapted to the axisymmetric problem statement in the command name *TYPE* by changing parameter NAXI (i.e., AXIsymmetric) to unity. Hence, the input file for Example 4.10 is the following minor modification to Study Problem 1.

```
TITL
***    CODE EXERCISE 4.6, M=2,AXISYMMETRIC ***
TYPE   [K      N      NNODEL     REFL  NPR  NTRAN   NAXI]
        1      1        2         1.0   1     0       1
PRIN   [NBUG(*)]
        1      0        0          3
GRID   [FOR N(XL,XR(*),PR(*)),NEM(PR*)]
                5.0     10.0      1.0
        2
MATL   [FOR NPR(KL(*),KR(*))]
        1     100.0   100.0
ROBN   [NROB,NODE(*),(H(*),TR(*))]
        1
        1
              40.0    1500.0
DIRI   [NDIR,NODE(*), QB(*)]
        1
        3
        306.85282
FORM
SOLV
NORM
STOP
```

Conduct some computational experiments for the axisymmetric problem description, for example by changing the data to $h = 200$ W/(cm · K) and/or replacing $T_b$ by a convection boundary condition. Now the exact solution is unknown; hence you'll have to execute a convergence study (recall Table 4.1) to verify the minimum $M$ needed for a selected accuracy. In doing this, stack your input files for various $M$ and execute the sequence by changing *STOP* to *EXIT* except for the last file.

## Finned-Tube Heat Exchanger

Now proceed to the very pertinent problem statement characterized by the finned-tube heat exchanger shown in Fig. 4.1. The engineer in us says that conduction effects through the fin thickness cannot be as important as radial conduction, but the convective heat transfer to the cooling air obviously occurs on the fin surface, not at its tip! Is it possible therefore to perform a meaningful one-dimensional, axisymmetric heat transfer analysis that can predict cooling fin performance? Figure 4.7 illustrates the essence of the matter; heat is transferred from the internal fluid ($h_f$, $T_f$) through the pipe wall into the fin, which is cooled by passing air ($h_a$, $T_a$). Hence, the problem statement is (4.97) and (4.98) without (4.99) since no fixed temperature $T_b$ exists.

The key formulation issue is the weak statement (4.100) boundary convection surface integral on $\partial\Omega_1$, hence the finite element algorithm (4.101). Figure 4.7*b* shows the nodes of a representative discretization $\Omega^h$. At the left end where $r = r_L$, the boundary convection term evaluation for $k = 1$ yields

$$(2\pi)^{-1}\int_{\partial\Omega_e}\{N_k\}h(\{N_k\}^T\{Q\}_e - T_r)r\,d\sigma \qquad \text{for } e = 1 \quad \text{and} \quad d\sigma = d\theta$$
$$= Z^*RL_1h_f\left(\begin{bmatrix}1 & 0\\ 0 & 0\end{bmatrix}\{Q\}_e - T_f\begin{Bmatrix}1\\0\end{Bmatrix}\right) \tag{4.104}$$

In (4.104), $Z^*$ is the ratio of the spacing between fins ($\Delta$) to the thickness of the fin, i.e., the axial extent of the pipe over which heat from $T_f$ will channel into one fin. Conversely, the boundary convection integral on the fin surface becomes

$$\frac{1}{2\pi}\int_{\partial\Omega_e}\{N_k\}h(\{N_k\}^T\{Q\}_e - T_r)r\,d\sigma$$
$$= (2\pi)^{-1}\int_{\partial\Omega_e}\{N_k\}h_a(\cdot)r\,dr\,d\theta$$
$$= \int_{\partial\Omega_e}\{N_k\}h_a(\{N_k\}^T\{Q\}_e - \{N_k\}^T\{TA\}_e)\{N_1\}^T\{R\}_e\,dr$$
$$= h_a\int_{\partial\Omega_e}\{R\}_e^T\{N_1\}\{N_k\}(\{N_k\}^T\{Q - TA\}_e)\,dr$$
$$= h_a\{R\}_e^T\int_0^{l_e}\{N_1\}\{N_k\}\{N_k\}^T\,d\bar{x}\{Q - TA\}_e \tag{4.105}$$

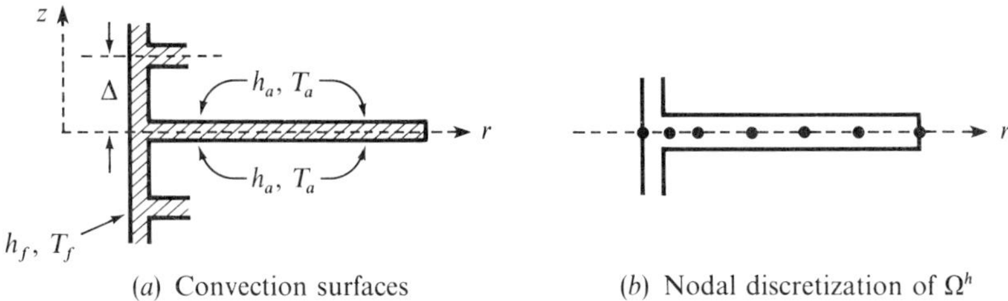


(a) Convection surfaces (b) Nodal discretization of $\Omega^h$

**FIGURE 4.7**
Essential aspects of the finned-tube heat transfer problem.

In establishing (4.105), for generality we chose $\{N_k\}$ to interpolate the cooling-air temperature $T_r$ using nodal values $\{TA\}_e$. As before, the interpolation of $r$ on $\Omega_e$ uses $\{N_1\}$ and the nodal radii $\{R\}_e$. Since $h_a$ is assumed constant, it can be extracted from the integral, as can the element data $\{R\}_e^T$ and $\{Q - TA\}_e$ for the same reason.

Hence, even though the finned-tube definition is one-dimensional, the convection boundary condition (4.105) produces an integral over $\Omega_e$ rather than an endpoint evaluation. For $k = 1$ in (4.59), and using the analytical integration formula (4.40), the fin convection element master matrix set for (4.105) then becomes

$$[H]_e = \frac{l_e h_a}{12}\{R\}_e^T \begin{bmatrix} \begin{Bmatrix} 3 \\ 1 \end{Bmatrix} & \begin{Bmatrix} 1 \\ 1 \end{Bmatrix} \\ \begin{Bmatrix} 1 \\ 1 \end{Bmatrix} & \begin{Bmatrix} 1 \\ 3 \end{Bmatrix} \end{bmatrix} \tag{4.106}$$

$$\{b\}_e = [H]_e\{TA\}_e \tag{4.107}$$

The hypermatrix construction developed for the $k = 2$ and $k = 3$ statements in Section 4.4 [recall (4.77) and (4.85)] remains highly efficient for the formulational detail necessary to evaluate (4.100), i.e., (4.106) and (4.107). Further, the element master matrix defined in (4.106) is identical to the source term contribution form given in (4.66*a*). The functional utility of the element library concept should be quite apparent.

The finite element algorithm formulation for the finned-tube problem is thus completed by (4.65*a*), (4.106), and (4.107). The following example illustrates the details.

**Example 4.11.** Symbolically set up and assemble the finite element global matrix statement

$$S_e(([K]_e + [H]_e)\{Q\}_e - \{b\}_e) = \{0\} \tag{4.70}$$

for an $M = 7$, $k = 1$ solution of the finned-tube heat exchanger problem (Fig. 4.2). Assume two elements span the tube of internal radius $r$, thickness $t$, and thermal conductivity $k_t$, through which fluid flows with $T_f$ and $h_f$. Employ a $p = 1.2$ nonuniform discretization of the fin of length $L$ and conductivity $k_f$, both sides of which are exposed to cooling airflow at $T_a$ and $h_a$. The ratio of fin spacing to fin thickness is $Z^*$.

Figure 4.8 sketches the details for this problem statement and the nodes of the $M = 7$ discretization. Elements are assumed ordered from left to right. The symbolic conductivity master matrix evaluations for the pipe elements $\Omega_{e=1}$ and $\Omega_{e=2}$ yield

$$[K]_e = \frac{k_e}{l_e}\left\{\frac{1}{2}, \frac{1}{2}\right\}\{R\}_e\begin{bmatrix} 1 & -1 \\ -1 & 1 \end{bmatrix} \tag{4.65a}$$

$$\Omega_{e=1}: \quad = \frac{k_t}{t/2}\left\{\frac{1}{2}, \frac{1}{2}\right\}\begin{Bmatrix} r \\ r + t/2 \end{Bmatrix}\begin{bmatrix} 1 & -1 \\ -1 & 1 \end{bmatrix} \equiv \begin{bmatrix} k_{11}^1 & k_{12}^1 \\ k_{21}^1 & k_{22}^1 \end{bmatrix} \tag{4.108}$$

$$\Omega_{e=1}: \quad = \frac{k_t}{t/2}\left\{\frac{1}{2}, \frac{1}{2}\right\}\begin{Bmatrix} r + t/2 \\ r + t \end{Bmatrix}\begin{bmatrix} 1 & -1 \\ -1 & 1 \end{bmatrix} \equiv \begin{bmatrix} k_{11}^2 & k_{12}^2 \\ k_{21}^2 & k_{22}^2 \end{bmatrix} \tag{4.109}$$

The internal flow convection master matrix evaluation on $\partial\Omega_{e=1}$ produces

$$[H]_e\{Q\}_e - \{b\}_e = Z^*RL_1h_f\left(\begin{bmatrix} 1 & 0 \\ 0 & 0 \end{bmatrix}\{Q\}_e - T_f\begin{Bmatrix} 1 \\ 0 \end{Bmatrix}\right)\delta_{e1} \tag{4.104}$$

$$\partial\Omega_{e=1}: \quad = Z^*rh_f\left(\begin{bmatrix} 1 & 0 \\ 0 & 0 \end{bmatrix}\{Q\}_e - T_f\begin{Bmatrix} 1 \\ 0 \end{Bmatrix}\right)$$

$$\equiv \begin{bmatrix} h_{11}^1 & 0 \\ 0 & 0 \end{bmatrix}\{Q\}_{e=1} - \begin{Bmatrix} b_1^1 \\ 0 \end{Bmatrix} \tag{4.110}$$

In (4.108) to (4.110), the superscripts on the symbolic matrix entries denote the element index $e$, while subscripts correspond to the usual matrix index convention $i, j$.

Proceeding to the fin, we need consider only the symbolic completion for the generic element $\Omega_e$. The conductivity master matrix evaluation

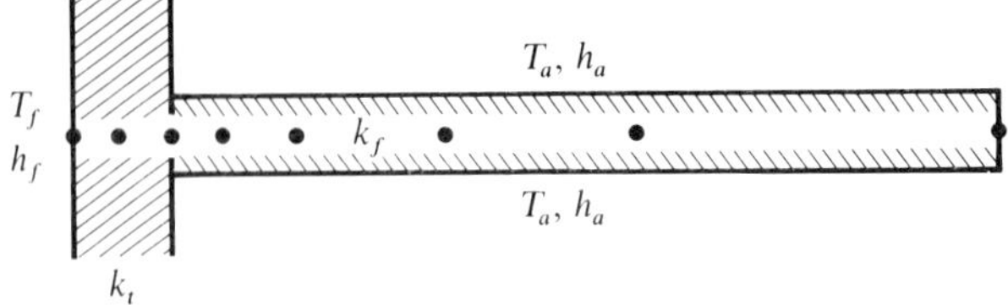


**FIGURE 4.8**
Finned-tube heat exchanger problem definition.

produces

$$[K]_e = \frac{k_e}{l_e}\left\{\frac{1}{2}, \frac{1}{2}\right\}\{R\}_e\begin{bmatrix} 1 & -1 \\ -1 & 1 \end{bmatrix} \tag{4.65a}$$

$$\Omega_{3-7}: \quad = \frac{k_f}{l_e}\left\{\frac{1}{2}, \frac{1}{2}\right\}\begin{Bmatrix} RL \\ RR \end{Bmatrix}_e\begin{bmatrix} 1 & -1 \\ -1 & 1 \end{bmatrix} \equiv \begin{bmatrix} k_{11}^e & k_{12}^e \\ k_{21}^e & k_{22}^e \end{bmatrix} \tag{4.111}$$

Completion of the fin surface convection master matrix produces

$$[H]_e\{Q\}_e - \{b\}_e = \frac{l_e h_e}{12}\{R\}_e^T\begin{bmatrix} \begin{Bmatrix} 3 \\ 1 \end{Bmatrix} & \begin{Bmatrix} 1 \\ 1 \end{Bmatrix} \\ \begin{Bmatrix} 1 \\ 1 \end{Bmatrix} & \begin{Bmatrix} 1 \\ 3 \end{Bmatrix} \end{bmatrix}\{Q - TA\}_e \tag{4.106}$$

$$\partial\Omega_{3-7}: \quad = \frac{2l_e h_a}{12}\{RL, RR\}_e[\cdot]\{\cdot\}_e$$

$$\equiv \begin{bmatrix} h_{11}^e & h_{12}^e \\ h_{21}^e & h_{22}^e \end{bmatrix}\{Q\}_e - \begin{Bmatrix} b_1^e \\ b_2^e \end{Bmatrix} \tag{4.112}$$

In (4.111) and (4.112), the superscript $e$ is the element index with range $3 \le e \le 7$, and the multiplier 2 in (4.112) results since boundary convection occurs on both surfaces of each fin element.

Equations (4.108) to (4.112) define the evaluations that would be completed by *LEARN.FE* in subroutine FORMIT, wherein also is executed the assembly of the global matrix statement. Using the column matrix storage format employed by *LEARN.FE* (recall the *various* code output reflections), we suggest that you verify that the symbolic assembled algorithm statement for this problem is

$$S_e(([K]_e + [H]_e)\{Q\}_e) = S_e(\{b\}_e)$$
$$1 \le e \le 7$$

$$\begin{bmatrix} 0 & k_{11}^1 + h_{11}^1 & k_{12}^1 \\ k_{21}^1 & k_{22}^1 + k_{11}^2 & k_{12}^2 \\ k_{21}^2 & k_{22}^2 + k_{11}^3 + h_{11}^3 & k_{12}^3 + h_{12}^3 \\ k_{21}^3 + h_{21}^3 & k_{22}^3 + h_{22}^3 + k_{11}^4 + h_{11}^4 & k_{12}^4 + h_{12}^4 \\ k_{21}^4 + h_{21}^4 & k_{22}^4 + \cdots & \\ \vdots & \vdots & \vdots \\ \vdots & k_{22}^7 + h_{22}^7 & 0 \end{bmatrix}\begin{Bmatrix} Q1 \\ Q2 \\ Q3 \\ Q4 \\ Q5 \\ Q6 \\ Q7 \\ Q8 \end{Bmatrix} = \begin{Bmatrix} b_1^1 \\ 0 \\ b_1^3 \\ b_2^3 + b_1^4 \\ b_2^4 + \cdots \\ \vdots \\ \\ b_2^7 \end{Bmatrix} \tag{4.113}$$

In (4.113), the "dot" column notation indicates a term that is a repetition of the previous term with only a change of element index. If the reader can "see into" the structure presented in (4.113), the role that the fin surface convection plays in the algorithm statement, and hence in the heat transfer process, is quite transparent. The cooling-air convection coefficient $h_a$ serves to enhance the fin radial thermal conductivity, while the product $h_a T_a$ appearing in $b_i$ acts as an equivalent "sink" for heat, in a manner identical to imposing a negative source term $s_e$. Specifically to this issue, note that the master matrix statements for internal source (4.66*a*) and surface convection (4.106) are *functionally identical*!

You probably were not privy to this interpretation prior to the example. The truly *important* fact is that the finite element algorithm statement (4.113) was completely constructed *without* your having to *devise* the somewhat intricate details. The finite element formulational "recipe book" is completely and entirely up to the challenge! Actually, the more complicated the mathematics or mechanics problem statement is, the more appreciative the computational analyst becomes of the true power represented by the finite element method.

**Code Exercise 4.7.** The input data file for this exercise contains the specifications for *LEARN.FE* to generate a numerical solution for Example 4.11. Access this file, and the following will appear on your screen.

```
TITL
**** CODE EXERCISE 4.7, M=7, FIN TUBE*****
TYPE  [K N NNODEL REFL    NPR    NTRAN NAXI NZSTR]
       1 1   2       1.0     2         0     1    1
PRIN  [NBUG(*)]
       1          1          1            3
GRID  [FOR N(XL, XR(*), PR(*)), NEM(PR*)]
                20.0      21.0          1.0    71.0          1.2
       2          7
MATL  [FOR NPR(KL(*), KR(*))]
       1        20.0      20.0
       2        30.0      30.0
ROBN  [NROB..ZSTR, NODE(*), (H(*), TR(*))]
       9   0    0
       1   3   4   5   6   7   8   9   10
            20.0   1000.0
            30.0     70.0
            30.0     70.0
            30.0     70.0
            30.0     70.0
            30.0     70.0
            30.0     70.0
            30.0     70.0
            30.0     70.0

FORM                       .
SOLV                       .
STOP                       .
```

The suggested geometric, material, and convection properties in any consistent set of units are

$$k_t = 20 \qquad k_f = 30 \qquad Z^* = 8 \qquad L = 50 \qquad r = 20$$

$$h_f = 20 \qquad T_f = 1000 \qquad h_a = 30 \qquad T_a = 70 \qquad t = 1$$

Note that entry for $Z^*$ is contained under command *ROBN* and indexed under *TYPE*. Execute this file, and the associated nodal temperature solution will appear in the output file on your screen. Now design and execute a computational study to give you confidence regarding the associated numerical solution accuracy.

## Heat Exchanger with Tapered Fins

One additional step will complete the axisymmetric heat transfer analysis. Recalling Fig. 4.1*b*, suppose the finned-tube heat exchanger is of cast construction with an associated taper to the fin. What generalization to the developed $k = 1$ basis master matrix library is then required? The taper certainly doesn't affect the tube conductivity characterization (4.65*a*), or the convection characterization (4.104), except for interpretation of $Z^*$ as the ratio of fin spacing to an average value of fin thickness (a requirement that falls on *your* shoulders as the computational experiment designer). Similarly, the surface convection formulation (4.106) and (4.107) for the fin is unaffected, since the difference between element span and surface length is negligible.

Therefore, only the fin element conductivity formulation must be generalized. Returning to (4.100), the form becomes

$$(2\pi)^{-1} \int_\Omega \phi_i \left[ -\frac{1}{r}\frac{d}{dr}\left( rk \frac{dT^N}{dr} \right) \right] d\tau$$

$$= (2\pi)^{-1} \int_\Omega \phi_i \left[ -\frac{1}{r}\frac{d}{dr}\left( kr \frac{dT^N}{dr} \right) \right] r\, dr\, d\theta\, dz \qquad 1 \le i \le N \qquad (4.114)$$

Thus, the domain $\Omega$ is now extended to include the fin vertical span $z$. The statement is still axisymmetric, hence the $2\pi$ from the integral over $\theta$ will cancel out. The integral on $dz$ can be effectively eliminated by assuming a fin thickness distribution $w(r)$ with radius. The finite element form for (4.114) then requires evaluation of

$$[K]_e\{Q\}_e = \int_{\Omega_e} \frac{d\{N_k\}}{dr}\, k_e\, \frac{d\{N_k\}^T}{dr}\, w_e(r) r\, dr \{Q\}_e \qquad (4.115)$$

The simplest evaluation occurs for the linear basis. Assuming that $k_e(r)$ and $w_e(r)$ are thus interpolated using $\{N_1\}$, as is the radius, (4.115) then becomes

$$[K]_e = \int_{\Omega_e} \frac{d\{N_1\}}{dr} k_e \frac{d\{N_1\}^T}{dr} w_e(r) r \, dr$$

$$= \{K\}_e^T \int_{\Omega_e} \{N_1\}\{N_1\}^T\{W\}\{N_1\}^T \, dr \{R\}_e \left( \frac{d\{N_1\}}{dr} \frac{d\{N_1\}^T}{dr} \right)$$

$$= \frac{1}{12l_e} \left( \{K\}_e^T \left( \{R\}_e^T \begin{bmatrix} \begin{Bmatrix} 3 \\ 1 \end{Bmatrix} & \begin{Bmatrix} 1 \\ 1 \end{Bmatrix} \\ \begin{Bmatrix} 1 \\ 1 \end{Bmatrix} & \begin{Bmatrix} 1 \\ 3 \end{Bmatrix} \end{bmatrix} \{W\}_e \right) \right) \begin{bmatrix} 1 & -1 \\ -1 & 1 \end{bmatrix} \tag{4.116}$$

The evaluation of the integrals in (4.116) is particularly easy, since they already occurred in forming the distributed source and/or the surface convection terms, [recall (4.105) and (4.106)]. This is a common occurrence; hence adding to the linear basis master matrix library is usually not time-consuming. A simplification occurs on assuming the element conductivity is a uniform constant, whereupon (4.116) reduces to the conventional matrix expression

$$[K]_e = \frac{k_e}{6l_e} \left( \{R\}_e^T \begin{bmatrix} 2 & 1 \\ 1 & 2 \end{bmatrix} \{W\}_e \right) \begin{bmatrix} 1 & -1 \\ -1 & 1 \end{bmatrix} \tag{4.116a}$$

Note in both (4.116) and (4.116*a*) that the term in parenthesis $(\cdot)$ is a scalar dependent on element data.

**Code Exercise 4.8.** To complete this exercise will require modifications to *LEARN.FE* to implement (4.116*a*). Should you care to, create an input data file for this exercise which would correspond in essence to that for Code Exercise 4.7. Under command names *TYPE* and *MATL* you can implement the data statements that feed the evaluation of (4.116*a*). After file creation and code modification, execute, view the output, and then proceed through a grid refinement study to quantize solution accuracy.

## 4.8 THE STRUCTURAL BEAM

The vehicle for exposition in this chapter has focused on heat conduction-convection as a highly suitable problem statement. The developed finite element methodology is applicable to field problems in *any* discipline, and the following section suggests a diverse range of continuum mechanics problems for computational examination. In closing this chapter, we consider a problem in structural mechanics to further expose the finite element method formulational versatility.

In structural beam theory, the differential equation governing the vertical deflection $y(x)$ of a homogeneous, initially horizontal beam is

$$L(y) = -\frac{d^2}{dx^2}\left(EI\frac{d^2y}{dx^2}\right) + p(x) = 0 \tag{4.117}$$

where $E$ is Young's modulus, $I(x)$ is the area moment of inertia (distribution) and $p(x)$ is the imposed transverse load distribution. In the special case of a beam vibrating about its static equilibrium under the action of its own weight, and assuming harmonic motion with frequency $\omega$, the expression for the load becomes

$$p(x) = \rho\omega^2 y(x) \tag{4.118}$$

where $\rho$ is the mass density per unit length. Hence, the equation governing the relative displacement of a self-induced harmonically oscillating horizontal beam is

$$L(y) = -\frac{d^2}{dx^2}\left(EI\frac{d^2y}{dx^2}\right) + \rho\omega^2 y = 0 \tag{4.119}$$

Equation (4.119) is a *Helmholtz* type fourth-order ordinary differential equation governing harmonic motion. Conversely, (4.117) is simply a fourth-order boundary value problem. Either statement requires that four boundary conditions be specified, to handle the four constants of integration that would result from a classical solution for $EI =$ constant. Applicable endpoint boundary conditions include fixed displacements, e.g., $y(x = 0) = 0 = y(x = L)$, a fixed slope constraint $dy/dx = b$ and/or an applied moment $M \equiv EI\, d^2y/dx^2$ or shear $V \equiv dM/dx$ at $x = 0$ or $x = L$.

The developed finite element methodology for a symmetric Galerkin weak statement is highly suited to developing a discrete approximate solution algorithm for (4.117), or the special case (4.119), and the range of boundary conditions. Since any approximation is of the form (3.6), we have

$$y^N(x) \equiv \sum_{i=1}^{N} a_i \phi_i(x) \tag{4.120}$$

The symmetric Galerkin weak statement construction is

$$WS \equiv \int_\Omega \phi_i L(y^N)\, d\tau \qquad \text{for } 1 \le i \le N$$

$$= \int_\Omega \phi_i \left[ -\frac{d^2}{dy^2}\left( EI \frac{d^2 y^N}{dx^2} \right) + p(x) \right] dx$$

$$= \int_\Omega \frac{d\phi_i}{dx}\frac{d}{dx}\left( EI \frac{d^2 y^N}{dx^2} \right) dx - \phi_i \frac{d}{dn}\left( EI \frac{d^2 y}{dx^2} \right)\Bigg|_{\partial\Omega} + \int_\Omega \phi_i p(x)\, dx$$

$$= \int_\Omega \left( -\frac{d^2\phi_i}{dx^2} EI \frac{d^2 y^N}{dx^2} + \phi_i p(x) \right) dx$$

$$+ \frac{d\phi_i}{dn} EI \frac{d^2 y^N}{dx^2}\Bigg|_{\partial\Omega} - \phi_i \frac{d}{dn}\left( EI \frac{d^2 y^N}{dx^2} \right)\Bigg|_{\partial\Omega} = 0 \qquad (4.121)$$

Note that the last line in (4.121) is indeed a symmetric form, since $\phi_i(x)$ must support as many derivatives as $y^N$. The use of two integration-by-parts has produced the indicated endpoint evaluations. For the given bending moment and shear definitions, the final form for (4.121) is

$$WS = \int_\Omega \left[ -\frac{d^2\phi_i}{dx^2} EI \frac{d^2 y^N}{dx^2} + \phi_i p(x) \right] dx$$

$$+ \frac{d\phi_i}{dn} M \Bigg|_{\partial\Omega} - \phi_i V_n \Bigg|_{\partial\Omega} = 0 \qquad \text{for } 1 \le i \le N \qquad (4.122)$$

In seeking a finite element discrete approximate evaluation of (4.122), the corresponding solution $y^h(x)$ remains the union of the element approximations (4.59):

$$y_e(x) = \{N_k(\zeta_i)\}^T \{Q\}_e \qquad (4.123)$$

Introducing the discretization $\Omega^h$ and inserting (4.123), (4.122) becomes, on multiplication by minus 1,

$$WS^h = S_e\left( \int_{\Omega_e} \frac{d^2\{N_k\}}{dx^2} EI \frac{d^2\{N_k\}^T}{dx^2} dx\{Q\}_e - \int_{\Omega_e} \{N_k\}\{N_k\}^T dx\{P\}_e \right.$$

$$\left. - \frac{d\{N_k\}}{dn} M \Bigg|_{\partial\Omega_e} + \{N_k\} V_n \Bigg|_{\partial\Omega_e} \right) = \{0\} \qquad (4.124)$$

Implementation of a specific finite element basis selection for (4.123) to (4.124) must address the issue of the number of applicable boundary conditions. Specifically, we must be able to independently define both the displacement $y$ and the slope $dy/dx$ at either endpoint of $\Omega$, in addition to imposing a moment and/or shear if required. One resolution is to redefine the beam equation (4.117) into a coupled *system* of two second-order differential equations. By introducing an intermediate dependent variable $g = g[y(x), x]$ we obtain

$$L(g) = -\frac{d^2}{dx^2}(EIg) + p(x) = 0 \tag{4.125a}$$

$$L(y) \equiv -\frac{d^2 y}{dx^2} + g = 0 \tag{4.125b}$$

as the new governing equation system. The moment and shear boundary conditions are then applied during solution of (4.125*a*), while the displacement and slope constraints are applicable to the solution statement for (4.125*b*).

The interested reader might proceed through derivation of the two appropriate weak statements for (4.125) and hence their finite element implementation using the developed basis family. To illustrate an added feature of finite element methodology, we elect instead to develop a *cubic Hermite* basis that meets the stated requirements for the primitive form (4.117). In developments to this point, the basis family $\{N_k(\zeta_i)\}$ possesses connections to Lagrange interpolation theory. The Hermite polynomial class is distinguished by using derivatives of the function as independent nodal degrees of freedom. In the finite element basis context, this means that certain elements in $\{Q\}_e$ in (4.123) are established by constraints involving the approximation slope $dy_e/dx$.

The simplest member of the Hermite basis family is the cubic. Hence, the basic element polynomial remains (4.17):

$$y_e(x) = a + b\left(\frac{\bar{x}}{l_e}\right) + c\left(\frac{\bar{x}}{l_e}\right)^2 + d\left(\frac{\bar{x}}{l_e}\right)^3 \tag{4.126}$$

In distinction to using the constraints (4.18) for determination of $a$, $b$, $c$, and $d$, the cubic Hermite defines that two degrees of freedom coexist at each geometric node which are the displacement and the slope. Hence, the resultant algebraic equation system determining $(a, b, c, d)$ in (4.126) is

$$
\begin{aligned}
y_e(\bar{x} = 0) &\equiv QL = a \\
y_e(\bar{x} = l_e) &\equiv QR = a + b + c + d \\
\frac{dy_e}{dx}(\bar{x} = 0) &\equiv QLP = \frac{1}{l_e}(b) \\
\frac{dy_e}{dx}(\bar{x} = l_e) &\equiv QRP = \frac{1}{l_e}(b + 2c + 3d)
\end{aligned}
\tag{4.127}
$$

where $QLP$ denotes "$Q$-left prime," the slope of $y_e$ at $\bar{x} = 0$, hence $QRP$ is "$Q$-right prime."

The solution of (4.127) for $\{Q\}_e$ yields the basis replacement of (4.126) as

$$
y_e(x) = \{N_3'(\zeta_i)\}^T\{Q\}_e \tag{4.128}
$$

For the ordering $\{Q\}_e = \{QL, QLP, QP, QRP\}_e^T$, the reader should verify that the cubic Hermite basis can be expressed in the form

$$
\{N_3'(\zeta_i)\} = \begin{Bmatrix} 1 - (\zeta_2)^2(1 + 2\zeta_1) \\ \zeta_2(\zeta_1)^2 l_e \\ (\zeta_2)^2(1 + 2\zeta_1) \\ -\zeta_1(\zeta_2)^2 l_e \end{Bmatrix} \tag{4.129}
$$

In (4.129) the $\zeta_i$, $1 \leq i \leq 2$, remain the one-dimensional *natural coordinate* system (4.4). Note that each entry is indeed a cubic polynomial, and that the set is distinctly different than the Lagrange cubic basis (4.19). Figure 4.9*a* illustrates the element nodal geometry $\{Q\}_e$ and Fig. 4.9*b* graphs the

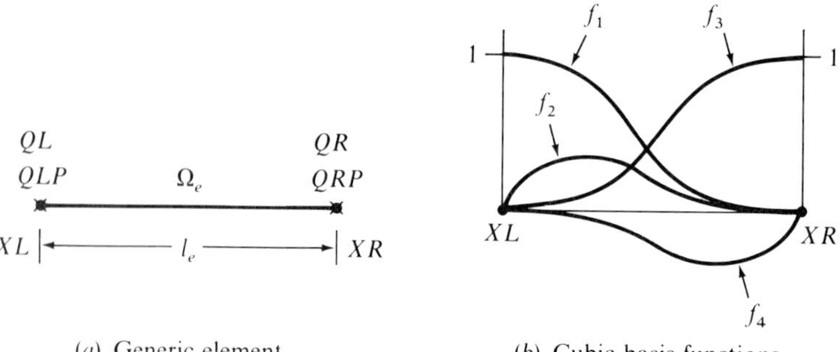


**FIGURE 4.9**
Hermite cubic basis finite element details.

entries in (4.129) with respective labeling $f_i$, $1 \le i \le 4$. Compare these data to the Lagrange basis (Fig. 4.5).

The determination (4.128) and (4.129) is the fundamental step to the finite element approximate evaluation of the weak statement (4.124) for the primitive statement (4.117). Both the first and second derivatives of (4.129) are required; a suggested exercise is to verify that

$$\frac{d\{N_3'(\zeta_i)\}}{dx} = \frac{1}{l_e}\begin{Bmatrix} -6\zeta_2(1-\zeta_2) \\ (1-4\zeta_2+3\zeta_2^2)l_e \\ 6\zeta_2(1-\zeta_2) \\ (-2\zeta_2+\zeta_2^2)l_e \end{Bmatrix} \tag{4.130}$$

$$\frac{d^2\{N_3'(\zeta_i)\}}{dx^2} = \frac{2}{l_e^2}\begin{Bmatrix} -3+6\zeta_2 \\ (-2+3\zeta_2)l_e \\ 3-6\zeta_2 \\ (-1+3\zeta_2)l_e \end{Bmatrix} \tag{4.131}$$

Equations (4.129) to (4.131) provide the definitions needed to evaluate the weak statement integrals. For the "stiffness" matrix $[K]_e$, and assuming that the beam is of nonuniform cross section, that is, the area moment of inertia is variable as $I_e(x) \simeq \{N_1(\zeta_i)\}^T\{I\}_e$, the matrix in the first term of (4.124) becomes

$$\begin{aligned}
[K]_e &\equiv \int_{\Omega_e} \frac{d^2\{N_3'\}}{dx^2}\, EI(x)\, \frac{d^2\{N_3\}^T}{dx^2}\, dx \\
&= E\{I\}_e^T \int_{\Omega_e} \begin{Bmatrix}\zeta_1\\ \zeta_2\end{Bmatrix} \frac{2}{l_e^2}\begin{Bmatrix} -3+6\zeta_2 \\ (-2+3\zeta_2)l_e \\ 3-6\zeta_2 \\ (-1+3\zeta_2)l_e \end{Bmatrix} \frac{2}{l_e^2}\{\cdot\}^T\, dx \\
&= \frac{4E}{l_l^4}\{I\}_e^T \int_0^{l_e} \begin{Bmatrix}\zeta_1\\ \zeta_2\end{Bmatrix} \\
&\quad\times \begin{bmatrix} (-3+6\zeta_2)^2 & \cdot & \cdot & \cdot \\ & (-2+3\zeta_2)^2 l_e^2 & \cdot & \cdot \\ & & (3-6\zeta_2)^2 & \cdot \\ (\text{sym}) & & & (-1+3\zeta_2)^2 l_e^2 \end{bmatrix} d\bar{x}
\end{aligned} \tag{4.132}$$

with the obvious fill-in of the $4 \times 4$ symmetric matrix. The integral evaluations in (4.132) again make ample use of the quadrature formula

(4.40). Note specifically that no difficulty is encountered in using mixed *Lagrange–Hermite* basis interpolations. A suggested reader exercise is to verify that the *stiffness hypermatrix* (4.132) is

$$[K]_e = \frac{E\{I\}_e^T}{l_e^3}\begin{bmatrix} \begin{Bmatrix}6\\6\end{Bmatrix} & \begin{Bmatrix}4\\2\end{Bmatrix}l_e & \begin{Bmatrix}-6\\-6\end{Bmatrix} & \begin{Bmatrix}2\\4\end{Bmatrix}l_e \\ \begin{Bmatrix}4\\2\end{Bmatrix}l_e & \begin{Bmatrix}3\\1\end{Bmatrix}l_e^2 & \begin{Bmatrix}-4\\-2\end{Bmatrix}l_e & \begin{Bmatrix}1\\1\end{Bmatrix}l_e^2 \\ \begin{Bmatrix}-6\\-6\end{Bmatrix} & \begin{Bmatrix}-4\\-2\end{Bmatrix}l_e & \begin{Bmatrix}6\\6\end{Bmatrix} & \begin{Bmatrix}-2\\-4\end{Bmatrix}l_e \\ \begin{Bmatrix}2\\4\end{Bmatrix}l_e & \begin{Bmatrix}1\\1\end{Bmatrix}l_e^2 & \begin{Bmatrix}-2\\-4\end{Bmatrix}l_e & \begin{Bmatrix}1\\3\end{Bmatrix}l_e^2 \end{bmatrix} \tag{4.133}$$

The second term in (4.124) involves the applied load $p = p(x)$, which is certainly variable with $x$. The selection for interpolation of these data would also not usually be Hermite, since this requires knowing the slope distribution of the load. Therefore, another mixture of Lagrange and Hermite bases is appropriate; selecting a linear interpolation for $p_e(x)$ yields

$$\begin{aligned}\{b\}_e &\equiv \int_{\Omega_e} \{N_3'\}\{N_1\}^T\,dx\{P\}_e \\ &= \int_0^{l_e} \begin{Bmatrix} 1-(\zeta_2)^2(1+2\zeta_1) \\ \zeta_2(\zeta_1)^2 l_e \\ (\zeta_2)^2(1+2\zeta_1) \\ -\zeta_1(\zeta_2)^2 l_e \end{Bmatrix}\{\zeta_1,\ \zeta_2\}\,d\bar{x}\{P\}_e \\ &= \frac{l_e}{a}[\,\cdot\,]\{P\}_e \end{aligned} \tag{4.134}$$

It is suggested you complete the entries in the $4 \times 2$ load matrix $[\,\cdot\,]$ in (4.134) and determine the common denominator $a$.

The resultant entries in $\{b\}_e$, as established by multiplication through by specific data $\{P\}_e$, are termed the *equivalent nodal loads* for the problem approximation statement. The matrix in (4.134) is nonsquare, as dictated by the choice of $\{N_1\}$ for $p_e(x)$. However, the resultant $4 \times 1$ column matrix $\{b\}_e$ correctly matches the order of $\{Q\}_e$. This requirement is met for any choice $\{N_k\}$ for $p_e(x)$; the reader should symbolically verify this for a $k = 2$ basis load interpolation.

The last two terms in (4.124) involve endpoint evaluations only, as has occurred many times. Recalling the endpoint Kronecker delta definitions $\delta_{e1}$ and $\delta_{eM}$, from (4.129) and (4.130) you may readily verify that

$$\{N'_3\}V_n\Big|_{\partial\Omega_e} = V_n\begin{Bmatrix}\delta_{e1}\\0\\\delta_{eM}\\0\end{Bmatrix} \tag{4.135a}$$

$$-\frac{d\{N'_3\}}{dn}M\Big|_{\partial\Omega_e} = M\begin{Bmatrix}0\\\delta_{e1}\\0\\-\delta_{eM}\end{Bmatrix} \tag{4.135b}$$

Note in (4.135) that an applied normal shear $V_n$ "loads" the displacement degrees of freedom in $\{Q\}_e$ while an applied moment $M$ becomes imposed on the slope degrees of freedom. Note also that the length scale $l_e$ cancels out in this enforcement.

*LEARN.FE* can be expanded to encompass solution of this class of beam problem. Similarly, the cubic Hermite basis can be applied to any second-order differential equation statement. For the heat transfer problem (4.55) to (4.57), *LEARN.FE* has in place the necessary element library contributions which are accessible via the command name *TYPE* specification K = 4 (as a close approximation to $k = 3'$) and NNODEL = 4.

## 4.9 SUMMARY

This chapter has thoroughly detailed implementation of the finite element discrete approximate evaluation of the Galerkin weak statement for a class of one-dimensional problem. The restriction to one dimension has permitted many hand operations that we hope gives the reader a feeling of dexterity with the method and its many details and attributes.

We now summarize the many informative developments stemming from the problem statement generalization and study problem analyses.

1. The choice of finite element basis $\{N_k\}$ is *fundamental* and is a decision left to the analyst.
2. Once the basis degree $k$ is defined, one can immediately generate the master matrix library valid for use with any discretization $\Omega^h$.
3. Once $\Omega^h$ is defined, creation of the algorithm global matrix statement is an algebraic detail, as is its solution.

4. Solution accuracy can be improved by refining the mesh $\Omega^h$ and/or increasing the basis degree $k$.
5. On a sufficiently refined mesh, the rate at which approximate solution accuracy improves is proportional to element measure $l_e$. For halving of the uniform mesh measure, the improvement factors are approximately 4, 16, and 64 for $k = 1$, 2 and 3 respectively. The mathematical statement of this observation is

$$\|\text{error}^h\| \simeq C_k l_e^{2k}(c_1\|\text{data}\|_\Omega + c_2\|\text{data}\|_{\partial\Omega}) \qquad (4.91a)$$

   where $C_k$ is a constant independent of $l_e$ but dependent on $k$, the finite element basis degree, and $\|\text{data}\|$ denotes the appropriate norm for problem definition data supplied on $\Omega$ and $\partial\Omega$. Equation (4.91*a*) confirms that the approximate solution error reduction will appear as a straight line of slope $2k$ on a log-log plot if the discretization $\Omega^h$ is adequate and if round-off error has not become significant.
6. The error estimate (4.91*a*) can be used to predict the exact solution values for the energy seminorm, and/or nodal values, without actually running the finest mesh solutions. The encroachment of round-off following verified convergence indicates that the finite element solution is essentially identical with the Lagrange interpolate of the exact solution to PC precision.
7. Implementation of nonhomogeneous Robin and Neumann boundary conditions is *independent* of the basis choice.
8. The vanishing normal derivative (adiabatic) boundary condition is intrinsic to the weak statement and hence comes for *free* in the absence of Dirichlet data.
9. Any distributed data are directly handled using the finite element basis for interpolation in terms of nodal values.
10. Data interpolation error can be minimized by prudent choice of the interpolation basis degree and using accurate integration.
11. Items 1 to 10 constitute confirmation that the finite element algorithm is an *exceedingly robust* approximate solution procedure exhibiting great versatility while leaving the fundamental decisions ($k$ and $\Omega^h$) to the user's discretion.

If you still have feelings of discomfort with these many details, we suggest that you reread this chapter and make sure you understand the examples and study problems. Carefully execute the *LEARN.FE* code exercises, and spend the time necessary to assimilate the output so that you feel comfortable with approximation error quantization and its control through $\{N_k\}$ and $\Omega^h$. Once this has been accomplished, you have the choice to proceed to Chap. 5, which extends the algorithm structures to two dimensions, or to skip ahead to Chap. 8, where the one-dimensional formulation is again employed for the unsteady problem class including fluid motion. We also suggest you consider constructing the finite element algorithm weak statement and set up and execute *LEARN.FE* for at least one of the following problem statements from continuum mechanics.

## 4.10 PROBLEMS

**1.** In structural mechanics, an Euler column is a slender vertical rod subject to equal and opposite compressive end loads $P$. For sufficiently large $P$, the column will deflect laterally as $y(z)$. For small deflections, the differential equation governing the lateral deflection distribution is

$$L(y) = -\frac{d^2y}{dz^2} + \frac{M}{EI} = 0 \tag{P.1}$$

where $M = -Py$ is the bending moment, $E$ is Young's modulus, and $I$ is the area moment of inertia. The applicable boundary conditions for (P.1) include $y(x = 0) = a$ and $y(x = L) = b$, for $a$ and $b$ offset data for simple pin connections, and/or $dy/dx = c$ for a clamped end.

(*a*) Following the development (4.55) to (4.58), set up the Galerkin weak statement for (P.1) and the applicable boundary conditions.
(*b*) Following (4.59) and (4.60), establish the corresponding finite element approximation $WS^h$.
(*c*) Drawing from the developments in Section 4.5, Linear Basis, establish the master matrix library for $WS^h$ for the $k = 1$ basis definition.
(*d*) Implement statement (*c*) into the FORMIT subroutine of *LEARN.FE*, using as a guide the setup for Study Problem 2.

For the case of pin connections, zero endpoint deflections ($a = 0 = b$), and uniform properties, there are restrictions on the load for which (P.1) has a solution. The analytical solution identifies specific numbers called *eigenvalues* which take the values $\lambda_n = n\pi/L$ where $L$ is the column length. The relationship to the load is $P = EI\lambda_n^2 \equiv P_n$; that is, only specific loads $P_n$ for $n = 0, 1, 2, \ldots,$ yield a solution, the normalized analytical form of which is $y_n = \sin(\lambda_n z)$.

In the approximate solution of (P.1) for this case, a numerical procedure must also be used to determine the approximation eigenvalues $\lambda_n^h$ that result from the discretization. In this process, the solution becomes degenerate, hence you can solve only for the normalized *mode shape* $y_n^h$. Computational structural mechanics codes handle all these details, which are beyond the scope of our development. Therefore, for this numerical experiment, assume that $\lambda_n^h = \lambda_n$ and pick the solution domain length as $L^* = L/(2n)$, which permits the normalized boundary condition $y(z = L^*) = 1$ to be applied. Then:

(*e*) Determine the uniform finite element discretization $\Omega^h$ of $L^*$ required for the linear basis solution $y_n^h$ to estimate the exact solution normalized mode shape $y_n$ to PC significance (six digits). The answer is $M = 16$ and round-off pollutes the $M = 64$ solution.
(*f*) Repeat step (*e*) for the $k = 2$ quadratic basis implementation. The answer is $M = 8$, and round-off corrupts the $M = 32$ solution.
(*g*) Repeat step (*e*) for the $k = 3$ cubic basis algorithm form. The answer is that all solutions for $M > 2$ are essentially Lagrange interpolates of the exact solution.

**2.** In electromagnetic field theory, the differential equation governing the propagation of a plane wave in an optically perfect medium is

$$L(\phi) = -\frac{d^2\phi}{dx^2} - \omega^2\phi = 0 \qquad \text{(P.2)}$$

where $\phi(x)$ is the electric potential and $\omega$ is the frequency of the plane wave. The applicable boundary conditions for (P.2) are typically $\phi(x = 0) = 0 = \phi(x = L)$.

(*a*) Following the development (4.55) to (4.58), set up the Galerkin weak statement for (P.1) and the pertinent boundary conditions.
(*b*) Following (4.59) and (4.60), establish the corresponding finite element approximation $WS^h$.
(*c*) Drawing from the developments in Section 4.5, Linear Basis, establish the master matrix library for $WS^h$ for the $k = 1$ basis definition.
(*d*) Implement the statement (*c*) into the FORMIT subroutine of *LEARN.FE*, using as a guide the setup for Study Problem 2.

As with Problem 1, there are a restricted set of frequencies for which (P.2) possesses a solution for the given boundary conditions. The analytical solution confirms that these are the eigenvalues $\omega_n = n\pi/L$, for all integers $n \geq 1$, and the corresponding modal solutions are $\phi_n = \sin(\omega_n x)$. In the approximate solution of (P.2), a numerical procedure must also be used to determine the approximation eigenvalues $\omega_n^h$ that result from the discretization. In this process, the matrix statement becomes degenerate, hence one can solve only for the normalized mode shape $\phi_n^h$. This detail is beyond our present scope; therefore, for the numerical experiment, assume that $\lambda_n^h = \lambda_n$ and apply the normalized modal boundary condition $\phi(x = L^*) = 1$ for the reduced wavetube length $L^* = L/2n$. Then:

(*e*) Determine the uniform finite element discretization $\Omega^h$ of $L^*$ required for the linear basis solution $\phi_n^h$ to estimate the exact solution normalized mode shape $\phi_n$ to PC significance (six digits). The answer is $M = 16$ and round-off pollutes the $M = 64$ solution.
(*f*) Repeat step (*e*) for the $k = 2$ quadratic basis implementation. The answer is $M = 8$ and round-off corrupts the $M = 32$ solution.
(*g*) Repeat step (*e*) for the $k = 3$ cubic basis algorithm form. The answer is that all solutions for $M > 2$ are essentially Lagrange interpolates of the exact solution.

**3.** An important fluid mechanics problem class concerns flow in ducts of smoothly varying cross-sectional area $A(x)$. For the assumption that the flow is isentropic and irrotational, the differential equation governing the velocity potential $\phi(x)$ is

$$L(\phi) = -\frac{d^2\phi}{dx^2} - \frac{d\phi}{dx}\frac{d(\ln A)}{dx} = 0 \tag{P.3}$$

The boundary conditions for $\phi$ are $\phi(x = L) = 0$ and $d\phi(x = 0)/dx = u_{in}$, the inlet velocity. Then:

(*a*) Following the development (4.55) to (4.58), set up the Galerkin weak statement for (P.3) and the pertinent boundary conditions.
(*b*) Following (4.59) and (4.60), establish the corresponding finite element approximation $WS^h$.
(*c*) Drawing from the developments in Section 4.5, Linear Basis, establish the master matrix form of $WS^h$ for the $k = 1$ basis definition.
(*d*) Implement the statement of (*c*) into the FORMIT subroutine of *LEARN.FE*, using as a guide the setup of Study Problem 2.
(*e*) For a specific area variation $A(x)$ of your choosing, determine the uniform finite element discretization $\Omega^h$ of $L$ required for the linear basis solution $\phi_n^h$ to approximate the exact solution to PC significance (six digits), knowing that (4.91*a*) remains valid as the expression for convergence rate.
(*f*) Repeat step (*e*) for the $k = 2$ quadratic basis implementation.
(*g*) Repeat step (*e*) for the $k = 3$ cubic basis algorithm form.

**4.** In the analysis of structural beam theory, the differential equation governing the vertical deflection $y(x)$ of a homogeneous, initially horizontal beam is

$$L(y) = -\frac{d^2}{dx^2}\left(EI\,\frac{d^2y}{dx^2}\right) + p(x) = 0 \tag{P.4}$$

where $E$ is Young's modulus, $I(x)$ is the area moment of inertia, and $p(x)$ is the imposed transverse load distribution. For a uniform beam vibrating about its static equilibrium under the action of its own weight, and assuming harmonic motion with frequency $\omega$, the load becomes

$$p(x) = \rho\omega^2 y(x) \tag{P.5}$$

where $\rho$ is the mass/length ratio. Combining (P.4) and (P.5) then yields

$$L(y) = -\frac{d^2}{dx^2}\left(EI\frac{d^2y}{dx^2}\right) + \rho\omega^2 y = 0 \tag{P.6}$$

as the Helmholtz equation governing the harmonic motion.

The natural frequencies of vibration for the case $EI$ = constant may be exactly determined (Thomson, 1988, p. 223) as

$$\omega_n = \beta_n^2\sqrt{\frac{EI}{\rho}} = (\beta_n l)^2\sqrt{\frac{EI}{\rho l^4}} \tag{P.7}$$

where $l$ is the length of the beam. The numbers $\beta_n$ depend on the boundary conditions for the problem, and for the simply supported case $(\beta_1 l)^2 = 9.87$ and $(\beta_2 l)^2 = 39.5$. In the approximate solution of (P.6), a numerical procedure must also be used to determine the approximate eigenvalues $\omega_n^h$ that result from the discretization. In this process, the matrix statement becomes degenerate, hence one can solve only for the normalized modal shape $y_n^h$. This detail is beyond our scope; therefore, for the numerical experiment, assume $\omega_n^h = \omega_n$ and apply the normalized modal boundary condition $y(x = l^*) = 1$ for the reduced beam length $l^* = l/2n$. Then determine the uniform finite element discretization $\Omega^h$ of $l^*$ required to predict the normalized mode shape $y_n^h$ to within a specified accuracy using the cubic Hermite basis implementation, Section 4.8.

5. Many practical heat transfer problems are characterized by transmission of a cold or hot fluid through a pipe covered with insulation. Rearrange the definition of Study Problem 1 to that of a two- or three-material conduction problem with boundary convection at both radial surfaces. Devise suitable material property jumps, such as metal, asbestos, and cork, and exercise *LEARN.FE* to analyze the problem definition, and then determine the mesh required to estimate the exact solution to PC precision.

## 4.11 ADDITIONAL REFERENCES

Babuska, I., O. C. Zienkiewicz, J. Gago and E. R. de Arantes Oliveira, (eds.) (1987): *Accuracy Estimates and Adaptive Refinements in Finite Element Computations*, Wiley, New York.

Demkowicz, L., J. T. Oden, and T. Strouboulis (1984): "Adaptive Finite Elements for Flow Problems with Moving Boundaries. Part I: Variational Principles and *a* Posteriori Estimates," *Comp. Mtd. Appl. Mech. Engrg.*, Vol. 46, pp. 217–251.

Demkowicz, L., J. T. Oden, and T. Strouboulis (1985): "An Adaptive *p*-Version Finite Element Method for Transient Flow Problems with Moving Boundaries," Chap. 15 in R. H. Gallagher, G. F. Carey, J. T. Oden, and O. C. Zienkiewicz (eds.), *Finite Elements in Fluids*, Vol. 6, Wiley, Chichester (U.K.)..

Oden, J. T., and J. N. Reddy (1976): *An Introduction to the Mathematical Theory of Finite Elements*, Wiley-Interscience, New York.

Thomson, W. T. (1988): *Theory of Vibration with Applications*, 3d ed., Prentice-Hall, Englewood Cliffs, N.J.

# CHAPTER 5

# TWO-DIMENSIONAL TRIANGULAR ELEMENTS

## 5.1 OVERVIEW

Discussions in Chap. 4 on one-dimensional problems, and the corresponding finite element bases, covered all fundamental aspects of algorithm construction. The restriction to one dimension allowed concentration on the essential topics, specifically, basis functions and the discretization $\Omega^h$. The extension to two- and three-dimensional problems presents no new fundamental concepts; thus the basic procedures are synonymous with the one-dimensional methodology. However, the two-dimensional problem does introduce one code practice aspect that was totally absent in the developments of Chap. 4.

Recall that the finite element approximate solution for the governing differential equation, via the corresponding weak statement, always yields the global matrix equation,

$$[K + H]\{Q\} = \{b\} \tag{5.1}$$

or equivalently,

$$S_e([K + H]_e\{Q\}_e - \{b\}_e) = \{0\} \tag{5.2}$$

This form results for all steady-state problems regardless of the specific governing equation. The global matrix equation (5.1) for two- or three-dimensional definitions does present a practical problem in that the matrix $[K + H]$ can become so large that its direct solution becomes impractical. This introduces the need for matrix iterative methods, which veritably opens a Pandora's box of opportunity. Our PC resolution in this and the succeeding chapters is to implement a Gauss-Seidel iteration technique which is simple to develop and is easily programmed. The general topic of stationary iteration is introduced in Section 5.5, and the Gauss-Seidel procedure is included in *LEARN.FE* and accessed through command name *GAUS*.

Triangular elements are the simplest of the available two-dimensional forms and can be used directly to model both structural as well as diffusion problems. The finite element method was originally based on the use of triangular elements to approximate displacement fields in aircraft structural analysis. The simplicity of the triangular element generally demands that a large number be used to accurately approximate a solution; this is particularly true when using the linear basis triangular element. This chapter develops and applies triangular element methodology for two-dimensional problems.

# A. METHODOLOGY

## 5.2 TRIANGULAR FINITE ELEMENT MESH

Discretization of a solution domain $\Omega$ is usually achieved by first subdividing it into one or more relatively large regions called *macro domains* which are denoted $\Omega^H$ as in Chap. 4. Each macro $\Omega^H$ is then subdivided into triangular elements to form a finite element mesh $\Omega^h$. A quadrilateral element always contains two triangular elements, as obtained by connecting the nodes from diagonally opposite corners of each quad. The general rule is that triangular elements which approximate an equilateral shape are preferred over those that are long and narrow. Figure 5.1 illustrates a generic quadrilateral macro domain $\Omega^H$ subdivided into triangular elements $\Omega_e$ forming a computational mesh $\Omega^h$.

A quadrilateral region $\Omega^H$ evenly divided into triangular elements contains $2(I - 1)(J - 1)$ elements, where $I$ and $J$ are the number of nodes

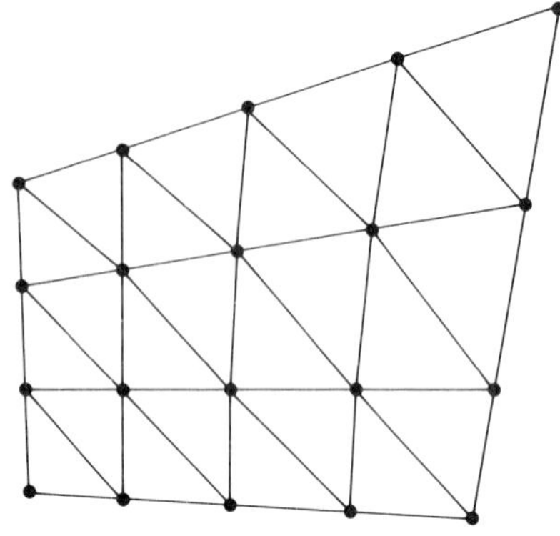

FIGURE 5.1
Quadrilateral region $\Omega$ subdivided into a mesh $\Omega^h$ of triangular elements $\Omega_e$.

on adjacent sides. A triangular macro $\Omega^H$, evenly divided, produces $(I - 1)^2$ elements, where $I$ is the number of nodes on one side. Since we will always be dealing with the *generic* element $\Omega_e$, and since nodes are regularly placed (as in the one-dimensional case), the geometric nodes defining a triangular element $\Omega_e$ are designated 1, 2, 3 and are conventionally ordered in a counterclockwise fashion. The straight-sided triangular element contains three vertex nodes, and the geometry and labeling convention is illustrated in Fig. 5.2.

As in the one-dimensional procedure, construction of an adequate computational mesh $\Omega^h$ is crucial to an accurate solution. For a simple geometry, a triangular array of elements can be easily hand-generated. However, for more complex geometries a "proper" mesh may require several tries before a suitable one is obtained, that is, yields an approximate solution with acceptable error. Obviously, generating a fairly dense multidimensional mesh by hand is tedious, since three vertex node coordinate

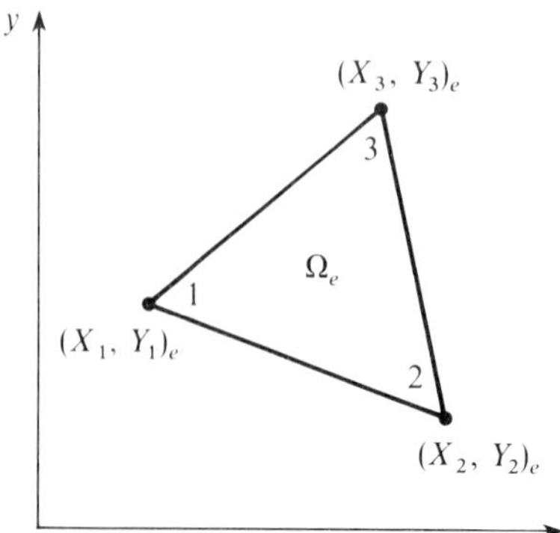


FIGURE 5.2
Generic two-dimensional triangular element $\Omega_e$.

pairs $(X_i, Y_i)_e$, $1 \leq i \leq 3$, are necessary in order to describe each element. [In three dimensions, the triangular element becomes a tetrahedron which is defined by four vertex node coordinate triples $(X_i, Y_i, Z_i)_e$, $1 \leq i \leq 4$.]

Automating grid generation via a computer code is essential when dealing with two- and three-dimensional geometries. *LEARN.FE* provides access to a generator for constructing two-dimensional meshes consisting of unions of either triangles or quadrilaterals. The code automatically generates the node numbering, the node-element connection table (MEL), and the geometric coordinate set $\{X, Y\}$ for all nodes after the user has input domain and boundary data. The grid generator is not sophisticated; it concisely illustrates the macro domain (block) method for algebraically creating computational meshes with uniform gradation of element size (measure) distributions.

This nonuniform grid-generation methodology is a direct extension of the geometric progression ratio procedure for one-dimensional macro domains $\Omega^H$ developed in Chap. 4. Recall that for node placement along the $x$-axis, according to progression ratio $p$, (4.2) expresses the right node coordinate for every element $\Omega_e$ created from $\Omega^H$. Generalizing notation slightly for $n = 2$

$$XR_{e+1} = XR_e + \frac{L_1 p_1^{e-1}}{\sum_{j=1}^{M_1} p_1^{j-1}} \qquad 1 \leq e \leq M_1 \tag{5.3}$$

where $p_1$ is now the progression ratio and $L_1$ the span of each macro domain $\Omega^H$ in the $x$ direction, i.e., $L_1 = x_R - x_L$ of $\Omega^H$. The extension to the second dimension is direct; i.e., (5.3) is reexpressed for defining the "upper" $y$ coordinate of finite element spans in the $y$-axis direction. Hence

$$YU_{e+1} = YU_e + \frac{L_2 p_2^{e-1}}{\sum_{j=1}^{M_2} p_2^{j-1}} \qquad 1 \leq e \leq M_2 \tag{5.4}$$

where $YU_e$ denotes the *upper* $y$ coordinate of each generated $\Omega_e$, $p_2$ is the corresponding progression ratio, and $L_2$ is the macro domain span in the $y$ direction, i.e., $L_2 = y_U - y_L$ of $\Omega^H$.

### Rectangular Macro Domains

The following *LEARN.FE* exercise illustrates use of (5.3) and (5.4) for triangular grid generation on a rectangular region $\Omega$, and then on the union of four rectangular macro domains $\Omega^H$ subdividing $\Omega$.

**Code Exercise 5.1.** Access the input file for this code exercise, and the following will appear on your monitor.

```
TITL
***** CODE EXERCISE 5.1, M=8, RECTANGULAR DOMAIN *****

TYPE    [K        N     NNODEL     REFL  NPRN  NTRAN   NAXI]
         1        2       3        1.0    1 1    0

PRIN    [NBUG(*)]
         1        0        1          3
GRID    [FOR N (XL XR(*)PR(*)),     NEM (PR*)]
                  2.0      3.0        1.0
         2
                  1.0      2.0        1.0
         2
STOP
```

Compare this file to that for Code Exercise 4.1; hence, note the changes under command name *TYPE*. The second entry is now a 2, denoting an $n = 2$ dimensional domain $\Omega$ is to be discretized, and NPR is generalized to NPRN with two entries for $n = 2$. Correspondingly, the input field under *GRID* now contains two sets of data entries. The first defines the discretization parallel to the $x$ axis, while the second is for the $y$-axis discretization. The former specification defines $x_L = 2.0$, $x_R = 3.0$, hence $L_1 = 1.0$. For the latter, $y_L = 1.0$ and $y_U = 2.0$; hence $L_2 = 1.0$ also. The progression ratios $p_i$ are unity in both directions, and $M_H = M_1 \times M_2 = 2 \times 2$ is the corresponding mesh generation request which will produce eight triangles $\Omega_e$ forming $\Omega^h$.

Execute *LEARN.FE* for this file, and the following output file will scroll onto your screen.

```
TITL
***** EXERCISE 5.1, M=8, RECTANGULAR DOMAIN *****
TYPE     K         N    NNODEL     REFL     NPRN   NTRAN    NAXI
         1         2      3         1.0     1 1      0        0
PRIN          NBUG(1)    (2)       (3)      (4)     (5)      (6)
                 1        0         1        3       0        0
GRID     NI       NPRI   NTRAN
         1         1      0
         XL       XR1    PR1       XR2       PR2     XR3      PR3
         2.0      3.0    1.0
         NEM1     NEM2   NEM3
         2
GRID     NI       NPRI   NTRAN
         2         1      0
         YL       YU1    PR1       YU2       PR2     YU3      PR3
         1.0      2.0    1.0
         NEM1     NEM2   NEM3
         2
```

```
ELEMENT NODE CONNECTION ARRAY(MEL)
ELEMENT NO.    NODE1    NODE2  NODE3    NODE4
         1       1        2      4
         2       5        4      2
         3       2        3      5
         4       6        5      3
         5       4        5      7
         6       8        7      5
         7       5        6      8
         8       9        8      6

NODE COORDINATES (X REAL VARIABLE)
         2.00    2.50     3.00
         2.00    2.50     3.00
         2.00    2.50     3.00

NODE COORDINATES (Y REAL VARIABLE)
         1.00    1.00     1.00
         1.50    1.50     1.50
         2.00    2.00     2.00
NODE COORDINATES (X,INTEGER MAP) EXPONENT E:-2
         200     250      300
         200     250      300
         200     250      300

NODE COORDINATES (Y,INTEGER MAP) EXPONENT E:-2
         200     250      200
         150     150      150
         100     100      100
STOP
```

For the *GRID* output entitled "NODE COORDINATES," *LEARN.FE* has generated a uniform 3 × 3 nodal mesh with $l_e = 0.5$ in each coordinate direction. *PRIN* command NBUG(4) = 3 produces both the real data and integer map output fields. Note that the integer display inverts the nodal *y*-coordinate array, such that the printfield appearance is "geometrically similar." *LEARN.FE* automatically constructs the MEL array for a triangular mesh, as shown, since NNODEL = 3 under *TYPE* keys that the *N*umber of *NOD*es per *EL*ement equals three, i.e., the triangle. The MEL array (debug) output is created when NBUG(3) = 1 as in one dimension. Request a graphic, and you will observe that indeed a uniform triangular mesh has been created. Experiment with generation of refined and/or nonuniform meshes ($p_i > 1$, $p_i < 1$) as you see fit by adapting this file.

*LEARN.FE* also generates two-dimensional rectangular meshes as unions of two macro domains $\Omega^H$ in each coordinate direction. The following exercise illustrates the use of this feature as directed by NPRN.

**Code Exercise 5.2** Access the input file for this exercise, and the following will appear on your monitor.

```
TITL
***** CODE EXERCISE 5.2, M=32, RECTANGULAR MACROS *****
TYPE    [K        N       NNODEL     REFL   NPRN   NTRAN   NAXI]
         1        2         3         1.0    2 2      0

PRIN    [NBUG(*)]
         1        0         0          2

GRID    [FOR N(XL XR(*)PR(*)), NEM(PR*)]
                 2.0        3.0         0.8      4.0      1.25
         2        2
                 1.0        2.0         0.8      3.0      1.25
         2        2
STOP
```

Compare this file to that of Code Exercise 4.1 for the third variation you created. Two macro domain spans $\Omega^H$ have been defined in each coordinate direction, which for $n = 2$ then yields a $2 \times 2 = 4$ $\Omega^H$ macro definition. Further, the progression ratio definitions will cluster smaller elements near the center of $\Omega$.

Execute this file, and the following will scroll onto your screen.

```
TITL
***** CODE EXERCISE 5.2, M=32, RECTANGULAR MACROS *****
TYPE     K        N       NNODEL     REFL   NPRN   NTRAN    NAXI
         1        2         3         1.0    2 2      0        0

PRINT      NBUG(1)       (2)       (3)       (4)     (5)      (6)
                 1        0         0         2       0        0

GRID    NI       NPRI     NTRAN
        1        2        0
        XL       XR1      PR1       XR2       PR2     XR3      PR3
        2.0      3.0      0.8       4.0       1.25
        NEM1     NEM2     NEM3
        2        2
GRID    NI       NPRI     NTRAN
        2        2        0
        YL       YU1      PR1       YU2       PR2     YU3      PR3
        1.0      2.0      0.8       3.0       1.25
        NEM1     NEM2     NEM3
        2        2

        NODE COORDINATES (X,INTEGER MAP) EXPONENT E:-2
                 200      255       300       344      400
                 200      255       300       344      400
                 200      255       300       344      400
                 200      255       300       344      400
                 200      255       300       344      400
```

```
        NODE COORDINATES (Y,INTEGER MAP) EXPONENT E:-2
                   300      300      300      300      300
                   244      244      244      244      244
                   200      200      200      200      200
                   155      155      155      155      155
                   100      100      100      100      100
STOP
```

Note that the elements with smaller spans $l_e$ are indeed located in the center region of $\Omega^h$ and hence $\Omega$, as readily viewable in the NODE COORDINATES integer map. Since NNODEL = 3, the code again creates a triangular mesh database (MEL) for these nodes. This intermediate output is now suppressed [NBUG(3) = 0]. Request a graphic, and observe the smooth nonuniformity of the created mesh. Experiment with this data file to gain familiarity with the two-dimensional grid generation capability that *LEARN.FE* possesses for rectangular domain discretizations constructed on unions of macro domains $\Omega^H$.

### Straight-Sided Quadrilateral Macro Domains

There are few really practical problems defined on regular rectangular domains. *LEARN.FE* possesses the versatility to generate smoothly nonuniform grids on both *straight-sided* and *curve-sided* general quadrilateral domains. Under command name *TYPE*, the integer variable NTRAN keys this feature, which was set to zero for the previous two exercise files. Setting NTRAN = 1 allows you to redefine the vertex nodes of a straight-sided quadrilateral macro domain $\Omega^H$ to match a corresponding geometrically genuine shape of $\Omega$. Figure 5.1 illustrates such a domain, and the following exercise illustrates accessing the *LEARN.FE* capability.

**Code Exercise 5.3.** Call up the input file for this exercise, and the following will appear on your monitor.

```
TITL
*** CODE EXERCISE 5.3, M=24, QUADRILATERAL MACRO ***
TYPE    [K        N      NNODEL    REFL  NPRN  NTRAN   NAXI]
         1        2        3        1.0   1 1    1

PRIN    [NBUG(*)]
         1        0        0        2

GRID    [FOR N (XL XR(*) PR(*)), NEM(PR*), XI(NTRAN)]
                  1.0     2.5      1.0
         4
                  1.0     2.0      1.0
         3
                  1.0     2.5      3.0        1.0
                  1.0     0.9      2.7        2.0
STOP
```

The appropriate entry under *TYPE* now defines NTRAN = 1. This invokes a bilinear *isoparametric* coordinate transformation that maps a general straight-sided quadrilateral, defined by the vertex node coordinates $XI$ and $YI$, $1 < I \leq 4$, of $\Omega^H$, to a rectangular domain. (The details of how this algebraic transformation operates are presented in the next chapter on the quadrilateral finite element.) As with the rectangle input procedure discussed, the data entries for $x_L - x_R$ with $p_1$ and $y_L - y_U$ with $p_2$ occur under *GRID*. However, the two subsequent data strings permit stretching the quadrilateral to a nonrectangular shape by defining the four vertex node coordinates of $\Omega^H$. Some duplication of data specification occurs, as illustrated, and the file requests a single macro 4 × 3 nodal mesh as shown in Fig. 5.1.

Execute this file, and the following will scroll onto your screen.

```
TITL
*** CODE EXERCISE 5.3, M=24, QUADRILATERAL MACRO ***
TYPE      K       N      NNODEL    REFL   NPRN   NTRAN   NAXI
          1       2        3        1.0    1 1     1       0

PRINT     NBUG(1)         (2)       (3)     (4)    (5)     (6)
              1            0         0       2      0       0

GRID      NI      NPRI    NTRAN
          1       1       1
          XL      XR1     PR1       XR2     PR2    XR3     PR3
          1.0     2.5     1.0
          NEM1    NEM2    NEM3
          4
GRID      NI      NPRI    NTRAN
          2       1       1
          YL      YU1     PR1       YU2     PR2    YU3     PR3
          1.0     2.0     1.0
          NEM1    NEM2    NEM3
          3

          NODE COORDINATES(X,INTEGER MAP) EXPONENT E:-2
                 100     150      200      250      300
                 100     146      192      238      283
                 100     142      183      225      266
                 100     138      175      213      250

          NODE COORDINATES(Y,INTEGER MAP) EXPONENT E:-2
                 200     217      235      252      270
                 166     177      188      199      210
                 133     137      141      145      150
                 100     97       95       92       90
STOP
```

The nodal coordinates are indeed linearly interpolated between the defined $\Omega^H$ vertex node definitions, as results for the unit progression ratio ($p_1 = 1 = p_2$) definitions. Request a graphic and note that the monitor figure appears identical to Fig. 5.1. Experiment with modifications to the input file to create meshes that are nonuniform using NPRN and XR(*), YU(*), PR(*), etc. defined on a union of $\Omega^H$. In all cases, *LEARN.FE* establishes the MEL array for a triangular element mesh, since NNODEL = 3 remains the basis definition.

## Curve-Sided Quadrilateral Macro Domains

The isoparametric coordinate transformation also permits establishment of smoothly varying nonuniform meshes $\Omega^h$ on macro domains $\Omega^H$ that possess either parabolic arc (NTRAN = 2) or circular arc (NTRAN = 3) curved sides. The resultant straight-sided triangle mesh geometrically approximates each curved line segment with the chords connecting the nodes of the generated discretization $\Omega^h$. Figure 5.3 illustrates this for a uniform $M = 32$ triangle discretization of a quarter circle macro $\Omega^H$ and the input procedure for *LEARN.FE* is as follows.

**Code Exercise 5.4.** Access this exercise input file, and the following will appear on your monitor.

```
TITL
*** CODE EXERCISE 5.4,CURVE SIDED MACRO ***
TYPE  [K         N     NNODEL     REFL  NPRN  NTRAN   NAXI]
       1         2       3         1.0   1 1     3

PRIN  [NBUG(*)]
       1         0          0          2

GRID  [FOR N (XL XR(*) PR(*)), NEM(PR*), XI(NTRAN)]
                1.0       2.0       1.0
       4
                0.0       1.0       1.0
       4

               1.0       2.0       0.0       0.0
               0.0       0.0       2.0       1.0
               1.5       1.4142    0.0       0.7071
               0.0       1.4142    1.5       0.7071

STOP
```

Note under *TYPE* that NTRAN = 3 keys the grid generator to the circular-arc curve-sided macro domain procedure. Four additional lines of nodal coordinate data for $\Omega^H$ are now required, as appear at the end of the *GRID* regular data set. The first two lines remain *XI* and *YI*, $1 \leq I \leq 4$, for the vertex nodes of $\Omega^H$. The last two lines are *XI* and *YI*, $5 \leq I \leq 8$, for the

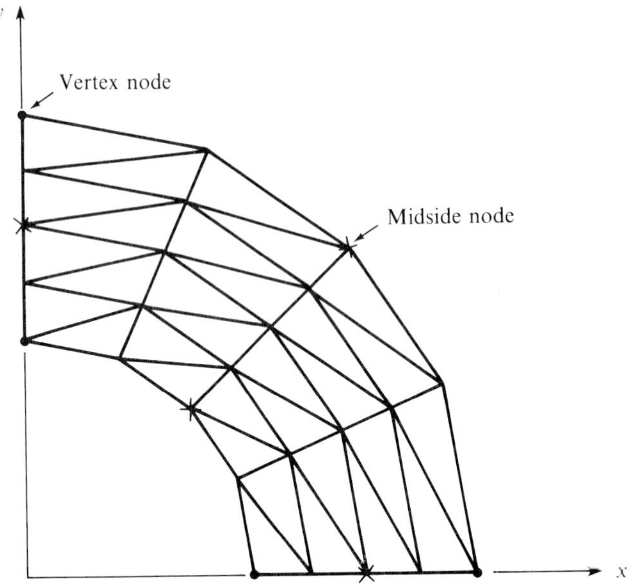


**FIGURE 5.3**
Curve-sided quadrilateral macro domain discretization.

nodes (denoted " × " in Fig. 5.3 located midway between the vertex nodes of $\Omega^H$. This extra node set permits geometric definition of curved boundaries for the macro. The specific example data yield two semicircular arcs and two straight line segments to produce the quarter-circle macro region shown in Fig. 5.3.

Execute this file, and the following will scroll onto your screen.

```
TITL
*** CODE EXERCISE 5.4, CURVE SIDED MACRO  ***
TYPE       K         N        NNODEL      REFL   NPRN   NTRAN   NAXI
           1         2          3          1.0    1 1      3       0
PRINT        NBUG(1)          (2)       (3)       (4)     (5)     (6)
                     1          0         0         2       0       0

GRID     NI        NPRI       NTRAN
         1         1          3
         XL        XR1        PR1       XR2       PR2     XR3     PR3
         1.0       2.0        1.0
         NEM1      NEM2       NEM3
         4

GRID     NI        NPRI       NTRAN
         2         1          3
         YL        YU1        PR1       YU2       PR2     YU3     PR3
         0.0       1.0        1.0
         NEM1      NEM2       NEM3
         4
```

```
X1        X2        X3        X4
1.0000    2.0000    0.0000    0.0000
Y1        Y2        Y3        Y4
0.0000    0.0000    2.0000    1.0000
X5        X6        X7        X8
1.5000    1.4142    0.0000    0.7071
Y5        Y6        Y7        Y8
0.0000    1.4142    1.5000    0.7071

NODE COORDINATES (X,INTEGER MAP) EXPONENT E:-2
       0          0          0          0          0
      38         47         57         66         76
      70         88        106        123        141
      92        115        138        161        184
     100        125        150        175        200

NODE COORDINATES (Y,INTEGER MAP) EXPONENT E:-2
     100        125        150        175        200
      92        115        138        161        184
      70         88        106        123        141
      38         47         57         66         76
       0          0          0          0          0

STOP
```

The integer map data reflection permits a good visual assimilation, while NBUG(4) = 2 again suppresses the real variable nodal coordinate output. Request a graphic; the output appears identical to Fig. 5.3. Then experiment with input file modifications to alter $\Omega^h$; for example, make it nonuniform, the union of two macros, more refined along the circular arc segments, or replace the circular arc sides with parabolic arcs (NTRAN = 2).

This completes the suggested exercises for the *LEARN.FE* two-dimensional grid generator. Hopefully you now have a feeling of familiarity with this important aspect of two-dimensional problem setup for a subsequent finite element analysis.

## 5.3 TWO-DIMENSIONAL TRIANGLE BASES

As always, the choice of the approximate solution *trial space* $\phi_i(x_j)$, $1 \le i \le N$ and $1 < j \le n = 2$, where $n$ denotes problem dimension, is *fundamental.* The necessary generalization to (3.6) for the two-dimensional problem class is

$$T^N(x, y) = \sum_{i=1}^{N} a_i \phi_i(x, y) \tag{5.5}$$

As occurred for $n = 1$ problems, the finite element method provides us with a direct translation from trial space set $\phi_i$ to the *basis function* $\{N_k(\zeta_i)\}$ appropriate for the two-dimensional triangular-shape finite element domain $\Omega_e$. This section develops this translation for the linear and quadratic basis functions.

### Linear Basis

Figure 5.2 illustrates the generic triangular finite element domain $\Omega_e$. The $k = 1$ basis $\{N_1\}$ must contain a linear polynomial expressed in the local *natural* coordinate system $\zeta_i$ spanning $\Omega_e$. The linear polynomial in global coordinates is

$$T_e(x, y) = a_1 + a_2 x + a_3 y \tag{5.6}$$

Since there are three vertex nodes for $\Omega_e$, and denoting the nodal variable array as $\{Q\}_e \equiv \{Q1, Q2, Q3\}_e^T$, the local natural coordinate expression must be

$$\begin{aligned} T_e(\zeta_i) &= \zeta_1 Q1_e + \zeta_2 Q2_e + \zeta_3 Q2_e \\ &= \{\zeta_1, \zeta_2, \zeta_3\} \begin{Bmatrix} Q1 \\ Q2 \\ Q3 \end{Bmatrix}_e \\ &= \{N_1(\zeta_i)\}^T \{Q\}_e \end{aligned} \tag{5.7}$$

The form of (5.7) firmly verifies that the $n = 2$ linear basis is the *direct extension* of the $n = 1$ basis [recall (4.6)]:

$$n = 1{:}\quad \{N_1(\zeta_i)\} = \begin{Bmatrix} \zeta_1 \\ \zeta_2 \end{Bmatrix} = \{\zeta_i\} \qquad 1 \le i \le 2 = n + 1 \tag{4.6}$$

$$n = 2{:}\quad \{N_1(\zeta_i)\} = \begin{Bmatrix} \zeta_1 \\ \zeta_2 \\ \zeta_3 \end{Bmatrix} = \{\zeta_i\} \qquad 1 \le i \le 3 = n + 1 \tag{5.8}$$

One thus can immediately state that the $n = 3$ linear basis (on the tetrahedra element $\Omega_e$) must be $\{N_1\} = \{\zeta_i\}$, $1 \le i \le 4 = n + 1$. This development is detailed in Chap. 7.

The fundamental step is to determine the functional form for the *natural coordinate system* $\zeta_i$ spanning a triangle. As in one dimension, the procedure is to evaluate (5.6) at the geometric nodes of $\Omega_e$ (recall Fig.

5.3), and then rearrange terms to the common multiplier $\{Q\}_e$. Thus

$$\begin{aligned} T_e(x = X_1, y = Y_1) &\equiv Q1 = a_1 + a_2 X_1 + a_3 Y_1 \\ T_e(x = X_2, y = Y_2) &\equiv Q2 = a_1 + a_2 X_2 + a_3 Y_2 \\ T_e(x = X_3, y = Y_3) &\equiv Q3 = a_1 + a_2 X_3 + a_3 Y_3 \end{aligned} \tag{5.9a}$$

which yields the $3 \times 3$ matrix statement

$$\begin{bmatrix} 1 & X_1 & Y_1 \\ 1 & X_2 & Y_2 \\ 1 & X_3 & Y_3 \end{bmatrix}_e \begin{Bmatrix} a_1 \\ a_2 \\ a_3 \end{Bmatrix} = \begin{Bmatrix} Q1 \\ Q2 \\ Q3 \end{Bmatrix}_e \tag{5.9b}$$

the solution of which is (a suggested exercise)

$$\begin{aligned} a_1 &= \frac{(X_2 Y_3 - X_3 Y_2)Q1 + (X_3 Y_1 - X_1 Y_3)Q2 + (X_1 Y_2 - X_2 Y_1)Q3}{2A_e} \\ a_2 &= \frac{(Y_2 - Y_3)Q1 + (Y_3 - Y_1)Q2 + (Y_1 - Y_2)Q3}{2A_e} \\ a_3 &= \frac{(X_3 - X_2)Q1 + (X_1 - X_3)Q2 + (X_2 - X_1)Q3}{2A_e} \end{aligned} \tag{5.10}$$

In (5.10), the divisor $2A_e$ is twice the triangle plane area. It is computed as the determinant of the matrix $[\cdot]$ in (5.9$b$).

$$\det[\cdot] = 2A_e = (X_1 Y_2 - X_2 Y_1) + (X_3 Y_1 - X_1 Y_3) + (X_2 Y_3 - X_3 Y_2) \tag{5.11}$$

Then, substituting (5.10) into (5.6), extracting the common multiplier $\{Q\}_e$, and equating this expression to (5.7) yields (verify this if you wish)

$$\begin{aligned} \zeta_1 &= \frac{(X_2 Y_3 - X_3 Y_2)_e + (Y_2 - Y_3)_e x + (X_3 - X_2)_e y}{2A_e} \\ \zeta_2 &= \frac{(X_3 Y_1 - X_1 Y_3)_e + (Y_3 - Y_1)_e x + (X_1 - X_3)_e y}{2A_e} \\ \zeta_3 &= \frac{(X_1 Y_2 - X_2 Y_1)_e + (Y_1 - Y_2)_e x + (X_2 - X_1)_e y}{2A_e} \end{aligned} \tag{5.12}$$

Thus, the natural coordinate system $\zeta_i$ spanning the triangle $\Omega_e$ is a linear function of the element geometric nodal coordinate set $(X_i, Y_i)_e$, $1 \le i \le 3$, the triangle plane area $A_e$, and the global coordinates $(x, y)$. It is instructive to compare (5.12) to a rearrangement of the $\zeta_i$, $1 \le i \le 2$, for the one-dimensional element $\Omega_e$. In current notation, $\bar{x} \equiv x - X_I$ in (4.4) would now be written as $\bar{x} = x - X_1$, yielding the expressions

$$\zeta_1 = \frac{X_2 - x}{l_e} \qquad \zeta_2 = \frac{-X_1 + x}{l_e} \tag{5.13}$$

where $l_e$ is the length of $\Omega_e$. Equation (5.12) is more complicated than (5.13), but (5.13) is certainly directly comparable in form. Importantly, both expressions are *explicit statements* for the elements of the linear basis $\{N_1(\zeta_i)\}$ for the appropriate-dimensional domain $\Omega_e$. This is the *sole requirement* to implement the finite element approximation of the Galerkin weak statement! Thus, the *LEARN.FE* grid generation procedures discussed in Sections 4.2 and 5.2, or any other consistent methodology that yields the global nodal array $\{X\}$ and $\{Y\}$, and the *MEL* array, provides *all data* needed to evaluate the $\zeta_i$ for any discretization formed as the "union" of finite element domains $\Omega_e$ for any problem statement of dimension $n$.

## Quadratic Basis

The derivation of the $k = 2$ basis for the triangle is straightforward. The quadratic polynomial in global coordinates is

$$T_e(x, y) = a_1 + a_2 x + a_3 y + a_4 xy + a_5 x^2 + a_6 y^2 \tag{5.14}$$

Rather than pursuing a direct algebraic assault on (5.14), however, it is easier to derive $\{N_2\}$ by analogy. Equations (4.6) and (5.8) express the $n$-dimensional generalization of the linear basis. For the additional quadratic nodal variable $QM$ located midway between the geometric nodes in one dimension (recall Section 4.3, Quadratic Basis), the $n = 1$, $k = 2$ basis was determined to be

$$\{N_2(\zeta_i)\} = \begin{Bmatrix} \zeta_1(2\zeta - 1) \\ 4\zeta_1\zeta_2 \\ \zeta_2(2\zeta_2 - 1) \end{Bmatrix} \tag{4.14}$$

For $n = 2$, correspondingly define three additional dependent variable nodes, each located midway between the geometric nodes of the generic triangle $\Omega_e$ (Fig. 5.2). Then, the quadratic triangle basis must be a cyclic permutation of the natural coordinate index $i$ in (4.14), extending its range to $n+1 = 3$. For the $\{Q\}_e$ node number convention shown in Fig.

5.4, you may easily verify that the $k = 2$ triangle basis must be

$$\{N_2(\zeta_i)\} = \begin{Bmatrix} \zeta_1(2\zeta_1 - 1) \\ \zeta_2(2\zeta_2 - 1) \\ \zeta_3(2\zeta_3 - 1) \\ 4\zeta_1\zeta_2 \\ 4\zeta_2\zeta_3 \\ 4\zeta_3\zeta_1 \end{Bmatrix} \tag{5.15}$$

In further analogy to one dimension, note that the three (vertex) geometric nodes (only) define the geometry of $\Omega_e$; hence the nodal variables $Q1$ to $Q3$ are colocated. The additional nodal dependent variables $Q4$ to $Q6$ are located at geometric midsides (denoted " × " in Fig. 5.4), hence no new geometry data are needed to specify $\Omega_e$.

In summary, (5.8) and (5.15) define the elements of the linear and quadratic *basis* for the triangular-shape finite element domain. Just as (5.13) defines the element *natural coordinate system* used for constructing *all* $n = 1$ bases, (5.12) with (5.11) specifies the corresponding $\zeta_i$ definition for all finite element bases on straight-sided triangles. The $n = 2$ cubic basis could thus be readily established by suitable extension on (4.19), the $n = 1$ form. We elect to not complete this step, however, since the geometric restriction to straight-sided triangles is severe and the resultant basis is of limited utility. The generalized bases for curve-sided triangle element domains can be developed (Zienkiewicz, 1978). We defer the associated isoparametric formulation methodology to Chap. 6. Herein, the focus is to thoroughly develop and illustrate the basic two-dimensional algorithm without undue geometric complications.

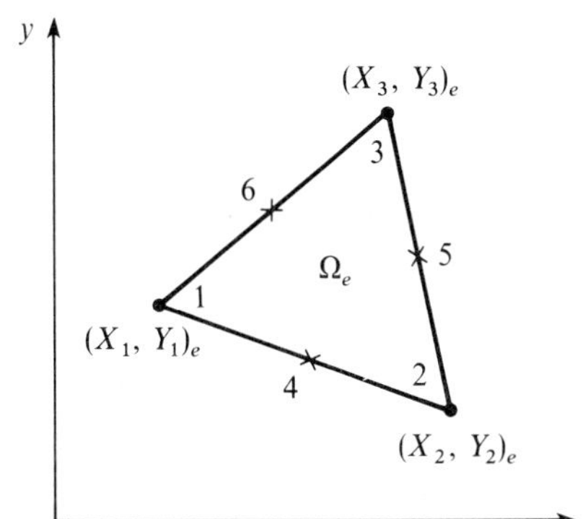


**FIGURE 5.4**
Nodal numbering for quadratic triangle finite element $\Omega_e$.

## 5.4 STEADY-STATE DIFFUSION EQUATION

We will develop the finite element algorithm for a fairly broad problem class in two-dimensional space. As in Chap. 4, the introductory development is for a linear steady-state diffusion equation with distributed source. The $n = 2$ generalization of (3.1) to (3.3), Section 4.4, is

$$\begin{aligned} L(T) &= -\frac{\partial}{\partial x}\left(k\frac{\partial T}{\partial x}\right) - \frac{\partial}{\partial y}\left(k\frac{\partial T}{\partial y}\right) - s \\ &\equiv -\nabla \cdot k\nabla T - s = 0 \qquad \text{on } \Omega \subset \mathbb{R}^2 \end{aligned} \tag{5.16}$$

$$l(T) = -k\nabla T \cdot \mathbf{n} - q_n = 0 \qquad \text{on } \partial\Omega_1 \tag{5.17}$$

$$T(\mathbf{x}_b) = T_b \qquad \text{on } \partial\Omega_2 \tag{5.18}$$

The second line in (5.16) introduces the gradient operator $\nabla \equiv \mathbf{i}\,\partial/\partial x + \mathbf{j}\,\partial/\partial y$, and the notation $\Omega \subset \mathbb{R}^2$ is read, "on the region $\Omega$ contained within the two-dimensional (Euclidean) space $\mathbb{R}^2$." For positive diffusion coefficient $k = k(x, y) > 0$, (5.16) is an *elliptic* equation; hence a linear combination of Dirichlet, Neumann, and/or Robin conditions must be defined all around the boundary $\partial\Omega$ of $\Omega$. The corresponding statements (5.17) and (5.18) meet the consistency requirement provided the "union" of $\partial\Omega_1$ and $\partial\Omega_2$ completely surrounds $\Omega$. By convention, the outward-pointing unit normal vector from $\Omega$ is denoted $\mathbf{n}$, and $\mathbf{i}$ and $\mathbf{j}$ are the unit vectors parallel to the $x$ and $y$ axes, respectively. Finally, $s = s(x, y)$ is the distributed internal source, $q_n$ is the outward-flowing fixed heat flux (Neumann data) on $\partial\Omega_1$, and $T_b = T_b(x, y \text{ on } \partial\Omega_2)$ is the fixed temperature (Dirichlet) boundary data.

The first step is to recast (5.16) to (5.18) into the symmetric Galerkin weak statement. Recalling the analogous development in Section 4.7, *any* approximation to the analytical solution $T(x, y)$ is defined as

$$T^N(x, y) \equiv \sum_{i=1}^{N} a_i \phi_i(x, y) \tag{5.19}$$

where $\phi_i(x, y)$ is the *trial function set*. We thus require that the symmetric weak form of (5.16) and (5.17) evaluated using (5.19) vanish:

$$\begin{aligned} WS &= \int_\Omega \phi_i(x, y) L(T^N)\, d\tau \equiv 0 \\ &= \int_\Omega \phi_i(-\nabla \cdot k\nabla T^N - s)\, d\tau \end{aligned}$$

$$= \int_{\Omega} (\nabla\phi_i \cdot k\nabla T^N - \phi_i s)\, d\tau - \int_{\partial\Omega} \phi_i k\nabla T^N \cdot \mathbf{n}\, d\sigma$$

$$= \int_{\Omega} (\nabla\phi_i \cdot k\nabla T^N - \phi_i s)\, d\tau + \int_{\partial\Omega_1} \phi_i q_n\, d\sigma$$

$$- \int_{\partial\Omega_2} \phi_i k\nabla T^N \cdot \mathbf{n}\, d\sigma \equiv 0 \qquad \text{for } 1 \leq i \leq N \tag{5.20}$$

The finite element implementation of (5.19) and (5.20) again replaces the (global) function set $\phi_i(x, y)$ with the non-overlapping sum (i.e., union denoted $\bigcup$) of finite element basis $\{N_k(\zeta_i)\}$, in concert with a discretization $\Omega^h$ of the domain $\Omega$ and its boundary $\partial\Omega$. The precise mathematical statement for the finite element approximation $T^h$ is

$$T^h(x, y) = \bigcup_e T_e(x, y) \tag{5.21}$$

$$T_e(x, y) = \{N_k(\zeta_i)\}^T\{Q\}_e \tag{5.22}$$

and the finite element discrete approximation to the Galerkin weak statement (5.20) is

$$WS^h = S_e\Bigg(\int_{\Omega_e} \nabla\{N_k\} \cdot k_e \nabla\{N_k\}^T\, d\tau\{Q\}_e$$

$$- \int_{\Omega_e} \{N_k\} s_e\, d\tau + \int_{\partial\Omega_1} \{N_k\} q_{n_e}\, d\sigma - \int_{\partial\Omega_2} \{N_k\} k_e \nabla T_e \cdot \mathbf{n}\, d\sigma\Bigg) = \{0\} \tag{5.23}$$

The discrete weak statement (5.23) thus requires specification of the *problem data* $k_e(x, y)$, $s_e(x, y)$, and $q_n(\mathbf{x}_b)$, and definition of the degree $k$ of the finite element basis. Without surprise, these requirements are identical to those encountered for the one-dimensional problems examined in Chap. 4. Note also that (5.23) again yields an estimation of the normal heat flux on all boundary segments $\partial\Omega_2$ where Dirichlet temperature data are specified. Finally, if we identify the first integral in (5.23) as the element conductivity matrix $[K]_e$, and combine all the remaining terms into the data matrix $\{b\}_e$, then (5.23) is the matrix statement

$$S_e([K]_e\{Q\}_e - \{b\}_e) = \{0\} \tag{5.24}$$

which is functionally identical to the one-dimensional expression (4.24). Hence, one concludes that the extension to the two-dimensional problem statement has indeed introduced no new fundamental concepts! We now proceed to fill in the algebraic detail.

**Linear Basis**

We need only to establish the conduction master matrix $[K]_e$ and the various data expressions in (5.23) and (5.24) to fully define the algorithm. For the linear basis selection, $k = 1$ in (5.22), the conduction matrix is

$$[K]_e \equiv \int_{\Omega_e} \nabla\{N_1\} \cdot k_e \nabla\{N_1\}^T \, d\tau$$
$$= \nabla\{N_1\} \cdot \nabla\{N_1\}^T \int_{\Omega_e} k_e \, d\tau \qquad (5.25)$$

Since $\{N_1\}$ contains only linear polynomials, the second equation above recognizes that first derivatives of $\{N_1\}$ are constants, hence are extractible. Specifically, and using the chain rule since $\{N_1\} = \{N_1(\zeta_i)\}$,

$$\nabla\{N_1\} \equiv \mathbf{i}\frac{\partial\{N_1\}}{\partial x} + \mathbf{j}\frac{\partial\{N_1\}}{\partial y}$$
$$= \mathbf{i}\frac{\partial\{N_1\}}{\partial \zeta_i}\frac{\partial \zeta_i}{\partial x} + \mathbf{j}\frac{\partial\{N_1\}}{\partial \zeta_i}\frac{\partial \zeta_i}{\partial y} \qquad \text{for } 1 \le i \le 3 \qquad (5.26)$$

Using (5.8), the lead expression in each term of (5.26) is trivial to evaluate as

$$\frac{\partial\{N_1\}}{\partial \zeta_i} = \begin{cases} \{1, 0, 0\}^T & \text{for } i = 1 \\ \{0, 1, 0\}^T & \text{for } i = 2 \\ \{0, 0, 1\}^T & \text{for } i = 3 \end{cases} \qquad (5.27)$$

For the expressions $\partial\zeta_i/\partial x$ and $\partial\zeta_i/\partial y$, (5.12) explicitly states the functional dependence $\zeta_i = \zeta_i(x, y) = \zeta_i(x_j)$, $1 \le j \le n = 2$, which is in fact the *coordinate transformation* from local to global coordinates. This transformation plays a central role in *LEARN.FE* for all multidimensional problem weak statement solutions. (In fact, the macro element grid generation procedure is really based on such coordinate transformations for quadrilateral macro elements $\Omega^H$.)

The *jacobian* of the element transformation $\zeta_i = \zeta_i(x_j)$ is defined as the matrix with entries $\partial\zeta_i/\partial x_j$:

$$[J]_e \equiv \left[\frac{\partial \zeta_i}{\partial x_j}\right]_e \qquad \text{for } 1 \le i \le n+1 \quad \text{and} \quad 1 \le j \le n \qquad (5.28)$$

For triangles on two-dimensional space [$n = 2$ in (5.28)], and using (5.12), you may readily determine that

$$\left[\frac{\partial \zeta_i}{\partial x_j}\right]_e = \frac{1}{2A_e}\begin{bmatrix} Y_2 - Y_3 & X_3 - X_2 \\ Y_3 - Y_1 & X_1 - X_3 \\ Y_1 - Y_2 & X_2 - X_1 \end{bmatrix}_e \tag{5.29}$$

Thus, the entries of the triangular finite element transformation jacobian are no more than algebraic differences among the nodal coordinate set $(X_i, Y_i)_e$ defining $\Omega_e$ divided by twice the element plane area $A_e$, which itself involves differences of products of the set [recall (5.11)].

With (5.27) and (5.29), we can now evaluate the matrix product $\nabla\{N_1\} \cdot \nabla\{N_1\}^T$ required to form $[K]_e$ in (5.25). Using (5.26), the matrix product is

$$\begin{aligned} &\nabla\{N_1\} \cdot \nabla\{N_1\}^T \\ &\quad = \left(\mathbf{i}\frac{\partial\{N_1\}}{\partial \zeta_i}\frac{\partial \zeta_i}{\partial x} + \mathbf{j}\frac{\partial\{N_1\}}{\partial \zeta_i}\frac{\partial \zeta_i}{\partial y}\right) \cdot \left(\mathbf{i}\frac{\partial\{N_1\}^T}{\partial \zeta_i}\frac{\partial \zeta_i}{\partial x} + \mathbf{j}\frac{\partial\{N_1\}^T}{\partial \zeta_i}\frac{\partial \zeta_i}{\partial y}\right) \\ &\quad = \mathbf{i}\cdot\mathbf{i}\left(\frac{\partial\{N_1\}}{\partial \zeta_i}\frac{\partial \zeta_i}{\partial x}\frac{\partial\{N_1\}^T}{\partial \zeta_k}\frac{\partial \zeta_k}{\partial x}\right) + \mathbf{j}\cdot\mathbf{j}\left(\frac{\partial\{N_1\}}{\partial \zeta_i}\frac{\partial \zeta_i}{\partial y}\frac{\partial\{N_1\}^T}{\partial \zeta_k}\frac{\partial \zeta_k}{\partial y}\right) \\ &\quad = \left(\frac{\partial\{N_1\}}{\partial \zeta_i}\frac{\partial\{N_1\}^T}{\partial \zeta_k}\right)\left(\frac{\partial \zeta_i}{\partial x_j}\frac{\partial \zeta_k}{\partial x_j}\right)_e \qquad \text{for } \begin{cases} 1 \le j \le n = 2 \\ 1 \le (i, k) \le n + 1 = 3 \end{cases} \end{aligned} \tag{5.30}$$

In writing (5.30), the new dummy index $k$ was introduced so that the summation rule on repeated indices (which cannot appear more than twice) would not be violated. Thereafter, the dummy index $j$ was introduced to allow combining the two scalar terms in the next to last line of (5.30) into the final single expression.

The term in the first parenthesis in the last line of (5.30) defines a $3 \times 3$ matrix with entries of either a zero or a one as formed using (5.27). For example, for $i \equiv 1$ and $k \equiv 3$

$$\frac{\partial\{N_1\}}{\partial \zeta_1}\frac{\partial\{N_1\}^T}{\partial \zeta_3} = \begin{Bmatrix} 1 \\ 0 \\ 0 \end{Bmatrix}\{0, 0, 1\} = \begin{bmatrix} 0 & 0 & 1 \\ 0 & 0 & 0 \\ 0 & 0 & 0 \end{bmatrix} \tag{5.31}$$

Experimenting with other specific index pairs, you soon find that this term in (5.30) can be compactly expressed as

$$\frac{\partial\{N_1\}}{\partial \zeta_i}\frac{\partial\{N_1\}^T}{\partial \zeta_k} = [\delta_{ik}] \qquad \text{for } 1 \le (i, k) \le 3 \tag{5.32}$$

where the matrix entry $\delta_{ik}$ is a *pseudo-Kronecker delta* with definition

$$\delta_{ik} \equiv \begin{cases} 1 \text{ for position } (i, k) \text{ in the matrix} \\ 0 \text{ for all other matrix locations} \end{cases} \tag{5.33}$$

With these developments, (5.30) now takes the form

$$\nabla\{N_1\} \cdot \nabla\{N_1\}^T = [\delta_{ik}]\left(\frac{\partial \zeta_i}{\partial x_j}\frac{\partial \zeta_k}{\partial x_j}\right)_e \qquad \text{for } \begin{cases} 1 \le j \le 2 \\ 1 \le (i, k) \le 3 \end{cases} \tag{5.34}$$

On substituting (5.34) into (5.25), the explicit statement of the element conduction master matrix for the $k = 1$ triangle basis is

$$\begin{aligned} [K]_e &= [\delta_{ik}]\left(\frac{\partial \zeta_i}{\partial x_j}\frac{\partial \zeta_k}{\partial x_j}\right)_e \int_{\Omega_e} k_e \, d\tau \\ &= \frac{1}{(2A_e)^2}\begin{bmatrix} \zeta_{11}^2 + \zeta_{12}^2 & \zeta_{11}\zeta_{21} + \zeta_{12}\zeta_{22} & \zeta_{11}\zeta_{31} + \zeta_{12}\zeta_{32} \\ \zeta_{21}\zeta_{11} + \zeta_{22}\zeta_{12} & \zeta_{21}^2 + \zeta_{22}^2 & \zeta_{21}\zeta_{31} + \zeta_{22}\zeta_{32} \\ \zeta_{31}\zeta_{11} + \zeta_{32}\zeta_{12} & \zeta_{31}\zeta_{21} + \zeta_{32}\zeta_{22} & \zeta_{31}^2 + \zeta_{32}^2 \end{bmatrix}_e \\ &\quad \times \int_{\Omega_e} k_e \, d\tau \end{aligned} \tag{5.35}$$

In (5.35), the variable $\zeta_{ij} \equiv \partial\zeta_i/\partial x_j$ is the corresponding term in the matrix (5.29), that is, a difference in a nodal coordinate pair. A suggested exercise is to verify one or two terms in (5.35), as established via (5.34) using the repeated index summation convention and (5.33).

The final step to establishing $[K]_e$ is to evaluate the integral remaining in (5.35). Thinking back to Section 4.5, Linear Basis, you must anticipate that this integral yields the average value. For the linear basis formulation, interpolating $k_e$ using $\{N_1\}$ is the natural choice. Hence

$$\int_{\Omega_e} k_e \, d\tau = \int_{\Omega_e} \{N_1\}^T\{K\}_e \, d\tau = \int_{\Omega_e} \{\zeta_1, \zeta_2, \zeta_3\} \, d\tau \{K\}_e \tag{5.36}$$

and the column matrix $\{K\}_e$ of (constant) nodal values of $k_e(x, y)$ for $\Omega_e$ is extractable. We are thus left with the requirement to integrate $\zeta_i$ over $\Omega_e$. The $n = 2$ generalization of the natural coordinate integration formula (4.40) is

$$\int_{\Omega_e} \zeta_1^p \zeta_2^q \zeta_3^r \, d\tau = 2A_e \frac{p!q!r!}{(2 + p + q + r)!} \tag{5.37}$$

where $A_e$ is the triangle plane area and $p$, $q$, $r$ remain integer exponents. You may thus readily verify that

$$\int_{\Omega_e} k_e \, d\tau = 2A_e\{\tfrac{1}{6}, \tfrac{1}{6}, \tfrac{1}{6}\}\{K\}_e = A_e\{\tfrac{1}{3}, \tfrac{1}{3}, \tfrac{1}{3}\}\{K\}_e = A_e\bar{k}_e \tag{5.38}$$

which confirms that the element-average value of $k_e$ as the consistent linear basis implementation.

The evaluation of (5.35) is now complete and

$$[K]_e = \frac{\bar{k}_e}{4A_e}\begin{bmatrix} \zeta_{11}^2 + \zeta_{12}^2 & \zeta_{11}\zeta_{21} + \zeta_{12}\zeta_{22} & \zeta_{11}\zeta_{31} + \zeta_{12}\zeta_{32} \\ \zeta_{21}\zeta_{11} + \zeta_{22}\zeta_{12} & \zeta_{21}^2 + \zeta_{22}^2 & \zeta_{21}\zeta_{31} + \zeta_{22}\zeta_{32} \\ \zeta_{31}\zeta_{11} + \zeta_{32}\zeta_{12} & \zeta_{31}\zeta_{21} + \zeta_{32}\zeta_{22} & \zeta_{31}^2 + \zeta_{32}^2 \end{bmatrix}_e \tag{5.39}$$

Equation (5.39) defines the element conductivity master matrix (5.25) for the choice $k = 1$ in (5.22) and (5.26). The two-dimensional matrix $[K]_e$ is certainly more detailed than the one-dimensional counterpart (4.26). Conversely, one cannot help noting that the functional forms (5.39) and (4.26) are intrinsically identical, that is, element-average conductivity $\bar{k}_e$ divided by the *element measure* (length $l_e$ or four times the plane area $A_e$) times a matrix of order equal to the number of geometric nodes of $\Omega_e$. In *LEARN.FE*, the subscripted Fortran name for the element jacobian matrix (5.29) is ZETA($I$, $J$). Look into subroutine FORMIT under linear triangle, and observe that the matrix elements in (5.39) are indeed formed via (5.29) and (5.11) as controlled by the MEL array created under *GRID*.

The remaining two completion steps are to evaluate the distributed source and flux boundary condition terms in $WS^h$, (5.23). Using the linear basis for interpolating all data, the former becomes

$$\int_{\Omega_e} \{N_1\} s_e \, d\tau = \int_{\Omega_e} \{N_1\}\{N_1\}^T d\tau \{S\}_e$$

$$= \frac{A_e}{12}\begin{bmatrix} 2 & 1 & 1 \\ 1 & 2 & 1 \\ 1 & 1 & 2 \end{bmatrix}\{S\}_e \tag{5.40}$$

The reader should verify (5.40) using (5.37). The specified heat flux will occur across any one (or two) boundary segments of any $\Omega_e$. Assuming for example that nodes 1 and 2 define the segment (see Fig. 5.2), the associated boundary length is

$$l_e = \sqrt{(X_2 - X_1)^2 + (Y_2 - Y_1)^2} \tag{5.41}$$

Then, using (4.40)

$$\int_{\partial\Omega_1} \{N_1\} q_{n,e}\, d\sigma = \int_{l_e} \{N_1\}\{N_1\}^T\, d\bar{x}\{QN\}_e$$

$$= \int_{l_e} \begin{Bmatrix} \zeta_1 \\ \zeta_2 \\ \zeta_3 \end{Bmatrix} \{\zeta_1, \zeta_2\}\, d\bar{x}\{QN\}_e$$

$$= \frac{l_e}{6} \begin{bmatrix} 2 & 1 \\ 1 & 2 \\ 0 & 0 \end{bmatrix} \{QN\}_e \tag{5.42}$$

since $\zeta_3 = 0$ on the line connecting nodes 1 and 2. The nodal values of the applied normal heat flux are denoted $\{QN\}_e$, and (5.42) really involves no more than a one-dimensional evaluation. Thus, the row of zeros in (5.42) are appropriately permuted should nodes 1 and 3 and/or 2 and 3 of $\Omega_e$ lie on $\partial\Omega_1$.

The master matrix library for the $k = 1$ triangular finite element algorithm is thus constituted of (5.39), (5.40), and (5.42). As in one-dimensional problems, the last term in (5.23) will vanish from (5.24) during imposition of nodal Dirichlet boundary conditions. The following example fully illustrates application to a simple problem.

**Example 5.1.** Determine the steady-state temperature distribution in a rectangular slab on $\Omega = (0 < x < 1, 0 < y < 2)$ subject to the boundary fixed-temperature distributions

$$T(x, y = 0) = 0.2x$$
$$T(x, y = 2) = 1 - 0.2x$$

while assuming the other two boundaries are adiabatic:

$$\left.\frac{\partial T}{\partial x}\right|_{x=0,y} = 0 = \left.\frac{\partial T}{\partial x}\right|_{x=1,y}$$

The slab is homogeneous with constant thermal conductivity $k = 2$.

Figure 5.5$a$ represents the solution domain $\Omega$ and the Dirichlet data distributions. The minimal macro mesh $\Omega^H$ that will yield a nodal solution is $M = M_1 \times M_2 = 1 \times 2$, which produces a $2 \times 3$ geometric nodal array. Figure 5.5$b$ shows one (out of four possible) triangular element discretizations $\Omega^h$ associated with this set of nodes.

This problem requires evaluation (only) of the element conductivity matrix $[K]_e$, $1 \le e \le 4$, using (5.39). Since $A_e = (1 \times 0.5) = 0.5$, and $k_e = 2$

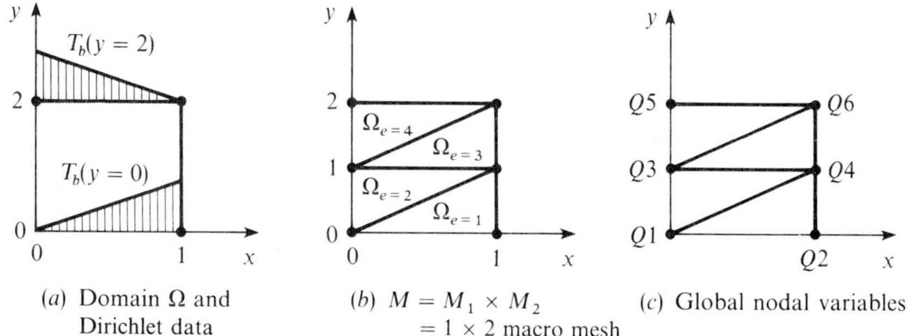


(a) Domain $\Omega$ and Dirichlet data

(b) $M = M_1 \times M_2 = 1 \times 2$ macro mesh

(c) Global nodal variables

**FIGURE 5.5**
Solution domain, triangular mesh, and nodal ordering for Example 5.1.

for all $\Omega_e$, the scalar multiplier $\bar{k}_e/4A_e$ is unity for each element. Assume the element geometric node numbering starts at the left lower corner of each $\Omega_e$ in Fig. 5.5*b*. Hence, for $\Omega_{e=1}$, (5.29) produces

$$\left[\frac{\partial \zeta_i}{\partial x_j}\right]_{e=1} = \begin{bmatrix} Y_2 - Y_3 & X_3 - X_2 \\ Y_3 - Y_1 & X_1 - X_3 \\ Y_1 - Y_2 & X_2 - X_1 \end{bmatrix}_e = \begin{bmatrix} -1 & 0 \\ 1 & -1 \\ 0 & 1 \end{bmatrix} \tag{5.43a}$$

Thus, (5.39) becomes

$$[K]_{e=1} = \frac{\bar{k}_e}{4A_e} \begin{bmatrix} \zeta_{11}^2 + \zeta_{12}^2 & \zeta_{11}\zeta_{21} + \zeta_{12}\zeta_{22} & \zeta_{11}\zeta_{31} + \zeta_{12}\zeta_{32} \\ \zeta_{21}\zeta_{11} + \zeta_{22}\zeta_{12} & \zeta_{21}^2 + \zeta_{22}^2 & \zeta_{21}\zeta_{31} + \zeta_{22}\zeta_{32} \\ \zeta_{32}\zeta_{11} + \zeta_{32}\zeta_{12} & \zeta_{31}\zeta_{21} + \zeta_{32}\zeta_{22} & \zeta_{31}^2 + \zeta_{32}^2 \end{bmatrix}_e$$

$$= 1 \begin{bmatrix} 1 & -1 & 0 \\ -1 & 2 & -1 \\ 0 & -1 & 1 \end{bmatrix} \tag{5.44a}$$

Proceeding directly, for $\Omega_{e=2}$ (5.29) produces

$$\left[\frac{\partial \zeta_i}{\partial x_j}\right]_{e=2} = \begin{bmatrix} Y_2 - Y_3 & X_3 - X_2 \\ Y_3 - Y_1 & X_1 - X_3 \\ Y_1 - Y_2 & X_2 - X_1 \end{bmatrix}_{e=2} = \begin{bmatrix} 0 & -1 \\ 1 & 0 \\ -1 & 1 \end{bmatrix} \tag{5.43b}$$

which yields for (5.39)

$$[K]_{e=1} = \frac{\bar{k}_e}{4A_e}[\cdot] = 1 \begin{bmatrix} 1 & 0 & -1 \\ 0 & 1 & -1 \\ -1 & -1 & 2 \end{bmatrix} \tag{5.44b}$$

Viewing Fig. 5.5*b*, you should anticipate that $[K]_{e=3} = [K]_{e=1}$ and $[K]_{e=4} = [K]_{e=2}$, since a translation of coordinates does not affect the element jacobian matrix (5.43). A suggested exercise is to verify that (5.43) and (5.44) are indeed appropriate for $\Omega_{e=3}$ and $\Omega_{e=4}$.

With (5.44) available for all $\Omega_e$, the next step is to assemble the algorithm matrix statement (5.24). Numbering the elements of $\{Q\}$ parallel to $x$ and in ascending $y$, (see Fig. 5.5*c*), the assembled matrix statement is

$$S_e([K]_e\{Q\}_e) = S_e(\{b\}_e)$$

As in one dimension, assembly of element matrices is keyed to the diagonal entries in $[K]_e$ regarding row placement in $[K]$. For example, for $\Omega_{e=1}$, the operation on (5.44*a*) is

$$S_e([K]_e\{Q\}_e) = [K]\{Q\}$$

$$S_{e=1}\left(\begin{bmatrix} 1 & -1 & 0 \\ -1 & 2 & -1 \\ 0 & -1 & 1 \end{bmatrix}_{e=1} \begin{Bmatrix} Q1 \\ Q2 \\ Q3 \end{Bmatrix}_{e=1}\right) = \begin{bmatrix} 1 & -1 & \cdots & 0 & \cdots \\ -1 & 2 & \cdots & -1 & \cdots \\ \cdots & \cdots & \cdots & \cdots & \cdots \\ 0 & -1 & \cdots & 1 & \cdots \\ \cdots & \cdots & \cdots & \cdots & \cdots \end{bmatrix} \begin{Bmatrix} Q1 \\ Q2 \\ Q3 \\ Q4 \\ \vdots \end{Bmatrix}$$

realizing that $[K]$ is empty when you start, hence the blank spaces. Advancing to element $\Omega_{e=2}$, and using (5.43*b*) and the above

$$S_{e=1,2}([K]_e\{Q\}_e) = S_e\left(\begin{bmatrix} 1 & -1 & 0 \\ -1 & 2 & -1 \\ 0 & -1 & 1 \end{bmatrix} \begin{Bmatrix} Q1 \\ Q2 \\ Q4 \end{Bmatrix}_{e=1}, \begin{bmatrix} 1 & 0 & -1 \\ 0 & 1 & -1 \\ -1 & -1 & 2 \end{bmatrix} \begin{Bmatrix} Q1 \\ Q4 \\ Q3 \end{Bmatrix}_{e=2}\right)$$

$$= \begin{bmatrix} 1+1 & -1 & -1 & 0+0 & \cdots \\ -1 & 2 & \cdots & -1 & \cdots \\ -1 & \cdots & 2 & -1 & \cdots \\ 0+0 & -1 & -1 & 1+1 & \cdots \\ \cdots & \cdots & \cdots & \cdots & \cdots \end{bmatrix} \begin{Bmatrix} Q1 \\ Q2 \\ Q3 \\ Q4 \\ \vdots \end{Bmatrix}$$

Proceeding on in this manner to include $\Omega_{e=3}$ and $\Omega_{e=4}$, and rearranging the diagonal presentation of $[K]$ to the column format used by

*LEARN.FE* yields for (5.24)

$$\begin{bmatrix} 0 & 0 & 1+1 & -1 & -1 \\ 0 & -1 & 2 & 0 & -1 \\ -1 & 0 & 2+1+1 & -1-1 & -1 \\ -1 & -1-1 & 1+1+2 & 0 & -1 \\ -1 & 0 & 2 & -1 & 0 \\ -1 & -1 & 1+1 & 0 & 0 \end{bmatrix} \begin{Bmatrix} Q1 \\ Q2 \\ Q3 \\ Q4 \\ Q5 \\ Q6 \end{Bmatrix} = \begin{Bmatrix} 0 \\ 0 \\ 0 \\ 0 \\ 0 \\ 0 \end{Bmatrix} \tag{5.45}$$

The reader should verify that (5.45) is indeed the assembly of (5.44) over $\Omega_e$, $1 \leq e \leq 4$, as defined and ordered in Fig. 5.5*b* and *c*.

The final step is to enforce the Dirichlet boundary data. For the mesh $\Omega^h$ in Fig. 5.5*b*, the resultant nodal constraints are

$$Q1 = 0 \qquad Q2 = 0.2 \qquad Q5 = 1.0 \qquad Q6 = 0.8$$

Thus, (5.45) is modified to the final global matrix statement $[K]\{Q\} = \{b\}$ as

$$\begin{bmatrix} 0 & 0 & 1 & 0 & 0 \\ 0 & 0 & 1 & 0 & 0 \\ -1 & 0 & 4 & -2 & -1 \\ -1 & -2 & 4 & 0 & -1 \\ 0 & 0 & 1 & 0 & 0 \\ 0 & 0 & 1 & 0 & 0 \end{bmatrix} \begin{Bmatrix} Q1 \\ Q2 \\ Q3 \\ Q4 \\ Q5 \\ Q6 \end{Bmatrix} = \begin{Bmatrix} 0 \\ 0.2 \\ 0 \\ 0 \\ 1.0 \\ 0.8 \end{Bmatrix} \tag{5.46}$$

*LEARN.FE* would solve (5.46) as presented. However, for a hand calculation, it is expedient to reduce it to a $2 \times 2$ matrix statement on the unknowns $Q3$ and $Q4$ by directly multiplying out the Dirichlet data. You should verify that the resultant matrix statement in standard (diagonal) form is,

$$\begin{bmatrix} 4 & -2 \\ -2 & 4 \end{bmatrix} \begin{Bmatrix} Q3 \\ Q4 \end{Bmatrix} = \begin{Bmatrix} 1 \\ 1 \end{Bmatrix} \tag{5.47}$$

the solution of which is

$$\begin{Bmatrix} Q3 \\ Q4 \end{Bmatrix} = \frac{1}{16-4} \begin{bmatrix} 4 & 2 \\ 2 & 4 \end{bmatrix} \begin{Bmatrix} 1 \\ 1 \end{Bmatrix} = \begin{Bmatrix} 0.5 \\ 0.5 \end{Bmatrix} \tag{5.48}$$

From the solution (5.48), the node line connecting $Q3$ and $Q4$ is indeed an isotherm. In fact, (5.48) is identical to the analytical solution on this coordinate line of $\Omega$. (Elsewhere on $\Omega$, this $M = 4$ solution is far from identical with the exact solution.) Of prime importance, this exercise illustrated *all details* of the $n = 2$ triangular finite element algorithm. It also verified that the *adiabatic* boundary condition again *comes for free*. If you feel at all uncomfortable with the details, we suggest you reread the material and make sure you understand each step. Analysis of practical problem statements follows in Section B.

## Quadratic Basis

We now complete the quadratic basis algorithm formulation. For $k = 2$ in (5.22), the element conduction matrix definition is

$$[K]_e = \int_{\Omega_e} \nabla\{N_2\} \cdot k_e \nabla\{N_2\}^T \, d\tau \tag{5.49}$$

and we may not extract $\nabla\{N_2\}$ from the integrand since it contains bilinear polynomials in the $\zeta_i$. Using the chain rule

$$\nabla\{N_2\} = \mathbf{i}\frac{\partial\{N_2\}}{\partial\zeta_i}\frac{\partial\zeta_i}{\partial x} + \mathbf{j}\frac{\partial\{N_2\}}{\partial\zeta_i}\frac{\partial\zeta_i}{\partial y} \qquad \text{for } 1 \le i \le 3 \tag{5.50}$$

and from (5.15), the lead expressions in each term in (5.50) are

$$\begin{aligned} \frac{\partial\{N_2\}}{\partial\zeta_1} &= \{4\zeta_1 - 1, 0, 0, 4\zeta_2, 0, 4\zeta_3\}^T \\ \frac{\partial\{N_2\}}{\partial\zeta_2} &= \{0, 4\zeta_2 - 1, 0, 4\zeta_1, 4\zeta_3, 0\}^T \\ \frac{\partial\{N_2\}}{\partial\zeta_3} &= \{0, 0, 4\zeta_3 - 1, 0, 4\zeta_2, 4\zeta_1\}^T \end{aligned} \tag{5.51}$$

Conversely, the second expression in each term in (5.50) is absolutely identical to the $k = 1$ basis development, since the $\zeta_i$ coordinate system is uniform for all basis degree $k$.

With (5.50) and (5.51), and assuming for simplicity that the conductivity $k_e$ is a constant on $\Omega_e$, (5.49) becomes

$$\begin{aligned} [K]_e &= k_e \int_{\Omega_e} \nabla\{N_2\} \cdot \nabla\{N_2\}^T \, d\tau \\ &= k_e \int_{\Omega_e} \left( \frac{\partial\{N_2\}}{\partial\zeta_i}\frac{\partial\{N_2\}^T}{\partial\zeta_k}\frac{\partial\zeta_i}{\partial x}\frac{\partial\zeta_k}{\partial x} + \frac{\partial\{N_2\}}{\partial\zeta_i}\frac{\partial\{N_2\}^T}{\partial\zeta_k}\frac{\partial\zeta_i}{\partial y}\frac{\partial\zeta_k}{\partial y} \right)_e d\tau \\ &= k_e \int_{\Omega_e} \frac{\partial\{N_2\}}{\partial\zeta_i}\frac{\partial\{N_2\}^T}{\partial\zeta_k} \, d\tau \left( \frac{\partial\zeta_i}{\partial x_j}\frac{\partial\zeta_k}{\partial x_j} \right)_e \qquad \text{for } \begin{cases} 1 \le (i, k) \le 3 \\ 1 \le j \le 2 \end{cases} \end{aligned} \tag{5.52}$$

Equation (5.52) is considerably more complicated than the corresponding linear basis expression (5.25). However, only nine integrals involving $6 \times 6$

matrices need to be evaluated. Further, there are many redundancies that significantly reduce the actual number of calculus operations required. For example, for $i = 1 = k$ in (5.52) and using (5.51)

$$\int_{\Omega_e} \frac{\partial\{N_2\}}{\partial\zeta_1}\frac{\partial\{N_2\}^T}{\partial\zeta_1}\,d\tau$$

$$= \int_{\Omega_e} \begin{Bmatrix} 4\zeta_1 - 1 \\ 0 \\ 0 \\ 4\zeta_2 \\ 0 \\ 4\zeta_3 \end{Bmatrix} \{4\zeta_1 - 1, 0, 0, 4\zeta_2, 0, 4\zeta_3\}\,d\tau$$

$$= \int_{\Omega_e} \begin{bmatrix} (4\zeta_1 - 1)^2 & 0 & 0 & 4\zeta_2(4\zeta_1 - 1) & 0 & 4\zeta_3(4\zeta_1 - 1) \\ & 0 & 0 & 0 & 0 & 0 \\ & & 0 & 0 & 0 & 0 \\ & & & 16\zeta_2^2 & 0 & 16\zeta_2\zeta_3 \\ & & & & 0 & 0 \\ (\text{sym}) & & & & & 16\zeta_3^2 \end{bmatrix} d\tau \tag{5.53}$$

The integral formula (5.37) remains available to evaluate the few nonzero terms in (5.53). The result is

$$\int_{\Omega_e} \frac{\partial\{N_2\}}{\partial\zeta_1}\frac{\partial\{N_2\}^T}{\partial\zeta_1}\,d\tau = \frac{A_e}{3}\begin{bmatrix} 3 & 0 & 0 & 0 & 0 & 0 \\ & 0 & 0 & 0 & 0 & 0 \\ & & 0 & 0 & 0 & 0 \\ & & & 8 & 0 & 4 \\ & & & & 0 & 0 \\ (\text{sym}) & & & & & 8 \end{bmatrix} \tag{5.54a}$$

Viewing (5.51), it should be obvious that corresponding expressions for $i = 2 = k$ and $i = 3 = k$ are simply row-column permutations of (5.54*a*). Thereby

$$\int_{\Omega_e} \frac{\partial\{N_2\}}{\partial\zeta_2}\frac{\partial\{N_2\}^T}{\partial\zeta_2}\,d\tau = \frac{A_e}{3}\begin{bmatrix} 0 & 0 & 0 & 0 & 0 & 0 \\ & 3 & 0 & 0 & 0 & 0 \\ & & 0 & 0 & 0 & 0 \\ & & & 8 & 4 & 0 \\ & & & & 8 & 0 \\ (\text{sym}) & & & & & 0 \end{bmatrix} \tag{5.54b}$$

and

$$\int_{\Omega_e} \frac{\partial\{N_2\}}{\partial\zeta_3}\frac{\partial\{N_2\}^T}{\partial\zeta_3}\,d\tau = \frac{A_e}{3}\begin{bmatrix} 0 & 0 & 0 & 0 & 0 & 0 \\ & 0 & 0 & 0 & 0 & 0 \\ & & 3 & 0 & 0 & 0 \\ & & & 0 & 0 & 0 \\ & & & & 8 & 4 \\ (\text{sym}) & & & & & 8 \end{bmatrix} \tag{5.54c}$$

The next step to completing the $k = 2$ master matrix library for $[K]_e$ is to complete these operations for $i \neq k$ in (5.52). For example, for $i = 1$ and $k = 2$

$$\int_{\Omega_e} \frac{\partial\{N_2\}}{\partial\zeta_1}\frac{\partial\{N_2\}^T}{\partial\zeta_2}\,d\tau$$

$$= \int_{\Omega_e} \begin{Bmatrix} 4\zeta_1 - 1 \\ 0 \\ 0 \\ 4\zeta_2 \\ 0 \\ 4\zeta_3 \end{Bmatrix} \{0,\, 4\zeta_2 - 1,\, 0,\, 4\zeta_1,\, 4\zeta_3,\, 0\}\,d\tau$$

$$= \int_{\Omega_e} \begin{bmatrix} 0 & (4\zeta_1 - 1)(4\zeta_2 - 1) & 0 & 4\zeta_1(4\zeta_1 - 1) & 4\zeta_3(4\zeta_1 - 1) & 0 \\ 0 & 0 & 0 & 0 & 0 & 0 \\ 0 & 0 & 0 & 0 & 0 & 0 \\ 0 & 4\zeta_2(4\zeta_2 - 1) & 0 & 16\zeta_1\zeta_2 & 16\zeta_2\zeta_3 & 0 \\ 0 & 0 & 0 & 0 & 0 & 0 \\ 0 & 4\zeta_3(4\zeta_2 - 1) & 0 & 16\zeta_1\zeta_3 & 16\zeta_3^2 & 0 \end{bmatrix} d\tau$$

$$= \frac{A_e}{3}\begin{bmatrix} 0 & -1 & 0 & 4 & 0 & 0 \\ 0 & 0 & 0 & 0 & 0 & 0 \\ 0 & 0 & 0 & 0 & 0 & 0 \\ 0 & 4 & 0 & 4 & 4 & 0 \\ 0 & 0 & 0 & 0 & 0 & 0 \\ 0 & 0 & 0 & 4 & 8 & 0 \end{bmatrix} \tag{5.55}$$

Then, since $i$ and $k$ are a symmetric index pair in (5.52), the sum of (5.55) plus the corresponding expression with $\zeta_1$ and $\zeta_2$ interchanged, which is the transpose of (5.55), produces the required (symmetric matrix) expression

$$\int_{\Omega_e}\left(\frac{\partial\{N_2\}}{\partial\zeta_1}\frac{\partial\{N_2\}^T}{\partial\zeta_2}+\frac{\partial\{N_2\}}{\partial\zeta_2}\frac{\partial\{N_2\}^T}{\partial\zeta_1}\right)d\tau=\frac{A_e}{3}\begin{bmatrix} 0 & -1 & 0 & 4 & 0 & 0 \\ & 0 & 0 & 4 & 0 & 0 \\ & & 0 & 0 & 0 & 0 \\ & & & 8 & 4 & 4 \\ & & & & 0 & 8 \\ (\text{sym}) & & & & & 0 \end{bmatrix} \tag{5.56a}$$

As before, the two other master matrix expressions generated for $i \neq k$ are again row-column permutations of (5.56*a*). Hence

$$\int_{\Omega_e}\left(\frac{\partial\{N_2\}}{\partial\zeta_1}\frac{\partial\{N_2\}^T}{\partial\zeta_3}+\frac{\partial\{N_2\}}{\partial\zeta_3}\frac{\partial\{N_2\}^T}{\partial\zeta_1}\right)d\tau=\frac{A_e}{3}\begin{bmatrix} 0 & 0 & -1 & 0 & 0 & 4 \\ & 0 & 0 & 0 & 0 & 0 \\ & & 0 & 0 & 0 & 4 \\ & & & 0 & 8 & 4 \\ & & & & 0 & 4 \\ (\text{sym}) & & & & & 8 \end{bmatrix} \tag{5.56b}$$

and

$$\int_{\Omega_e}\left(\frac{\partial\{N_2\}}{\partial\zeta_2}\frac{\partial\{N_2\}^T}{\partial\zeta_3}+\frac{\partial\{N_2\}}{\partial\zeta_3}\frac{\partial\{N_2\}^T}{\partial\zeta_2}\right)d\tau=\frac{A_e}{3}\begin{bmatrix} 0 & 0 & 0 & 0 & 0 & 0 \\ & 0 & -1 & 0 & 4 & 0 \\ & & 0 & 0 & 4 & 0 \\ & & & 0 & 4 & 8 \\ & & & & 8 & 4 \\ (\text{sym}) & & & & & 0 \end{bmatrix} \tag{5.56c}$$

Thus, the six matrices given in (5.54) and (5.56) plus (5.29) constitute the quadratic basis library elements for the conductivity master matrix $[K]_e$ defined in (5.52). Note that while $A_e$ is a common multiplier in (5.54) and (5.56), it will partially cancel with the $(2A_e)^{-2}$ generated in products of (5.29) [recall (5.39)]. Finally, the variable set ZETA($I$, $J$) established for the $k = 1$ basis formulation is directly utilized for the $k = 2$ algorithm computation for $[K]_e$ [i.e., (5.52)].

Two steps remain to full completion of the quadratic basis master matrix library. The source term becomes

$$\int_{\Omega_e} \{N_2\} s_e \, d\tau = \int_{\Omega_e} \{N_2\}\{N_2\}^T \, d\tau \{S\}_e$$

$$= \frac{A_e}{180} \begin{bmatrix} 6 & -1 & -1 & 0 & -4 & 0 \\ & 6 & -1 & 0 & 0 & -4 \\ & & 6 & -4 & 0 & 0 \\ & & & 32 & 16 & 16 \\ & & & & 32 & 16 \\ (\text{sym}) & & & & & 32 \end{bmatrix} \{S\}_e \tag{5.57}$$

Evaluating the flux boundary condition for (5.17) involves no more than a one-dimensional quadratic basis integral, with row-column entry adjusted for the specific triangle element boundary nodes lying on $\partial\Omega_1$. Assuming, for example, that these nodes are 1, 4, 2, recall Fig. 5.4, and for the corresponding boundary segment length $l_e$ determined from (5.41)

$$\int_{\partial\Omega_1} \{N_2\} q_{n,e} \, d\sigma = \int_{\partial\Omega_1} \{N_2\}\{N_2\}^T \, dx \{QN\}_e$$

$$= \int_{l_e} \begin{Bmatrix} \zeta_1(2\zeta_1 - 1) \\ \zeta_2(2\zeta_2 - 1) \\ \zeta_3(2\zeta_3 - 1) \\ 4\zeta_1\zeta_2 \\ 4\zeta_2\zeta_3 \\ 4\zeta_3\zeta_1 \end{Bmatrix} \{\zeta_1(2\zeta_1 - 1), 4\zeta_1\zeta_2, \zeta_2(2\zeta_2 - 1)\} \, d\bar{x} \{QN\}_e$$

$$= \frac{l_e}{30} \begin{bmatrix} 4 & 2 & -1 \\ -1 & 2 & 4 \\ 0 & 0 & 0 \\ 2 & 16 & 2 \\ 0 & 0 & 0 \\ 0 & 0 & 0 \end{bmatrix} \{QN\}_e \tag{5.58}$$

Equations (5.54) and (5.56) to (5.58) define the complete triangluar element quadratic basis master matrix library.

**Example 5.2.** Set up a solution for the steady-state temperature distribution in the rectangular slab problem of Example 5.1 using the quadratic basis formulation on an $M = M_1 \times M_2 = 1 \times 1$ $\Omega^H$ discretization. Figure 5.6*a* shows the mesh $\Omega^h$, and Fig. 5.6*b* labels the associated element (local) nodal degrees of freedom. This problem requires evaluation (only) of the element conduction matrix $[K]_e$, $1 \le e \le 2$, as defined in (5.52) and using (5.29), (5.54), and (5.56).

For the counterclockwise vertex node numbering convention (see Fig. 5.4), and noting the $A_e = (1 \times 2)/2 = 1$ for each $\Omega_e$, (5.29) yields

$$\left[\frac{\partial \zeta_i}{\partial x_j}\right]_{e=1} = \frac{1}{2A_e}\begin{bmatrix} Y_2 - Y_3 & X_3 - X_2 \\ Y_3 - Y_1 & X_1 - X_3 \\ Y_1 - Y_2 & X_2 - X_1 \end{bmatrix}_{e=1} = \frac{1}{2}\begin{bmatrix} -2 & 0 \\ 2 & -1 \\ 0 & 1 \end{bmatrix}_{e=1} \tag{5.59a}$$

$$\left[\frac{\partial \zeta_i}{\partial x_j}\right]_{e=2} = \frac{1}{2}\begin{bmatrix} 0 & -1 \\ 2 & 0 \\ -2 & 1 \end{bmatrix}_{e=1} \tag{5.59b}$$

As defined in (5.52), formation of $[K]_e$ requires summation over the repeated indices $i$ and $k$. For notational simplicity, label the integrals contained in the master matrix library (5.54) and (5.56) as

$$[K_{ik}]_e \equiv \int_{\Omega_e} \frac{\partial\{N_2\}}{\partial \zeta_i}\frac{\partial\{N_2\}^T}{\partial \zeta_k}\, d\tau \tag{5.60}$$

Hence, the DO loop forming $[K]_e$ is

$$[K]_e \equiv \sum_{i=1}^{3}\sum_{k=1}^{3}\sum_{j=1}^{3} k_e[K_{ik}]\left(\frac{\partial \zeta_i}{\partial \zeta_j}\frac{\partial \zeta_k}{\partial \zeta_j}\right)_e \tag{5.61}$$

For $\Omega_{e=1}$, we then have

$$[K]_{e=1} = k_e\left\{[K_{11}]\left[\left(\frac{\partial \zeta_1}{\partial x}\right)_e^2 + \left(\frac{\partial \zeta_1}{\partial y}\right)_e^2\right] + [K_{22}]\left[\left(\frac{\partial \zeta_2}{\partial x}\right)_e^2 + \left(\frac{\partial \zeta_2}{\partial y}\right)_e^2\right]\right.$$

$$\left. + \cdots + ([K_{23}] + [K_{32}])\left(\frac{\partial \zeta_3}{\partial x}\frac{\partial \zeta_2}{\partial x} + \frac{\partial \zeta_3}{\partial y}\frac{\partial \zeta_2}{\partial y}\right)_e\right\}$$

$$= \frac{A_e k_e}{(2)^2 3}\left(\begin{bmatrix} 3 & 0 & 0 & 0 & 0 & 0 \\ & 0 & 0 & 0 & 0 & 0 \\ & & 0 & 0 & 0 & 0 \\ & & & 8 & 0 & 0 \\ & & & & 0 & 0 \\ (\text{sym}) & & & & & 8 \end{bmatrix}((-2)^2 + (0)^2)\right.$$

$$\left. + \cdots + \begin{bmatrix} 0 & 0 & 0 & 0 & 0 & 0 \\ & 0 & -1 & 0 & 4 & 0 \\ & & 0 & 0 & 4 & 0 \\ & & & 0 & 4 & 8 \\ & & & & 8 & 4 \\ (\text{sym}) & & & & & 8 \end{bmatrix}(0(2) + 1(-1))\right)$$

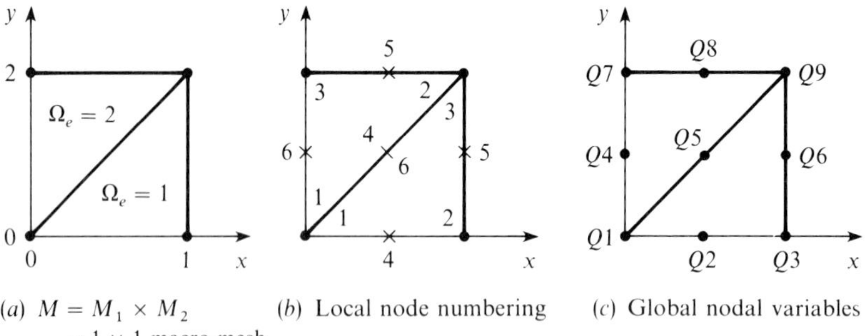


(a) $M = M_1 \times M_2$ $= 1 \times 1$ macro mesh (b) Local node numbering (c) Global nodal variables

**FIGURE 5.6**
Quadratic element discretization and nodal definition for Example 5.2.

using (5.57*a*) and extracting the $(1/2)^2$ as a common multiplier. Noting that $A_e k_e/4 \times 3 = 1 \times 1/4 \times 3 = 1/12$, the completion of the DO loop (5.61) for $e = 1$ yields

$$[K]_{e=1} = \frac{1}{12}\begin{bmatrix} 12 & 4 & 0 & -16 & 0 & 0 \\ & 15 & 1 & -16 & -4 & 0 \\ & & 3 & 0 & -4 & 0 \\ & & & 40 & 0 & -8 \\ & & & & 40 & -32 \\ (\text{sym}) & & & & & 40 \end{bmatrix} \tag{5.62a}$$

Proceeding through the same detail for $e = 2$ produces

$$[K]_{e=2} = \frac{1}{12}\begin{bmatrix} 3 & 0 & 1 & 0 & 0 & -4 \\ & 12 & 4 & 0 & -16 & 0 \\ & & 15 & 0 & -16 & -4 \\ & & & 40 & -8 & -32 \\ & & & & 40 & 0 \\ (\text{sym}) & & & & & 40 \end{bmatrix} \tag{5.62b}$$

Equation (5.62) defines the two finite element conduction matrices. The next step is to form the global statement (5.1) as the assembly over the elements of the mesh. Each matrix (5.62) is ordered according to the element local node convention defined in Fig 5.6*b*. The assembly operation produces the system matrix ordered according to the global node numbering defined in Fig. 5.6*c*. Hence, in the conventional (diagonal) presentation format the global statement is

$$[K]\{Q\} = S_e([K]_e\{Q\}_e) = \{b\} \qquad \text{for } 1 \le e \le 2$$

$$= \frac{1}{12}\begin{bmatrix} \begin{pmatrix} 3 \\ +12 \end{pmatrix} & -16 & 4 & -4 & 0 & 0 & 1 & 0 & 0 \\ -16 & 40 & -16 & 0 & -8 & 0 & 0 & 0 & 0 \\ 4 & -16 & 16 & 0 & 0 & -4 & 0 & 0 & 1 \\ -4 & 0 & 0 & 40 & -32 & 0 & -4 & 0 & 0 \\ 0 & -8 & 0 & -32 & \begin{pmatrix} 40 \\ +40 \end{pmatrix} & -32 & 0 & -8 & 0 \\ 0 & 0 & -4 & 0 & -32 & 40 & 0 & 0 & -4 \\ 1 & 0 & 0 & -4 & 0 & 0 & 15 & -16 & 4 \\ 0 & 0 & 0 & 0 & -8 & 0 & -16 & 40 & -16 \\ 0 & 0 & 1 & 0 & 0 & -4 & 4 & -16 & \begin{pmatrix} 3 \\ +12 \end{pmatrix} \end{bmatrix} \begin{Bmatrix} Q1 \\ Q2 \\ Q3 \\ Q4 \\ Q5 \\ Q6 \\ Q7 \\ Q8 \\ Q9 \end{Bmatrix}$$

$$= \{0\} \tag{5.63}$$

In (5.63), the assembled entries for global nodes $Q1$, $Q5$, and $Q9$ are presented as individual contributions to help illustrate the assembly process. Note that $[K]$ is indeed symmetric, banded about the diagonal and relatively sparse. The final operation is imposition of the Dirichlet data, which for this discretization requires

$$Q1 = 0.0 \qquad Q2 = 0.1 \qquad Q3 = 0.2$$
$$Q7 = 1.0 \qquad Q8 = 0.9 \qquad Q9 = 0.8$$

(recall Fig. 5.5$a$). For a hand computation, the first three and last three rows in (5.63) are deleted and the data moved to the right side in the middle three row locations in $\{b\}$. The resultant rank 3 global matrix statement is

$$\begin{bmatrix} 40 & -32 & 0 \\ -32 & 80 & -32 \\ 0 & -32 & 40 \end{bmatrix} \begin{Bmatrix} Q4 \\ Q5 \\ Q6 \end{Bmatrix} = \begin{Bmatrix} 4.0 \\ 8.0 \\ 4.0 \end{Bmatrix} \tag{5.64}$$

By inspection, the solution of (5.64) is

$$\begin{Bmatrix} Q4 \\ Q5 \\ Q6 \end{Bmatrix} = \begin{Bmatrix} 0.5 \\ 0.5 \\ 0.5 \end{Bmatrix} \tag{5.65}$$

hence the middle line of nodes lies on an isotherm.

The nodal solution (5.65) should not be a surprise. It is in agreement with the exact solution and with the linear basis nodal solution (5.48). (Again, this is the only location on $\Omega$ where either approximate solution agrees with the exact solution. Elsewhere, both of these *coarse grid* solutions are in substantial error!) The exercise value is a complete exposure to the triangular element $k = 2$ basis algorithm operational sequence for an $n = 2$ problem.

*LEARN.FE* does not contain the quadratic basis master matrix library, but you certainly can install it by using the transparent sequences in subroutines BLOCK DATA and ELMAT. Subroutine FORMIT is flagged at the location where the element DO loop (5.61) should be installed. It is left as a reader exercise to add (5.54) and (5.56) to (5.58) to the capabilities of *LEARN.FE.* Before doing so, we suggest you read the following section to become cognizant of the moderating issue.

## 5.5 NUMERICAL LINEAR ALGEBRA METHODS

This is the final and very important topic considered under the scope of "methodology." As commented on in Section 5.1, multidimensional finite element methods produce a matrix statement (5.1) for which the global rank of $[K + H]$ can become large compared to current PC memory capacity. While (5.1) is a linear equation, and hence can be solved directly in theory, in PC practice it may be preferable and perhaps mandatory to use a numerically approximate solution procedure wherein the large global matrix is not equation solved. There are numerous candidates about which the reader may be knowledgeable such as *Jacobi* iteration, *Gauss-Seidel*, successive overrelaxation (SOR), strongly implicit (SIP), and alternating direction (ADI), to name only a few. The design and construction of methods in numerical linear algebra is a mature but expanding field (Varga, 1965; Ames, 1977). Our purpose here is to expose the basic concept only, hence to introduce a representative method as implemented in the *LEARN.FE* code.

For the linear field problem class addressed herein, our choice is a stationary iteration procedure. To make notation consistent with the field of linear algebra, the requirment is to solve the matrix statement

$$Au = s \tag{5.66}$$

where a lowercase letter denotes a column matrix while a capital letter signifies a square matrix. Thus, comparing (5.66) and (5.1), $A$ notationally replaces $[K + H]$, $u$ replaces $\{Q\}$, and $s$ replaces $\{b\}$.

Provided $A$ is not singular, i.e., $\det A \neq 0$, the formal solution statement for (5.66) is

$$u = A^{-1}s \tag{5.67}$$

Of course, (5.67) is never exercised in practice since the operation count for forming the inverse $A^{-1}$ of $A$ is practically prohibitive. Instead, a *Gauss elimination* direct solution of (5.66), as accessed in *LEARN.FE* under command *SOLV*, achieves (5.67) in an operational sequence to within machine round-off error. However, such a procedure becomes difficult to execute on a PC when $A$ is large. Therefore, letting superscript $k$ denote the $k$th estimate (called an *iterate*) of a variable, a numerical linear algebra procedure replaces (5.67) with the operation

$$u^k = f^k(A, s, u^{k-1}, u^{k-2}, \ldots) \tag{5.68}$$

In (5.68), $u^k$ is the unknown and $f^k(\cdot)$ denotes a function of the indicated arguments. If $f^k(\cdot)$ is independent of $k$, then (5.68) defines a *stationary iteration*. Further, if $f^k(\cdot)$ is linear in $u^{k-1}$, the stationary iteration is called *linear* and the general form is

$$u^k = B^k u^{k-1} + r^k \tag{5.69}$$

Hence, using (5.67) we obtain

$$r^k = (I - B^k)A^{-1}s \tag{5.70}$$

where $I$ is the identity matrix. Combining (5.69) and (5.70), the general *linear stationary iteration* formula is

$$u^k = B^k u^{k-1} + M^k s \tag{5.71}$$

where

$$M^k A + B^k = I \tag{5.72}$$

The principal interest in any numerical linear algebra procedure is efficiency, in particular determination of how rapidly $u^k$ approaches the true solution $u$ to (5.66). The error in any iterate is

$$\begin{aligned} e^k &\equiv u^k - u \\ &= u^k - A^{-1}s \\ &= B^k u^{k-1} + M^k s - A^{-1}s \\ &= B^k e^{k-1} \end{aligned} \tag{5.73}$$

Thus, the size of $e^k$ depends directly on the choice of the *iteration matrix* $B^k$. The theoretical requirement is that $e^k \to 0$ as $k$ becomes sufficiently large. Equation (5.73) indicates that

$$e^k = B^k e^{k-1} = B^k B^{k-1} e^{k-2} = \cdots = B^k B^{k-1} \cdots B^1 e^0 \tag{5.74}$$

where $e^0$ is the error in the initial guess $u^0$ for $u$. Thus, provided the matrix product $B^k B^{k-1} \cdots B^1$ has a magnitude of less than one, that is, the matrix spectral radius is less than unity, then (5.71) defines a *convergent* iteration procedure.

The variety of *point iterative* methods are distinguished by the choice of the iteration matrix $B^k$. The best known methods construct $B$ around the partition of $A$ into

$$A = L + D + U \tag{5.75}$$

where $L$, $D$, and $U$ are square matrices having the same elements as $A$ below the main diagonal ($L$), on the main diagonal ($D$) and above it ($U$). In particular, $L$ is a lower triangular matrix and $U$ is an upper triangular matrix. The Jacobi iteration is the simplest, wherein for the indicated partitioning of $A$, (5.69) takes the form

$$u^k = -D^{-1}(L + U)u^{k-1} + D^{-1}s \tag{5.76}$$

Comparing to (5.71), the following definitions result

$$\begin{aligned} B^k &\to B = -D^{-1}(L + U) \\ M^k &\to M = D^{-1} \end{aligned} \tag{5.77}$$

Thus, for the Jacobi stationary iteration, both $B^k$ and $M^k$ are independent of $k$ and are very easy to form. A code implementation of (5.76) is thus very straightforward since the inverse of the diagonal matrix $D$ (containing the diagonal elements of $A$) is trivial to form. Conversely, the Jacobi method is not in significant use today since the rate of convergence of $u^k$ to $u$ is ponderously slow.

The Gauss-Seidel iteration procedure differs from the Jacobi iteration in that each new iterate value is used immediately in calculating successive solution estimate entries (in the column matrix $u^k$). It enjoys an improved convergence rate at the cost of tracking the sequential ordering in $u^k$, hence the structure of $A$. Denoting the elements of $B$ as $b_{ij}$, with the corresponding notation for $M$, the *algebraic form* of the Gauss-Seidel definition for (5.71) is

$$u_i^k = \sum_{j=1}^{i-1} b_{ij} u_j^k + \sum_{j=i+1}^{N} b_{ij} u_j^{k-1} + \sum_{j=1}^{N} m_{ij} s_j \tag{5.78}$$

where $N$ is the number of entries in $u^k$ which is equal to the rank of $A$. The corresponding matrix statement form of (5.71) is

$$Iu^k = -D^{-1}Lu^k - D^{-1}Uu^{k-1} + D^{-1}s \tag{5.79}$$

Multiplying (5.79) through by $D$ and then solving for $u^k$ yields the matrix statement form of Gauss-Seidel iteration as

$$u^k = -(L - D)^{-1}Uu^{k-1} + (L + D)^{-1}s \tag{5.80}$$

The distinction between Jacobi and Gauss-Seidel iteration is clearly seen by comparing (5.76) and (5.80). The improved convergence rate of the latter comes at the expense of forming $(L + D)^{-1}$, which is not particularly difficult since $(L + D)$ is lower triangular. Command name *GAUS*, when used to replace *SOLV*, invokes a Gauss-Seidel solution process of the form (5.80) in *LEARN.FE*.

For finite element algorithm matrix statements [recall (5.1)] of moderate size, as created for linear basis two-dimensional problems, for example, either direct or iterative numerical methods may be applied. Gauss elimination and/or Cholesky decomposition are the most widely used direct methods (*SOLV* employs the former), while Gauss-Seidel and successive overrelaxation (SOR) are the most popular iterative schemes. For comparison to (5.80) and/or (5.76), the matrix statement of SOR is

$$u^k = (D + \omega L)^{-1}\{[(1 - \omega)D - \omega U]u^{k-1} + \omega b\} \tag{5.81}$$

where $1 \leq \omega \leq 2$ is the *relaxation factor* that must be determined for each problem statement (5.66), i.e., (5.1), for optimal efficiency.

Regarding direct methods, no algorithm for equation solving involves fewer operation counts than Gauss elimination (Akin, 1982), which is a two-step process. The first involves factorization of the matrix $A$ into the partition (5.75). To eliminate the matrix terms below the main diagonal in column $i$, one multiplies row $i$ by $(-a_{ij}/a_{ii})$ and adds the result to row $j$ where $(i + 1) \leq j \leq n$ and $a_{ji}$ denotes the matrix coefficients of $A$. Once the factorization is complete, the last matrix row is of the form $\bar{a}_{nn}u_n = \bar{s}_n$, where the superscript bar denotes the elimination-modified terms. The second step of Gauss elimination is back-substitution, which starts with the solution $u_n = \bar{s}_n/\bar{a}_{nn}$. Once determined, $u_n$ is substituted into the next-to-last equation, hence $u_{n-1}$ is determined and so on until $u_1$ becomes established.

Most commercial finite element codes replace Gaussian elimination with variations of the *Cholesky* method (Conte 1968), which yields more efficient storage and faster computational speed. For many problems, less than half of the global matrix coefficients $a_{ij}$ are needed because of the symmetric and banded nature of $A$ [recall (5.46) and/or (5.63)]. For a symmetric matrix $A$ with half-bandwidth $c$, the upper triangular portion of the submatrix of order $c \times c$ must be available in memory at all times and the storage requirement is $c(c+1)/2$. The magnitude of $c$ depends directly on the global node ordering which is critical when using banded solution techniques. The grid generator in *LEARN.FE* uses a simple method to minimize $c$ during construction of the MEL array.

## B. PRACTICAL PROBLEMS

### 5.6 STEADY HEAT CONDUCTION WITH BOUNDARY CONVECTION

Following Chap. 4, the steady conduction equation is generalized for variable properties and boundary surface thermal convection. Recalling (5.16) to (5.18), the two-dimensional generalization is

$$L(T) = -\nabla \cdot k(x_i)\nabla T - s(x_i) = 0 \qquad \text{on } \Omega \subset \mathbb{R}^2 \tag{5.82}$$

$$l(T) = k\nabla T \cdot \mathbf{n} + h(T - T_r) = 0 \qquad \text{on } \partial\Omega_1 \tag{5.83}$$

$$l(T) = k\nabla T \cdot \mathbf{n} + q_n = 0 \qquad \text{on } \partial\Omega_2 \tag{5.84}$$

$$T(\mathbf{x}_b) = T_b \qquad \text{on } \partial\Omega_3 \tag{5.85}$$

and the union of $\partial\Omega_1$, $\partial\Omega_2$, and $\partial\Omega_3$ forms the complete boundary $\partial\Omega$ of $\Omega$.

The weak statement (5.20) remains appropriate with the obvious generalization for (5.83). Further, (5.21) and (5.22) remain the expression of the finite element approximation in terms of basis degree $k$. The finite element approximate evaluation of $WS$ for (5.82) to (5.85) is then

$$\begin{aligned} WS^h = S_e\Bigg(&\int_{\Omega_e} \nabla\{N_k\} \cdot k_e \nabla\{N_k\}^T \, d\tau \{Q\}_e \\ &- \int_{\Omega_e} \{N_k\} s_e \, d\tau + \int_{\partial\Omega_1} \{N_k\} h_e \{N_k\}^T \{Q - TR\}_e \, d\sigma \\ &+ \int_{\partial\Omega_2} \{N_k\} q_{n_e} \, d\sigma - \int_{\partial\Omega_3} \{N_k\} k_e \nabla T_e \cdot \mathbf{n} \, d\sigma\Bigg) \\ &= S_e([K_e + H_e]\{Q_e\} - \{SC\}_e - [H]_e\{TR\}_e + \{QF\}_e - \{FF\}_e) \\ &= S_e([K_e + H_e]\{Q\}_e - \{b\}_e) = \{0\} \end{aligned} \tag{5.86}$$

which is a modest generalization of (5.23) and (5.24). As before, the boundary flux term nodal values $\{FF\}_e$ on $\partial\Omega_3$ never enter the global matrix statement as a result of Dirichlet data imposition.

All master matrices required for (5.86) have been established except for $[H]_e$ and $\{QF\}_e$, the boundary condition terms corresponding to convective flux and imposed normal flux, respectively. Actually, since $\partial\Omega_1$ and $\partial\Omega_2$ each correspond to a one-dimensional line segment, these two-dimensional master matrix expressions were determined in Chap. 4. Recalling the development in Section 4.7:

$k = 1$:

$$[H]_e \equiv \int_{\partial\Omega_e} \{N_1\}h_e\{N_1\}T d\sigma$$

$$= \frac{l_e}{12}\{H\}_e^T \begin{bmatrix} \begin{Bmatrix} 3 \\ 1 \end{Bmatrix} & \begin{Bmatrix} 1 \\ 1 \end{Bmatrix} \\ \begin{Bmatrix} 1 \\ 1 \end{Bmatrix} & \begin{Bmatrix} 1 \\ 3 \end{Bmatrix} \end{bmatrix} \tag{5.87a}$$

$$\{QF\}_e = \int_{\partial\Omega_e} \{N_1\}q_{n_e}\,d\sigma$$

$$= \frac{l_e}{6}\begin{bmatrix} 2 & 1 \\ 1 & 2 \end{bmatrix}\{QN\}_e \tag{5.87b}$$

$k = 2$:

$$[H]_e = \int_{\partial\Omega_e} \{N_2\}h_e\{N_2\}^T\,d\sigma$$

$$= \frac{l_e}{420}\{H\}_e^t \begin{bmatrix} \begin{Bmatrix} 39 \\ 20 \\ -3 \end{Bmatrix} & \begin{Bmatrix} 20 \\ 16 \\ -8 \end{Bmatrix} & \begin{Bmatrix} -3 \\ -8 \\ -3 \end{Bmatrix} \\ \begin{Bmatrix} 20 \\ 16 \\ -8 \end{Bmatrix} & \begin{Bmatrix} 16 \\ 192 \\ 16 \end{Bmatrix} & \begin{Bmatrix} -8 \\ 16 \\ 20 \end{Bmatrix} \\ \begin{Bmatrix} -3 \\ -8 \\ -3 \end{Bmatrix} & \begin{Bmatrix} -8 \\ 16 \\ 20 \end{Bmatrix} & \begin{Bmatrix} -3 \\ 20 \\ 39 \end{Bmatrix} \end{bmatrix} \tag{5.87c}$$

$$\{QF\}_e = \int_{\partial\Omega_e} \{N_2\} q_{n_e} \, d\sigma$$

$$= \frac{l_e}{30} \begin{bmatrix} 4 & 2 & -1 \\ 2 & 16 & 2 \\ -1 & 2 & 4 \end{bmatrix} \{QN\}_e \qquad (5.87d)$$

The order of the element arrays $\{H\}_e$ and $\{QN\}_e$ is appropriate for $k$ and each correspondingly contains the nodal values of convection coefficient $h_e$ and imposed normal flux $q_{n_e}$ on boundary segment $\partial\Omega_e$. Equation (5.87) completes the triangle basis master matrix library.

**STUDY PROBLEM 3.** Determine the steady-state temperature distribution through the wall of an axisymmetric pipe through which hot fluid is flowing and for which the exterior surface is first maintained at a fixed temperature $T_b$, and then subject to a convection boundary condition. Thereafter, cover the pipe with insulation and recompute the temperature distribution, enforcing the same exterior convection boundary condition. Hence, estimate the savings in thermal energy as a function of insulation thickness. The thermal conductivity of the pipe material is $k_p = 10$ Btu/(h · ft · °F) while that of the insulation material is $k_i = 0.1$ Btu/(h · ft · °F). The internal flow convection data are $h_f = 20$ Btu/(h · ft$^2$ · °F) and $T_f = 1500$°F. For the uninsulated pipe, $T_b = 306.85$°F, while for the addition of insulation assume the exterior data for the surrounding air is $h_a = 5$ Btu/(h · ft$^2$ · °F) and $T_a = 70$°F.

**Code Exercise 5.5.** Generate a $k = 1$ basis solution for the first part of Study Problem 3 for the specific geometry of a (very) thick-walled pipe of internal radius $r_1 = 1$ ft and external radius $r_2 = 2$ ft using a 5 × 5 nodal uniform discretization of a (symmetric) quarter-circle region of the pipe. The resultant $M = 32$ triangle element discretization is illustrated in Fig. 5.3 and the associated creation file is listed in Code Exercise 5.4. The *LEARN.FE* input file completion for this problem constitutes extension to include the command instructions *MATL*, *ROBN*, *FORM*, and *SOLV* or *GAUS*. Therefore, the input file for this code exercise is

```
TITL
*** CODE EXERCISE 5.5, PIPE WITH TB FIXED ***
TYPE    [K       N       NNODEL       REFL      NPRN    NTRAN    NAXI]
         1       2         3           1.0       1 1      3
PRIN    [NBUG(*)]
         1          0          0          2
```

```
GRID    [FOR N(XL XR(*) PR(*)), NEM(PR*), XI(NTRAN)]
                  1.0      2.0      1.0
         4
                  0.0      1.0      1.0
         4

        1.0     2.0       0.0      0.0
        0.0     0.0       2.0      1.0
        1.5     1.41421   0.0      0.70711
        0.0     1.41421   1.5      0.70711

MATL    [FOR MACRO (I), KI(*)]
        1    10.0
ROBN    [(NROB, REPEAT), NODE(*), (H(*), TR(*))]
        5    1      1
        5
        1    6      11      16       21
        5     20.0    1500.0

DIRI    [N(NDIR, REPEAT), NODE(*), (NRPT, QB(*))]
        5     1    0
        5          10          15          20          25
        5        306.8528
FORM
GAUS    [CONVERGENCE]
        0.001
STOP
```

Note that nodes 1, 6, 11, 16, and 21 lie on the internal convection surface while nodes 5, 10, 15, 20, and 25 lie on the external fixed-temperature surface.

The definition $T_b = 306.85°F$ may signal that you have encountered this problem statement before. Indeed, the solution should be identical to the one-dimensional variable conductivity problem, and/or the axisymmetric counterpart with uniform conductivity developed in Chap. 4. Execute the input file, and the following will scroll onto your monitor.

```
TITL
*** CODE EXERCISE 5.5, PIPE WITH TB FIXED ***
TYPE    [K       N        NNODEL       REFL       NPRN   NTRAN   NAXI]
        1        2        3            1.0        1 1    3       0
PRIN    [NBUG(1)    (2)       (3)        (4)        (5)     (6)]
           1        0         0          2          0       0
GRID    NI       NPRI     NTRAN
        1        1        3
        XL       XR1      PR1      XR2      PR2      XR3      PR3
        1.0      2.0      1.0
        NEM1     NEM2     NEM3
        4
GRID    NI       NPRI     NTRAN
        2        1        3
        YL       YU1      PR1      YU2      PR2      YU3      PR3
        0.0      1.0      1.0
        NEM1     NEM2     NEM3
        4
```

```
              X1       X2         X3         X4
              1.00     2.00       0.00       0.00
              Y1       Y2         Y3         Y4
              0.00     0.00       2.00       1.00
              X5       X6         X7         X8
              1.50     1.41       0.00       0.71
              Y5       Y6         Y7         Y8
              0.00     1.41       1.50       0.71

              NODE COORDINATES (X, INTEGER MAP) EXPONENT E:-2
                     0          0          0         0        0
                    38         47         57        66       76
                    70         88        106       123      141
                    92        115        138       161      184
                   100        125        150       175      200

              NODE COORDINATES (Y, INTEGER MAP) EXPONENT E:-2
                   100        125        150       175      200
                    92        115        138       161      184
                    70         88        106       123      141
                    38         47         57        66       76
                     0          0          0         0        0

MATL           MACRO CONDUCTIVITIES
               1       10.0

ROBN          NROB:    5
              JROB:    1          6          11          15          21
                       H          TREF
                       20.0     1500.0
                       20.0     1500.0
                       20.0     1500.0
                       20.0     1500.0
                       20.0     1500.0
DIRI        NFIX:      5
            JFIX:      5          10           15           20           25
                306.8528    306.8528    306.8528    306.8528    306.8528
FORM
GAUS        NODAL SOLUTION (INTEGER MAP) EXPONENT E:0, TOL=0.001
                   993        772         592        439        306
                   993        772         592        439        306
                   993        772         592        439        306
                   993        772         592        439        306
                   993        772         592        439        306
STOP
```

Listed following *GAUS* is the integer map of the approximate solution nodal temperature distribution in transform space. Note it is indeed axisymmetric, that is, the temperature is uniformly distributed along each radial node line in the mesh. However, these nodal data are only modestly close to the comparative $M = 4$, $n = 1$, $k = 1$ solution created by *LEARN.FE* in Section 4.7. Specifically, the axisymmetric $n = 1$ solution yielded a surface temperature of 999.19°F, which is much closer to the exact solution (1000°F) than are these data.

You must therefore inquire, "What is the cause of this discrepancy, since the problem data are certainly specified correctly?" The answer lies in the approximation to the boundary geometry, hence the resultant discrete approximate convection data load on $\partial\Omega_1$. Specifically, for the straight-sided triangular element, the surface on which $hT_r$ is imposed is a chord rather than a circular arc segment, recall Fig. 5.3. The following code exercise creates the verification computational database.

**Code Exercise 5.6.** Modify the input file for Code Exercise 5.5 to increase the mesh refinement in the azimuthal direction by setting NEM equal to 9 and then 18 in the *GRID* data following the $y_L$ and $y_U$ specification. Correspondingly, expand the data arrays under *ROBN* and *DIRI* to account for the finer azimuthal mesh. Also add command *NORM* after *GAUS*. After executing *LEARN.FE*, tabulate the convection surface nodal solution and the energy seminorm in the form of Tables 4.2 and 4.3. Then verify the quadratic convergence rate for surface temperature (to 999.19°F) as the cylinder surface geometry is progressively redefined by measure factors of 2.

The results from these two code exercises confirm the consequence of boundary geometry compromises when dealing with flux boundary conditions. The use of curve-sided finite elements $\Omega_e$ would significantly moderate this error mechanism at the expense of a more CPU-demanding solution procedure. In either instance, the (isoparametric) coordinate transformation for $\Omega^H$, and its refinement to $\Omega^h$, is an effective method for replacing specialty coordinate systems (e.g., polar coordinates) and for handling absolutely arbitrarily nonregular boundaries of problem domains $\Omega$. This boundary shape approximation error mechanism is essentially negligible when Dirichlet data are imposed.

Now proceed with the suggested study problem sequence. Since the problem domain and boundary conditions remain axisymmetric, to improve accuracy with decent efficiency you should reduce the quarter-circle domain span to a much smaller arc segment and keep the azimuthal discretization at (say) four finite elements. Hence, reexecute Code Exercise 5.6 for $M_2 = 18$ after setting NBUG(4) = 1 (to get E format output) and moving *STOP* to follow *GRID* (to avoid executing a solution). Then record the vertex and midside nodal coordinates of a macro domain $\Omega^H$ defined by the first pair of resultant azimuthal elements in $\Omega^h$. The resultant angular span of $\Omega^H$ will be $\Delta\theta = 10°$, which would yield a corresponding uniform $\Omega^h$ span of $\Delta\theta = 2.5°$ for $M_2 = 4$. Hence, for Study Problem 3 continuation,

**Code Exercise 5.7.** Assume the pipe inner radius remains $r_1 = 1$ ft and that the wall thickness is 0.1 ft; hence $r_2 = 1.1$ ft. Compute the temperature distribution for the uninsulated pipe with the external Dirichlet $T_b$ constraint on an $M_H = 5 \times 5$ nodal uniform $\Omega^h$ with $\Omega_e$ azimuthal span $\Delta\theta = 2.5°$ (following the suggested procedure above). The resultant *LEARN.FE* input file you will create should appear as

```
TITL
*** CODE EXERCISE 5.7, STUDY PROBLEM 3 STEP 2 ***
TYPE    [K      N       NNODEL      REFL   NPRN      NTRAN   NAXI]
        1       2          3         1.0    1 1         3
PRIN    [NBUG(*)]
               0         0          3
GRID    [FOR N(XLXR(*) PR(*)), NEM(PR*), XI(NTRAN)]
              1.0       1.1        1.0
        4
              0.0   0.17365       1.0
        4
        1.0     1.1         1.08329      0.98481
        0.0     0.0         0.19101      0.17365
        1.05    1.09581 1.03405          0.99619
        0.0     0.09587 0.09151          0.08716
MATL    [FOR MACRO(I), KI(*)]
        1       10.0
ROBN    [NROB, NODE(*), (H(*), TR(*))]
        5       0
        1       6    11         16         21
                20.0    1500.0
                20.0    1500.0
                20.0    1500.0
                20.0    1500.0
                20.0    1500.0
DIRI    [(NDIR, REPEAT), NODE(*), (NRPT, QB(*))]
        5     1     0
        5          10          15           20           25
        5        306.8528
FORM
GAUS    [CONVERGENCE]
        0.001
NORM
STOP
```

Execute an $\Omega^h$ radial span convergence study to (say) 32 × 4 nodes; hence estimate the exact solution surface temperature as illustrated in Table 4.2 in Chap. 4. Then compute the estimated heat flux (loss) $q_n = 2\pi r_1 h(T - T_f)$ from the fluid, which completes this step.

Next, exchange the exterior Dirichlet constraint $T_b$ for $h_a$ and $T_a$ as defined for Study Problem 3. The associated modifications to the input file above are

```
TITL
*** CODE EXERCISE 5.8, PIPE WITH EXTERIOR HA,TA ***
ROBN    [(NROB, REPEAT) NODE(*), (NRPT, H(*), TR(*))]
        10   2      1
        5    10
        1    6     11     16     21
         5   20.0    1500.0
        5    10    15     20     25
         5    5.0       70.0
```

Create this new file, which utilizes the *ROBN* data repeat option in *LEARN.FE*, and execute the convergence study to (say) $M_H = 32 \times 4$. Hence, estimate the exact solution internal and external surface temperatures, and then compute the resultant heat flux (loss) $q_n = 2\pi r_1 h(T - T_f)$. Compare the improvement over the $T_b$ fixed specification. Also verify that the heat-flux loss from the fluid equals (to PC significance) the heat flux into the surrounding air at $h_a$ and $T_a$.

Now continue the Study Problem 3 analysis by adding a 6-in-thick layer of insulation to the pipe. Use a two-macro-element definition to produce an $M_H = 2 \times 1$ discretization for the pipe and insulation. Further, you may wish to define it nonuniform within the insulation, anticipating that the steepest temperature gradient therein will occur adjacent to the pipe. The specification of a distinct material in each macro domain is facilitated under *MATL*. The resultant Code Exercise 5.9 input file you will create should appear as follows.

```
TITL
*** CODE EXERCISE 5.9, PIPE WITH EXTERIOR HA,TA ***
TYPE    [K      N      NNODEL      REFL    NPRN    NTRAN    NAXI]
        1       2        3          1.0     2 1      3
PRIN    [NBUG(*)]
         1        0        0          3
GRID    [FOR N(XL XR(*) PR(*)), NEM(PR*), XI(NTRAN)]
                 1.0      1.1      1.0      1.6      1.2
         2    8
                 0.0     0.17365   1.0
         4

        1.0    1.6        1.57569     0.98481
        0.0    0.0        0.27784     0.17365
        1.3    1.59391    1.29505     0.99619
        0.0    0.13945    0.11330     0.08716
MATL    [FOR MACRO(I), KI(*)]
        1     10.0
        2      0.1
```

```
ROBN     [NROB, NODE(*), (H(*), TR(*))]
          10   1   2
          5    10
          1    6    11     16        21
            5   20.0  1500.0
          5      10         15            20        25
            5    5.0    70.0
FORM
GAUS
    0.001
NORM
STOP
```

Construct and execute this file on *LEARN.FE*; hence verify that the resultant discrete approximate solution nodal temperature distribution is

```
GAUS: NODAL SOLUTION (INTEGER MAP) EXPONENT E: 1, TOL = 0.001
       148  148  147  128  108  87  64  38 11
       148  147  147  128  108  87  64  38 11
       148  147  147  128  108  87  64  38 11
       148  147  147  128  108  87  64  38 11
       148  147  147  128  108  87  64  38 11
NORM: F.E. SOLN. ENERGY SEMI-NORM: E(QH)=0.10765993E+06
```

Note the discontinuity in the slope of the temperature distribution at the pipe-insulation interface. Estimate the resultant heat loss $q_n$ from the fluid and compare the significant improvement over the data from the previous study. Then refine the radial mesh taking care to avoid a significant jump in element span at the pipe-insulation interface. Execute the resultant modified input file, and refine the estimate of the heat loss savings. Then reflect back on how Code Exercises 5.5 to 5.9 have led you through a *computational experiment* that has yielded an *engineering data set* to help you characterize this elementary but fully illustrative design problem.

## 5.7 AXISYMMETRIC CONDUCTION WITH BOUNDARY CONVECTION

Many practical two-dimensional problems exist for which the geometry is axisymmetric. This generalization, which is a modest extension on the material developed in the preceding sections, and in Chap. 4, totally eliminates the flux boundary condition error that we laboriously examined in the previous study. Further, since axisymmetric "conduction with boundary convection" occurs over a wide range of problem descriptions in mechanics, such as electromagnetics, multiphase biological and other species diffusion, fluid potential flows, and elasticity, this generalization is an important step.

Changing the dependent variable symbol from $T$ to $q(x, y)$, to dissociate our mind-set with the thermal problem, the generalized axisymmetric elliptic boundary value problem statement for which a finite element algorithm is sought is

$$\begin{aligned} L(q) &= -\nabla \cdot k(x_i)\nabla q - s(x_i) \\ &= -\frac{1}{r}\frac{\partial}{\partial r}\left(kr\frac{\partial q}{\partial r}\right) - \frac{\partial}{\partial z}\left(k\frac{\partial q}{\partial z}\right) - s(r, z) = 0 \qquad \text{on } \Omega \subset \mathbb{R}^2 \end{aligned} \tag{5.88}$$

$$l(q) = aq + k\nabla q \cdot \mathbf{n} - b = 0 \qquad \text{on } \partial\Omega_1 \tag{5.89}$$

$$q(\mathbf{x}_b) = q_b(\mathbf{x}_b) \qquad \text{on } \partial\Omega_2 \tag{5.90}$$

The gradient differential operator $\nabla$ in the $r$, $z$ coordinate system has been expanded in (5.88) for clarity. Further, by introducing the flux boundary condition parameters $a(\mathbf{x}_b)$ and $b(\mathbf{x}_b)$, for $\mathbf{x}_b$ on $\partial\Omega_1$, (5.89) combines the previously separate Neumann and Robin expressions [recall (5.83) and (5.84)]. As before, (5.90) expresses the Dirichlet constraint $q_b(\mathbf{x}_b)$ for $\mathbf{x}_b$ on $\partial\Omega_2$.

Derivation of the finite element algorithm for (5.88) to (5.90) follows the developed recipe. Combining (5.19), (5.21), and (5.22), the discrete approximate solution to be determined is

$$q(r, z) \simeq q^h(r, z) = \sum_{j=1} a_j \phi_j(x_i) \equiv \bigcup_e \{N_k(\zeta_i)\}^T \{Q\}_e \tag{5.91}$$

on the discretization $\Omega^h$ of the domain of definition of (5.88) to (5.90). As always, $\Omega^h$ is formed as the nonoverlapping sum (union) of the finite element domains $\Omega_e$. Figure 5.7 illustrates a prototypical domain $\Omega \bigcup \partial\Omega$ for an L-shaped axisymmetric geometry, including a "hand-generated" triangular element discretization.

The symmetric weak statement remains (5.23) with a modest modification for (5.89); hence

$$\begin{aligned} WS^h = S_e\Bigg(&\int_{\Omega_e} \nabla\{N_k\} \cdot k_e \nabla\{N_k\}^T \, d\tau \{Q\}_e - \int_{\Omega_e} \{N_k\} s_e \, d\tau \\ &+ \int_{\partial\Omega_1} \{N_k\}(a_e q_e - b_e) \, d\sigma - \int_{\partial\Omega_2} \{N_k\} k_e \nabla q_e \cdot \mathbf{n} \, d\sigma\Bigg) = \{0\} \end{aligned} \tag{5.92}$$

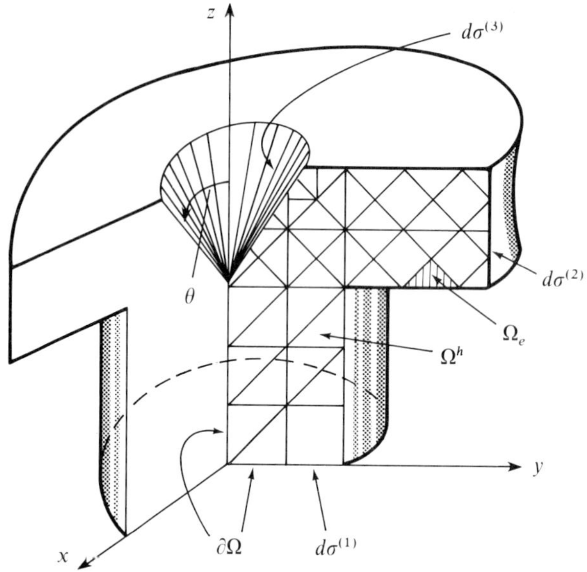


**FIGURE 5.7**
An axisymmetric problem geometry with $\Omega^h \equiv \bigcup \Omega_e$.

The differential element for axisymmetric coordinates is $d\tau = r\,dr\,d\theta\,dz$. As introduced in Section 4.7, the integration over $d\theta$ yields a uniform multiplier $2\pi$ which is ultimately canceled out, but retaining $d\theta$ is conceptually important for forming surface integrals with $d\sigma^{(a)}$ (see Fig. 5.7). For the $k = 1$ triangle basis, assuming variable conductivity $k_e$, and dividing through by $2\pi$, the first term in (5.92) becomes

$$
\begin{aligned}
[K]_e &\equiv (2\pi)^{-1} \int_{\Omega_e} \nabla\{N_1\} \cdot k_e \nabla\{N_1\}^T r\,dr\,d\theta\,dz\{Q\}_e \\
&= \nabla\{N_1\} \cdot \nabla\{N_1\}^T \int_{\Omega_e} k_e r_e\,dr\,dz\{Q\}_e \\
&= \nabla\{N_1\} \cdot \nabla\{N_1\}^T \left( \{K\}_e^T \int_{\Omega_e} \{N_1\}\{N_1\}^T\,dr\,dz\{R\}_e \right) \{Q\}_e \\
&= \frac{1}{4A_e} \begin{bmatrix} \zeta_{11}^2 + \zeta_{12}^2 & \zeta_{11}\zeta_{21} + \zeta_{12}\zeta_{22} & \zeta_{11}\zeta_{31} + \zeta_{12}\zeta_{32} \\ \zeta_{21}\zeta_{11} + \zeta_{22}\zeta_{12} & \zeta_{21}^2 + \zeta_{22}^2 & \zeta_{21}\zeta_{31} + \zeta_{22}\zeta_{32} \\ \zeta_{31}\zeta_{11} + \zeta_{32}\zeta_{12} & \zeta_{31}\zeta_{21} + \zeta_{32}\zeta_{22} & \zeta_{31}^2 + \zeta_{32}^2 \end{bmatrix} \\
&\quad \times \left( \{K\}_e^T \begin{bmatrix} 2 & 1 & 1 \\ 1 & 2 & 1 \\ 1 & 1 & 2 \end{bmatrix} \{R\}_e \right) \{Q\}_e
\end{aligned} \tag{5.93}
$$

The second line in (5.93) recognizes that $\nabla\{N_1\}$ is constant on $\Omega_e$ [recall (5.30)]. The third line has substituted the linear basis interpolation for $k_e$ and $r_e$ and extracted the nodal arrays $\{K\}_e$ and $\{R\}_e$ [recall (5.36)]. The last line combines the previously established matrix expression (5.35) with the integral evaluation (5.40), and hence establishes the final form. Thus, all components necessary to form the axisymmetric conductivity element matrix for the $k = 1$ triangular basis are available for a *LEARN.FE* code implementation, which would be keyed by NAXI = 1 for N = 2 under *TYPE*.

Continuing, the source term evaluation yields

$$\begin{aligned}\{b\}_e &\equiv (2\pi)^{-1}\int_{\Omega_e}\{N_1\}s_e\,d\tau = (2\pi)^{-1}\int_{\Omega_e}\{N_1\}\{N_1\}^T r\,dr\,d\theta\,dz\{S\}_e \\ &= \{R\}_e^T\int_{\Omega_e}\{N_1\}\{N_1\}\{N_1\}^T\,dr\,dz\{S\}_e \\ &= \frac{A_e}{60}\{R\}_e^T\begin{bmatrix}\begin{Bmatrix}6\\3\\3\end{Bmatrix} & \begin{Bmatrix}3\\3\\1\end{Bmatrix} & \begin{Bmatrix}3\\1\\3\end{Bmatrix} \\ & \begin{Bmatrix}3\\6\\3\end{Bmatrix} & \begin{Bmatrix}1\\3\\3\end{Bmatrix} \\ \text{sym} & & \begin{Bmatrix}3\\3\\6\end{Bmatrix}\end{bmatrix}\{S\}_e\end{aligned} \tag{5.94}$$

As before, $\{S\}_e$ contains the nodal values of the source term distribution $s_e(x_i)$. The element master matrix defined in (5.94) is new and hence must be added to the *LEARN.FE* library to complete the two-dimensional axisymmetric problem statement capability.

The boundary condition terms in (5.92) also yield new element master matrices. The boundary data $a_e$ and $b_e$ are interpolated by an appropriate $n = 1$ element basis. The respective differential elements $d\sigma^{(a)}$ take specific forms dependent on the orientation of the boundary surface segment. From Fig. 5.7, the following definitions occur

$$\begin{aligned}d\sigma^{(1)} &= r\,dr\,d\theta && \partial\Omega_1 \text{ in } x, y \text{ plane} \\ d\sigma^{(2)} &= r\,dz\,d\theta && \partial\Omega_1 \text{ parallel to the } z \text{ axis} \\ d\sigma^{(3)} &= r(z)\,dl\,d\theta && \partial\Omega_1 \text{ not aligned parallel to the } z \text{ axis}\end{aligned} \tag{5.95}$$

where $dl = \sqrt{dr^2 + dz^2}$ on the arbitrarily oriented surface. In each instance in (5.95), $r_e(x_i)$ on $\partial\Omega_e$ is interpolated in terms of the appropriate nodal values. Hence, the boundary condition (5.92) expression is

$$(2\pi)^{-1} \int_{\partial\Omega_1} (N_k\}(a_e q_e - b_e)\, d\sigma^{(a)} = \int_{\Omega_e} \{N_k\}(a_e q_e - b_e)\{N_1\}^T\{R\}_e\, ds \quad (5.96)$$

where $ds$ is taken as $dr$, $dz$, or $dl$ as appropriate.

The first term in (5.96), evaluated using the linear triangle-compatible $k = 1$ basis, and interchanging scalar terms as convenient, becomes

$$\begin{aligned} \int_{\partial\Omega_e} \{N_1\}a_e q_e\{N_1\}^T\{R\}_e\, ds &= \{A\}_e^T \int_{\partial\Omega_e} \{N_1\}\{N_1\}\{N_1\}^T\{R\}_e\{N_1\}^T\, ds\{Q\}_e \\ &= l_e(\{A\}_e^T[BC1]\{R\}_e)\{Q\}_e \end{aligned} \quad (5.97a)$$

In (5.97$a$), $l_e$ is the element boundary segment length and the master matrix $[BC1]$ is the $B$oundary $C$ondition $2 \times 2$ matrix with elements which are themselves $2 \times 2$ matrices, that is, a hypermatrix of degree 2. The evaluation of the integrals in (5.97$a$) is direct using (4.40) and yields

$$[BC1] = \frac{1}{60}\begin{bmatrix} \begin{bmatrix} 12 & 3 \\ 3 & 2 \end{bmatrix} & \begin{bmatrix} 3 & 2 \\ 2 & 3 \end{bmatrix} \\ \begin{bmatrix} 3 & 2 \\ 2 & 3 \end{bmatrix} & \begin{bmatrix} 2 & 3 \\ 3 & 12 \end{bmatrix} \end{bmatrix} \quad (5.97b)$$

As with all hypermatrix expressions, the inner products within the parentheses in (5.97$a$) must be completed prior to any "regular" matrix operations.

The second term in (5.96) is also formed directly as

$$\begin{aligned} \int_{\partial\Omega} \{N_1\}b_e\{N_1\}^T\{R\}_e\, ds &= \{B\}_e^T \int_{\partial\Omega_e} \{N_1\}\{N_1\}\{N_1\}^T\, ds\{R\}_e \\ &= l_e\{B\}_e^T[BC2]\{R\}_e \end{aligned} \quad (5.98a)$$

The master matrix $[BC2]$ is a degree-1 $2 \times 2$ hypermatrix with $2 \times 1$ column matrix elements. The values can be determined using (4.40);

however, it is easier to recall that (5.98*a*) is identical to (4.110), hence

$$[BC2] = \frac{1}{12}\begin{bmatrix} \begin{Bmatrix} 3 \\ 1 \end{Bmatrix} & \begin{Bmatrix} 1 \\ 1 \end{Bmatrix} \\ \begin{Bmatrix} 1 \\ 1 \end{Bmatrix} & \begin{Bmatrix} 1 \\ 3 \end{Bmatrix} \end{bmatrix} \tag{5.98b}$$

Equations (5.93) to (5.98) complete the linear triangle master matrix library for the finite element algorithm (5.92) for the axisymmetric boundary value problem (5.88) to (5.90). In review, the additions for the *LEARN.FE* library are the boundary condition matrix $[BC1]$ and the source term matrix in (5.94). This serves to further substantiate a major attribute of the finite element algorithm and its code implementation. The generalization of problem class typically constitutes no more than expansion of the *library*, with the weak statement theory precisely defining the exact form of every created term for any problem dimensionality and basis degree. The library concept is formalized in the next section, yielding a notational structure that facilitates logical Fortran expressions for $[K]_e$, $[H]_e$, and all other terms. You have already seen this structure if you have viewed the Appendix listings for subroutines ELMAT, BLOCK DATA, and FORMIT.

## 5.8 MASTER MATRIX LIBRARY ORGANIZATION

Looking back through Chaps. 4 and 5, much "detail" certainly exists in converting the finite element weak statement approximation into an operational code. However, the *master matrix library* concept, coupled with a rational nomenclature, brings order. The weak statement approximation always produces the computational form

$$WS^h = S_e(([K]_e + [H]_e)\{Q\}_e - \{b\}_e) = \{0\} \tag{5.99}$$

In (5.99), $[K]_e$ is the element matrix equivalent of the diffusion term, $[H]_e$ corresponds to the homogeneous portion of a flux boundary condiion, and $\{b\}_e$ contains all data given on $\Omega$ and $\partial\Omega$. The specific form for each term is tied to the dimension $n$ of the problem statement, with extension for axisymmetric geometries, and the degree $k$ of the finite element basis. Further, many one-dimensional expressions have appeared as boundary condition master matrices for $n = 2$ problems.

The concept of the *finite element library* capitalizes on the ordered structure that pervades the algorithm. A method proven informative for code organization is a Fortran-compatible matrix nomenclature keyed to $n$, $k$, and the number of trial space basis functions $\{N_k\}$, and their derivatives, appearing in any term. This is a generalization of the original development by Baker (1983, Chap. 3), wherein any specific finite element master matrix is given a Fortran name of the form

$$\begin{aligned} [M] &\equiv [a\, b\, c\, c\, c \cdots d], \textit{ and/or} \\ \{M\} &\equiv \{a\, b\, c\, d\} \end{aligned} \tag{5.100}$$

The variable set ($a$, $b$, $c$, $d$) in (5.100) is defined as follows.

$$a \equiv \begin{cases} A & \text{for an } n = 1 \text{ matrix} \\ B & \text{for an } n = 2 \text{ matrix} \\ C & \text{for an } n = 3 \text{ matrix} \end{cases}$$

$$b \equiv \begin{cases} \text{an integer equal to the number of trial space bases } \{N_k\} \\ \text{appearing in the matrix definition} \end{cases}$$

$$c \equiv \begin{cases} 0 & \text{appropriate basis } \{N_k\} \text{ not differentiated} \\ 1 & \text{appropriate basis } \{N_k\} \text{ differentiated, } n = 1 \text{ only} \\ J & \text{appropriate basis } \{N_k\} \text{ differentiated parallel to } \eta_j,\ 1 < J \le n \end{cases}$$

$d \equiv$ the label for basis degree $k$

All the developed master matrix expressions can be assigned a usable Fortran name using this convention. It is appropriate to return first to Chap. 4 and develop the $n = 1$ library nomenclature.

### One-Dimensional Library

The steady-state diffusion problem was first examined in Section 4.4, and the matrix $[K]_e$ was developed for bases $\{N_k\}$ for $k = 1$, 2, and 3. For example, then, recalling (4.25) for constant conductivity $k_e$

$$\begin{aligned} [K]_e &= \int_{\Omega_e} k_e \frac{d\{N_1\}}{dx} \frac{d\{N_1\}^T}{dx}\, dx \\ &= \frac{k_e}{l_e} \begin{bmatrix} 1 & -1 \\ -1 & 1 \end{bmatrix} \\ &\equiv \frac{k_e}{l_e} [A211L] \end{aligned} \tag{5.101}$$

The last line in (5.101) states that $[K]_e$, for $n = 1$, $k = 1$, and uniform thermal conductivity $k_e$, involves a master matrix formed on one dimension, ($a \equiv A$), contains two bases $\{N_k\}$, ($b = 2$), both of which are differentiated, ($c = 1$, $c = 1$), and is formed using the $k = 1$ linear basis $\{N_k\}$, ($d = L$). Thus, for the convention (5.100),

$$[A211L] \equiv \begin{bmatrix} 1 & -1 \\ -1 & 1 \end{bmatrix} \tag{5.102}$$

which is easy to place in a library data statement and to use as a Fortran variable.

In the same manner, the uniform source element matrix (4.28) becomes

$$\{b\}_e = s_e \int_{\Omega_e} dx = s_e l_e \{A10L\} \tag{5.103}$$

Hence,

$$\{A10L\} \equiv \begin{Bmatrix} \frac{1}{2} \\ \frac{1}{2} \end{Bmatrix} = \frac{1}{2} \begin{Bmatrix} 1 \\ 1 \end{Bmatrix} \tag{5.104}$$

is the one-dimensional $k = 1$ singular basis matrix. Combining (5.101) to (5.104), the linear basis steady conduction algorithm (4.29) can be re-expressed as the Fortran statement

$$\begin{aligned} S_e([K]_e\{Q\}_e - \{b\}_e) &= S_e\left(\frac{k_e}{l_e}\begin{bmatrix} 1 & -1 \\ -1 & 1 \end{bmatrix}\{Q\}_e - \frac{s_e l_e}{2}\begin{Bmatrix} 1 \\ 1 \end{Bmatrix} - k_c \frac{dT}{dx}\begin{Bmatrix} -\delta_{e1} \\ \delta_{eM} \end{Bmatrix}\right) \\ &= S_e\left(\frac{k_e}{l_e}[A211L]\{Q\}_e - s_e l_e\{A10L\} - q_n\{\delta_{eE}\}\right) \end{aligned} \tag{5.105}$$

The pointer master matrix $\{\delta_{eE}\}$ contains $\{1, 0\}^T$ or $\{0, 1\}^T$ dependent on $E = 1$ or $E = M$.

By direct extension, the quadratic basis algorithm Fortran statement definition is

$$S_e([K]_e\{Q\}_e - \{b\}_e) = S_e\left(\frac{k_e}{l_e}[A211Q]\{Q\}_e - s_e l_e\{A10Q\} - q_n\{\delta_{eE}\}\right) \tag{5.106}$$

where index $d \to Q$ signifies the $k = 2$ (Quadratic) basis matrices and $\{\delta_{eE}\}$ is extended to the appropriate rank. Comparing (5.106) to (4.43) and

(4.44) yields the library data definitions

$$[A211Q] \equiv \frac{1}{3}\begin{bmatrix} 7 & -8 & 1 \\ -8 & 16 & -8 \\ 1 & -8 & 7 \end{bmatrix} \tag{5.107}$$

$$\{A10Q\} \equiv \frac{1}{6}\begin{Bmatrix} 1 \\ 4 \\ 1 \end{Bmatrix} \tag{5.108}$$

The Lagrange cubic basis algorithm statement amounts to a minor notational change in (5.106) with $[A211C]$ replacing $[A211Q]$, and so on. Comparing to (4.53) and (4.54)

$$[A211C] = \frac{1}{40}\begin{bmatrix} 148 & -189 & 54 & -13 \\ & 432 & -297 & 54 \\ & & 432 & -189 \\ (\text{sym}) & & & 148 \end{bmatrix} \tag{5.109}$$

$$\{A10C\} = \frac{1}{8}\{1, 3, 3, 1\}^T \tag{5.110}$$

The *LEARN.FE* code utilizes this nomenclature as Fortran variable names for all element master matrices. The integer data arrays in (5.102), (5.104), and (5.107) to (5.110) are stored in subroutine BLOCK DATA. Prior to any solution, the call to subroutine ELMAT from READIT normalizes these arrays by the appropriate common integer divisor. Subroutine FORMIT then uses the resultant real arrays as Fortran variables to complete algorithm statements. Thus, *LEARN.FE* typifies an orderly use of the master matrix library concept.

Recall in Section 4.5, Quadratic Basis, that the $k = 2$ basis algorithm was extended for variable thermal conductivity; see (4.74) and (4.75). The resultant element conduction matrix is defined in (4.77), which would now be reexpressed using (5.100) as

$$[K]_e = \frac{1}{l_e}\{K\}_e^T[A3011Q] \tag{5.111}$$

In (5.111), $[A3011Q]$ states that the matrix is for $n = 1$ dimension, $(a = A)$, it contains three bases, $(b = 3)$, the first of which involves the interpolation of element conductivity in terms of nodal values $\{K\}_e$, $(c = 0)$, the other two are differentiated, $(c = 1,\ c = 1)$, and it is for the $k = 2$ basis $(d = Q)$.

Comparing (5.111) and (4.77) yields the library statement

$$[A3011Q] \equiv \frac{1}{6}\begin{bmatrix} \begin{Bmatrix} 11 \\ 3 \end{Bmatrix} & \begin{Bmatrix} -12 \\ -4 \end{Bmatrix} & \begin{Bmatrix} 1 \\ 1 \end{Bmatrix} \\ & \begin{Bmatrix} 16 \\ 16 \end{Bmatrix} & \begin{Bmatrix} -4 \\ -12 \end{Bmatrix} \\ \text{(sym)} & & \begin{Bmatrix} 3 \\ 11 \end{Bmatrix} \end{bmatrix} \tag{5.112}$$

Take a look into subroutine BLOCK DATA, and you will indeed observe identity (5.112).

Any library hypermatrix of rank $p$ can be used to generate appropriate hypermatrices and/or matrices of rank less than $p$. For example, recalling Exercise 4.1, it is easy to verify that

$$\{1\}^T[A3011Q] = [A211Q] \tag{5.113}$$

by inspection of (5.107) and (5.111). For completeness, the $k = 2$ basis master matrix statement for distributed source (4.78) is now

$$\{b\}_e = l_e[A200Q]\{S\}_e \tag{5.114}$$

and

$$[A200Q] \equiv \frac{1}{30}\begin{bmatrix} 4 & 2 & -1 \\ 2 & 16 & 2 \\ -1 & 2 & 4 \end{bmatrix} \tag{5.115}$$

is the library entry.

Section 4.6 generalized the $n = 1$ algorithm statement to axisymmetric geometries which yielded the element node radius array as the premultiplier for the thermal conduction matrix. Thus, for assumed constant conductivity on $\Omega_e$ we can now write

$$[K]_e = k_e \int_{\Omega_e} \frac{d\{N_k\}}{dr}\frac{d\{N_k\}^T}{dr} r_e\,dr = \frac{k_e}{l_e}\{R\}_e^T[A3011d] \tag{5.116}$$

as the replacement for (4.65*a*), (4.77*a*), and (4.85*a*) with $d$ taking the label $L$, $Q$, $C$ for $k = 1, 2, 3$, respectively. Subsequent extension to the finned tube heat-exchanger geometry introduced the generalization of the boundary thermal convection matrix $[H]_e$. Recalling (4.105) and (4.106), the new master matrix definition is

$$[H]_e = l_e h_e \{R\}_e^T[A3000d] \tag{5.117}$$

and, for example,

$$[A3000L] \equiv \frac{1}{12}\begin{bmatrix} \begin{Bmatrix} 3 \\ 1 \end{Bmatrix} & \begin{Bmatrix} 1 \\ 1 \end{Bmatrix} \\ \begin{Bmatrix} 1 \\ 1 \end{Bmatrix} & \begin{Bmatrix} 1 \\ 3 \end{Bmatrix} \end{bmatrix} \tag{5.118}$$

for the $k = 1$ basis implementation, $(d = L)$ in (5.117).

## Two-Dimensional Triangle Library

We have expressed the master matrix library for the range of $n = 1$ problem statements, and for $1 \leq k \leq 3$ basis implementations, in $A$-matrix terminology. As defined in (5.100), the library is now extended to the $n = 2$ algorithm statements using the lead prefix $B$ as appropriate.

Equations (5.16) to (5.18) define the introductory steady-state diffusion equation system, and (5.23) expresses the weak statement finite element approximation. Converting the resultant matrix statement (5.24), to the Fortran library syntax as a function of $k$, is direct. Section 5.4, Linear Basis, details the $k = 1$ element diffusion matrix with final form (5.39). The Fortran nomenclature development starts with (5.25)

$$\begin{aligned} [K]_e &= \int_{\Omega_e} \nabla\{N_1\} \cdot k_e \nabla\{N_1\}^T \, d\tau \\ &= \nabla\{N_1\} \cdot \nabla\{N_1\}^T \int_{\Omega_e} k_e \, d\tau \end{aligned} \tag{5.25}$$

The integral remaining in (5.25) becomes, by definition

$$\int_{\Omega_e} k_e \, d\tau = \int_{\Omega_e} \{N_1\}^T \, d\tau \{K\}_e \equiv A_e \{K\}_e^T \{B10L\} \tag{5.119}$$

and from (5.38),

$$\{B10L\} \equiv \frac{1}{3}\begin{Bmatrix} 1 \\ 1 \\ 1 \end{Bmatrix} \tag{5.120}$$

The derivatives in the vector matrix product in (5.25) must be transformed to the $\zeta_i$ coordinate system. Recalling (5.30), the resultant library expression is

$$\nabla\{N_1\}\cdot\nabla\{N_1\}^T = \left(\frac{\partial\zeta_i}{\partial x_j}\right)_e\left(\frac{\partial\zeta_k}{\partial x_j}\right)_e \frac{\partial\{N_1\}}{\partial\zeta_i}\frac{\partial\{N_1\}^T}{\partial\zeta_k}$$

$$\equiv ZETAIJ_e\ ZETAKJ_e[B2IKL]/(2A_e)^2$$

$$\text{for } 1 \leq (I, K) \leq 3 \quad \text{and} \quad 1 \leq J \leq 2 \qquad (5.121)$$

In (5.121), the metric transformation components $(\partial\zeta_i/\partial x_j)_e$ constitute the array $ZETAIJ_e$ ( pronounced "zeta-*i*-*j*" ), which is defined in (5.29) without the divisor $2A_e$. $ZETAKJ_e$ contains the same data; thus, each is simply the Fortran subscripted variable ZETA($I, J$) to be formed on each $\Omega_e$. The two-dimensional ($a = B$) master matrix $[B2IKL]$ results from the column-row product of two bases ($b = 2$), both of which are (vectorally) differentiated ($c = I$, $c = K$, $1 \leq (I, K) \leq n + 1 = 3$). The linear basis ($d = L$) form is the pseudo-Kronecker delta matrix

$$[B2IKL] \equiv [DELTAIK] \qquad (5.122)$$

as derived and defined in (5.32).

The remaining two algorithm master matrices are created from the distributed source and convection boundary condition [recall (5.40) to (5.42)]. In library terminology, these expressions are

$$\int_{\Omega_e}\{N_1\}s_e\,d\tau = \int_{\Omega_e}\{N_1\}\{N_1\}^T\,d\tau\{S\}_e$$

$$\equiv A_e[B200L]\{S\}_e \qquad (5.123)$$

$$\int_{\partial\Omega_1}\{N_1\}q_{n_e}\,d\sigma = \int_{\Omega_e}\{N_1\}\{N_1\}^T\,dx\{QN\}_e$$

$$\equiv l_e[A200L]\{QN\}_e \qquad (5.124)$$

The library arrays are

$$[B200L] = \frac{1}{12}\begin{bmatrix} 2 & 1 & 1 \\ 1 & 2 & 1 \\ 1 & 1 & 2 \end{bmatrix} \qquad (5.125)$$

$$[A200L] = \frac{1}{6}\begin{bmatrix} 2 & 1 \\ 1 & 2 \end{bmatrix} \qquad (5.126)$$

Note indeed that the $n = 1$ master matrix $[A200L]$ defines the $n = 2$ convection boundary condition statement.

Equations (5.119) to (5.126) thus constitute the library definitions for the triangular element linear basis algorithm for (5.16) to (5.18). We can immediately write the finite element weak statement (5.23) in the equivalent Fortran form as

$$
\begin{aligned}
WS^h &= S_e\left(\int_{\Omega_e} \nabla\{N_1\}\cdot k_e \nabla T_e\, d\tau - \int_{\Omega_e}\{N_1\}s_e\, d\tau + \int_{\partial\Omega_1}\{N_1\}q_{n_e}\, d\sigma\right)\\
&= S_e((4A_e)^{-1}\{K\}_e^T\{B10L\}ZETAIJ_e ZETAKJ_e[B2IKL]\{Q\}_e\\
&\qquad - A_e[B200L]\{S\}_e + l_e[A200L]\{QN\}_e)\\
&\equiv S_e([K]_e\{Q\}_e - \{b\}_e) = \{0\}\\
&\qquad \text{for } 1 \le J \le n = 2 \quad \text{and} \quad 1 \le (I, K) \le n + 1 = 3
\end{aligned}
\tag{5.127}
$$

The repeated (tensor) indices $I$, $J$, $K$ in (5.127) define the range of the DO loops in subroutine FORMIT, and the presence (absence) of $e$ subscripts explicitly keys element-dependent (-independent) data. Figure 5.8 is an excerpt from the listing of FORMIT in Section 3.1, the linear triangle procedure. It clearly illustrates the element metric data handling and the associated DO loop for the $[K]_e$ term in (5.127). The element working matrix is called EMAT($I$, $J$), and element-dependent data are logically ordered by the MEL array keys created during grid generation.

The syntax of (5.127) can be immediately expressed for the quadratic basis algorithm for uniform element conductivity; see (5.16) to (5.18). The Fortran statement is

$$
\begin{aligned}
WS^h &= S_e(k_e(2A_e)^{-1}ZETAIJ_e ZETAKJ_e[B2IKQ]\{Q\}_e\\
&\qquad - A_e[B200Q]\{S\}_e + l_e[A200Q]\{QN\}_e)\\
&= \{0\} \qquad \text{for } 1 \le J \le n = 2 \quad \text{and} \quad 1 \le (I, K) \le n + 1 = 3
\end{aligned}
\tag{5.128}
$$

There are nine element-independent quadratic basis master matrices $[B2IKQ]$. Each is a $6 \times 6$ array, as given in (5.54) and (5.56), in distinction to the single linear basis $3 \times 3$ pseudo-Kronecker delta matrix (5.122). Note that since $(I, K)$ is a symmetric index pair in (5.128), only the sum of $[B2IKQ]$ for $I \neq K$ is ever needed, as developed and illustrated in (5.55) and (5.56). Hence, all library arrays for the $k = 2$ basis algorithm are symmetric, yielding a savings in memory and computer execution requirement.

Should the thermal conductivity not be uniform on $\Omega_e$, the Fortran generalization for (5.128) would be

$$
k_e[B2IKQ] \rightarrow \{K\}_e^T[B30IKQ] \tag{5.129}
$$

```
C       ************************************************************
C       SUBROUTINE FORMIT
C
C       FORM THE ALGORITHM MATRIX STATEMENT
C
C       ************************************************************
        SUBROUTINE FORMIT(IENERQ)
         .
         .
         .
C       T R I A N G L E S

C       3.1) LINEAR TRIANGLES

C       SET UP ZETAS

        ZETA(1,1) = YG(MEL2) - YG(MEL3)
        ZETA(2,1) = YG(MEL3) - YG(MEL1)
        ZETA(3,1) = YG(MEL1) - YG(MEL2)
        ZETA(1,2) = XG(MEL3) - XG(MEL2)
        ZETA(2,2) = XG(MEL1) - XG(MEL3)
        ZETA(3,2) = XG(MEL2) - XG(MEL1)

        DET = XG(MEL2)*YG(MEL3) + XG(MEL1)*
              YG(MEL(2))+
              XG(MEL3)*YG(MEL1) - XG(MEL2)*
              YG(MEL(1))-
              XG(MEL3)*YG(MEL2)-XG(MEL1)*
              YG(MEL3)
        AREA=DET/2.
        ECOND=COND(MEL)
C       SET UP ELEMENT DIFFUSION MATRIX

        DO 160I=1,NDOFE
           DO 170 J=1,NDOFE

           EMAT(I,J)=ECOND*(ZETA(I,1)*ZETA(J,1) +
                                ZETA(I,2)*ZETA(J,2))/(4.*AREA)
              .
              .
              .
```

**FIGURE 5.8**
Excerpt from *LEARN.FE* subroutine FORMIT for linear triangle $[K]_e$ element matrix formation.

which significantly increases master matrix complexity and storage requirements. One resolution is to employ a numerical integration procedure which exchanges use of (5.37) and memory for additional DO loops. This methodology is introduced and developed in Chap. 6.

The remaining generalizations in this chapter involve thermal convection boundary conditions and the axisymmetric geometry. The former introduces $[H]_e$ into $WS^h$ as presented in (5.86) and defined in (5.87). For distributed convection coefficient $h_e$ and specified normal flux $q_n$, the library expressions are

$$\begin{aligned}[H]_e &\equiv l_e\{H\}_e^T[A3000d] \\ \{QF\}_e &\equiv l_e[A200d]\{QN\}_e\end{aligned} \tag{5.130}$$

where $d = L, Q$ as keyed to basis degree $k$. For the axisymmetric, variable conductivity $k = 1$ basis algorithm, the element conductivity matrix $[K]_e$ expressed in (5.93) becomes the Fortran statement

$$[K]_e = (4A_e)^{-1}\{K\}_e^T[A200L]\{R\}_e ZETAIJ_e ZETAKJ_e[B2IKL] \quad (5.131)$$

Equation (5.131) is a truly modest generalization on the nonaxisymmetric statement (5.127), that can be readily implemented into FORMIT using the established library base.

The corresponding extensions for distributed source and convection boundary condition terms (5.94) to (5.97) are

$$\int_{\Omega_e} \{N_1\}s_e r\, dr\, dz = A_e\{R\}_e^T[B3000L]\{S\}_e \quad (5.132)$$

$$\int_{\partial\Omega_e} \{N_1\}(a_e q_e - b_e)\, d\sigma_{(i)} = l_e\{A\}_e^T[A40000L]\{R\}_e\{Q\}_e - l_e\{B\}_e^T[A3000L]\{R\}_e \quad (5.133)$$

The specialty matrix names $[BC1]$ and $[BC2]$ introduced in (5.97) and (5.98) are thus replaced by Fortran variables that clearly specify the intrinsic DO loop structure for FORMIT.

## 5.9 AERODYNAMIC POTENTIAL FLOW

Potential flow aerodynamics is an important problem class that enjoys widespread application in aerospace, automotive, and air movement industry sectors, to name only a few. Any fluid flowfield is described by its velocity vector distribution $\mathbf{u}(x_i)$, which in cartesian scalar component form for $n = 2$ is

$$\mathbf{u}(x_i) = u(x, y)\mathbf{i} + v(x, y)\mathbf{j} \quad (5.134)$$

The *potential* simplification assumes that $\mathbf{u}$ in (5.134) is *irrotational* and that it occurs in an isentropic environment, specifically, no viscous effects or heat transfer. The *aerodynamic* simplification further assumes that $\mathbf{u}$ is a (modest) modification to a uniform onset flowfield defined by the reference velocity vector $\mathbf{U}_\infty$. Combining these assumptions allows the analyst to express $\mathbf{u}$ as,

$$\mathbf{u}(x_i) \equiv U_\infty(\mathbf{e} - \nabla\phi) \quad (5.135)$$

where $\mathbf{e}$ is the unit vector defining the direction of $\mathbf{U}_\infty$, $\nabla$ is the gradient differential operator [recall (5.82)], and $\phi(x_i)$ is the *perturbation* potential function. For example, if we assume that $\mathbf{U}_\infty$ is parallel to the $x$ axis, i.e., $\mathbf{U}_\infty = U_\infty \mathbf{i}$, then (5.135) can be expanded as

$$\begin{aligned}\mathbf{u}(x_i) &= u(x, y)\mathbf{i} + v(x, y)\mathbf{j} \\ &= U_\infty\left[\left(1 - \frac{\partial \phi}{\partial x}\right)\mathbf{i} + \left(-\frac{\partial \phi}{\partial y}\right)\mathbf{j}\right]\end{aligned} \tag{5.136}$$

Hence, $-\partial\phi/\partial x$ is the $x$-perturbation distribution to the imposed free-stream flow and $-\partial\phi/\partial y$ is the total lateral velocity component distribution.

The differential equation system governing the perturbation potential $\phi(x, y)$ is derived from conservation of mass, the *continuity equation*, and the requirement that $\mathbf{u}$ be tangent to select boundary surfaces. The continuity equation states that $\mathbf{u}$ must be divergence-free, i.e., $\nabla \cdot \mathbf{u} \equiv 0$. Substituting (5.135) into this requirement yields

$$\begin{aligned}L(\phi) &= \nabla \cdot \mathbf{u} \\ &= \nabla \cdot U_\infty(\mathbf{e} - \nabla\phi) \\ &= -\nabla^2\phi = 0\end{aligned} \tag{5.137}$$

since $U_\infty \mathbf{e}$ is a constant and hence possesses no spatial derivative. Thus, $\phi$ is the solution to a homogeneous Laplace equation on a two-dimensional region $\Omega$ for which the weak statement finite element algorithm is highly appropriate.

Certain boundary conditions for (5.137) are derived from the flow tangency requirement. For any solid aerodynamic surface, using (5.135), we obtain

$$\mathbf{u} \cdot \mathbf{n} \equiv 0 = U_\infty(\mathbf{e} - \nabla\phi) \cdot \mathbf{n} \tag{5.138}$$

Actually, the form of (5.138) is also valid on any boundary segment $\partial\Omega$ on which the normal flux of fluid is known, for instance, at inflow or in the farfield where $\mathbf{u} = U_\infty \mathbf{e}$. In the general case, therefore, letting $\mathbf{u}_b$ denote any boundary segment velocity distribution, from (5.135)

$$l(\phi) = -\nabla\phi \cdot \mathbf{n} + \mathbf{e} \cdot \mathbf{n} - \frac{\mathbf{u}_b \cdot \mathbf{n}}{U_\infty} = 0 \tag{5.139}$$

Equation (5.139) constitutes the nonhomogeneous Neumann boundary condition constraint on boundary segments $\partial\Omega_1$ where knowledge exists regarding the fluid normal flux. The sole Dirichlet constraint is to fix the level of $\phi$ to an arbitrary constant as admitted by the definition (5.135).

Thus, the aerodynamic perturbation potential flow differential equation statement is

$$L(\phi) = -\nabla^2\phi = 0 \qquad \text{on } \Omega \subset \mathbb{R}^2 \tag{5.140}$$

$$l(\phi) = -\nabla\phi \cdot \mathbf{n} + (\mathbf{e} - \mathbf{u}_b/U_\infty) \cdot \mathbf{n} = 0 \qquad \text{on } \partial\Omega_1 \tag{5.141}$$

$$\phi(\mathbf{x}_b) = \phi_b \qquad \text{on } \partial\Omega_2 \tag{5.142}$$

which is functionally identical to the heat transfer problem description considered to this point; see (5.82) to (5.85) or (5.88) to (5.90). Therefore, (5.91) defines the finite element solution while (5.92) clearly states the weak statement approximation for the definitions $k_e = 1$, $a_e = 0$ and $b_e = (\mathbf{e} - \mathbf{u}_b/U_\infty) \cdot \mathbf{n}$. Hence

$$\begin{aligned} WS^h &= S_e\left(\int_{\Omega_e} \nabla\{N_k\} \cdot \nabla\{N_k\}^T \, d\tau \{Q\}_e - \int_{\partial\Omega_1} \{N_k\}(\mathbf{u}_b/U_\infty - \mathbf{e})_e \cdot \mathbf{n} \, d\sigma\right) \\ &= S_e([K]_e\{Q\}_e - \{b\}_e) \end{aligned} \tag{5.143}$$

Viewing (5.143), the finite element statement for aerodynamic potential flow is simply a special case of (5.99), the general statement. Recalling (5.127) as the master matrix library statement for (5.99), we need only define the boundary data interpolation to transform (5.143) into the library syntax. Hence

$$\left(\frac{\mathbf{u}}{U_\infty} - \mathbf{e}\right)_e \cdot \mathbf{n} \equiv \{N_k\}^T\{UBDOTN - EDOTN\}_e \tag{5.144}$$

where $\{UBDOTN\}_e$, pronounced "U-B-dot-N," contains the boundary nodal values of $\mathbf{u}_b/U_\infty \cdot \mathbf{n}$, with the corresponding interpretation for $\{EDOTN\}_e$.

Then, the Fortran statement replacing (5.143) for arbitrary basis degree $k$ is

$$\begin{aligned} WS^h = S_e(&MEAS_e ZETAIJ_e ZETAKJ_e[B2IKd]\{Q\}_e \\ &- l_e[A200d]\{UBDOTN - EDOTN\}_e) = \{0\} \end{aligned} \tag{5.145}$$

In (5.145), $MEAS_e$ is the measure of $\Omega_e$ which equals $(4A_e)^{-1}$ or $(2A_e)^{-1}$ for $k = 1$ or 2; see (5.127) and (5.128). As before, $l_e$ is the length of the triangular element boundary segment coincident with $\partial\Omega_1$. Finally, $d$ in $[B2IKd]$ and $[A200d]$ takes the value $L$ or $Q$ for the linear or quadratic basis implementation.

**STUDY PROBLEM 4.** Determine the velocity perturbation potential field $\phi^h(\mathbf{x})$ associated with steady external flow about an elliptic cylinder of major and minor axes $a$ and $b$. The imposed farfield flowfield is parallel to the $x$ axis with magnitude $U_\infty$.

This problem statement is challenging since the domain of definition is infinite in extent. Therefore, the first requirement is to establish a suitable solution domain $\Omega$, that is one for which boundary conditions (5.141) and (5.142) can be readily defined. Figure 5.9$a$ illustrates the essential aspect, and it is immediately apparent that both the $x$ and $y$ axes are lines of symmetry for $\mathbf{U}_\infty$ parallel to the $x$ axis. Therefore, since efficiency is a guiding influence, any single quadrant is adequate, so let's choose the second, Fig. 5.9$b$.

The next step is to determine how far away from the elliptic cylinder the inflow and top boundaries in Fig. 5.9$b$ must be to admit defining the appropriate "farfield" boundary condition. Guidance for this determination comes from the analytical solution for the special case of a circular cylinder ($b = a$) of radius $A$. The exact solution in polar coordinate form is

$$\phi(r, \theta) = -\left(\frac{A}{r}\right)^2 \cos\theta \tag{5.146}$$

where $r = \sqrt{x^2 + y^2}$ and $\theta$ is measured counterclockwise from the positive $x$ axis. From (5.146), $\phi$ will decrease to 1 percent of its maximum value at $r = 10A$, which is probably the minimum allowable displacement for $\partial\Omega$ to be specified as farfield. Therefore, assuming this is appropriate for the elliptic cylinder, define the two farfield boundaries at $x = -10a/2$ and $y = 5b$.

With this determination, the succeeding step is to define the precise boundary condition form on $\partial\Omega$ of Fig. 5.9$b$. At the far left, the imposed inflow is $\mathbf{u}_b = U_\infty \mathbf{i}$; hence $\mathbf{e} = \mathbf{i}$ and (5.141) yields $-\nabla\phi \cdot \mathbf{n} = 0$ (which you should verify). For the top boundary of $\Omega$ sufficiently remote from the cylinder, $\mathbf{u}_b$ will be essentially unperturbed from $U_\infty \mathbf{i}$. Since $\mathbf{n} = \mathbf{j}$ on this segment, (5.142) again yields $-\nabla\phi \cdot \mathbf{n} = 0$, which you should also verify. The lower $x$-axis boundary segment is a symmetry line, and no flow crosses over the cylinder boundary, either, so $-\nabla\phi \cdot \mathbf{n} = 0$ is appropriate on both boundary segments. Finally, the right $y$-axis boundary segment is also a symmetry line, but an unknown distribution of velocity efflux exists thereon, so (5.139) cannot be used. However, flow symmetry requires there be no component of the efflux velocity vector parallel to this boundary segment; therefore, $\phi$ can be set to *any* constant $\phi_b$ along the entire length, which constitutes (5.142).

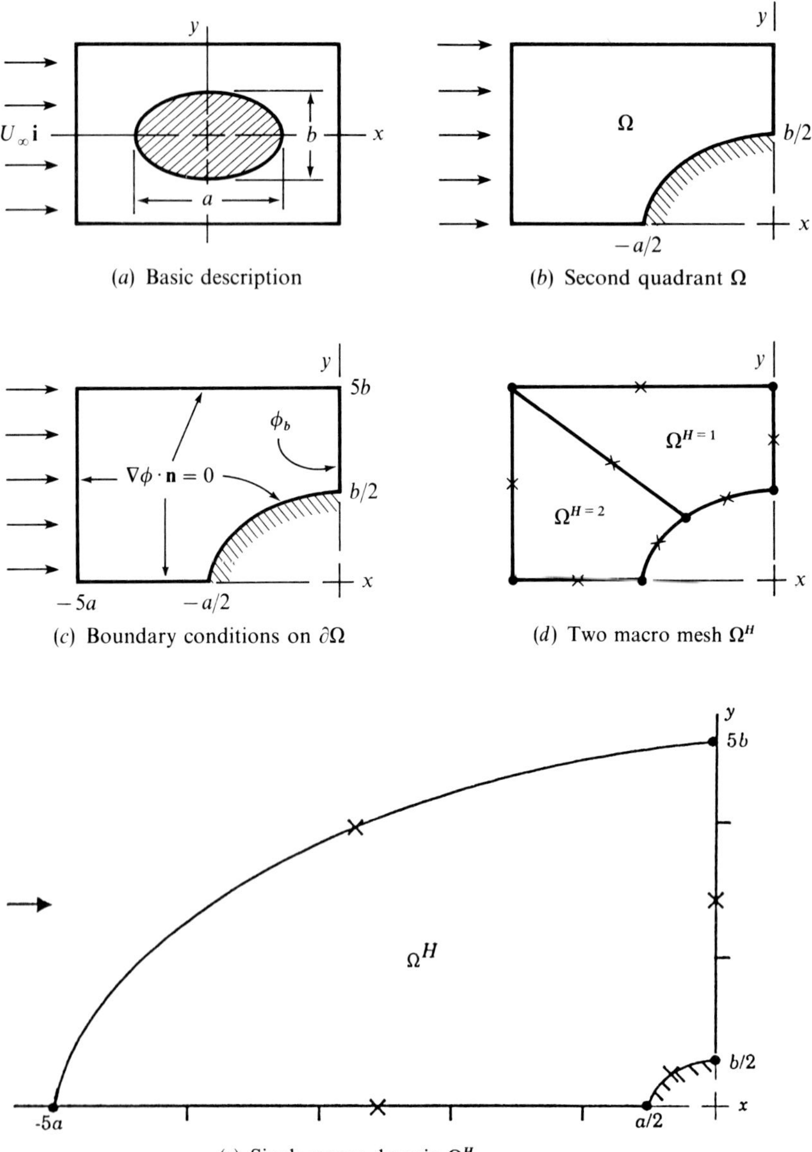


(e) Single macro domain $\Omega^H$

**FIGURE 5.9**
Geometric descriptions for aerodynamic potential flow around an elliptic cylinder.

Figure 5.9*c* summarizes the selected domain $\Omega$ and boundary condition distribution on $\partial\Omega$. The problem statement is well posed, since either Neumann or Dirichlet data are specified everywhere on $\partial\Omega = \partial\Omega_1 \bigcup \partial\Omega_2$. (At this point in problem synthesis, the analyst again appreciates that the finite element method directly handles Neumann constraints.) The next step is to generate an appropriate discretization $\Omega^h$ on which the (first) computational experiment can be conducted. Figure 5.9*d* shows a double macro domain $\Omega^H$ discretization for which the grid procedures utilized by *LEARN.FE* could create a nonuniform computational mesh $\Omega^h$. However, on further insight drawn from the more correct geometric scale, Fig. 5.9*e*, a single macro mesh is easily identified as a direct extension on past experience. The wasted mesh in the upper left farfield of Fig. 5.9*d* is eliminated, and the required vanishing Neumann boundary condition is readily implemented on any curved domain boundary segment.

The preceding discussion illustrates the steps through which the computational experimentalist must proceed to convert an engineering problem statement into computable form. We suggest you set up an input file for *LEARN.FE* to generate one or more finite element approximate solutions for Study Problem 4. The basic ingredients of input file construction are highlighted below, but the details are left to you. We suggest you start off with an $M_H = M_1 \times M_2 = 9 \times 9$ nonuniform nodal mesh on $\Omega^H$, recall Code Exercise 5.9, and then proceed to a mesh refinement study to assess solution error reduction. Insert a STOP command after GRID until you achieve a "decent" first mesh. The resultant nodal coordinates created on the solid surface will require correction, as in NTRAN = 3, if the genuine ellipse simulation is sought. Otherwise the aerodynamic surface will be a parabola. For guidance, Fig. 5.10 illustrates a typical $\phi^h$ solution

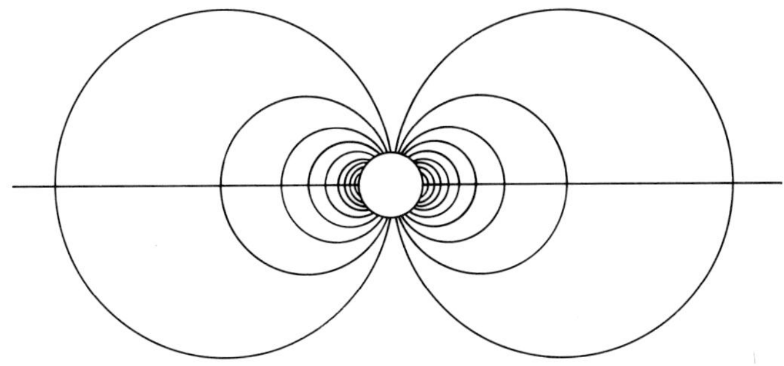

**FIGURE 5.10**
Isopotential distribution for cylinder flow perturbation potential solution $\phi^h$.

as computed on an adequate mesh $\Omega^h$ for the circular cylinder problem (Baker 1983, Chap. 3).

```
TITL
***** AERODYNAMICS STUDY PROBLEM # 4 *****
TYPE      [K      N      NNODEL     REFL   NPRN     NTRAN NAXI]
           1      2        3         1.0    1 1        2     0
PRIN     [NBUG(*)]
           1      0        0         2
GRID     [FOR N (XL XR(*) PR(*)), NEM(PR*), XI(NTRAN)]
          -5A     -A/2     0.8
           8
           0       5B      1.0
           8
          -5A     -A/2     0.        0.
           0.      0.      B/2       5B
        -2.5A      .        .        .
           0.      .        .        .
      STOP
MATL     [FOR MACRO (I), KI(*)]
           1      1.0
DIRI     [(NDIR, REPEAT), NODE(*), (NRPT, QB(*))]
           9       1       0
           9       18      .   .   .
           9       0.0
FORM
GAUS     [CONVERGENCE]
          0.001
NORM
STOP
```

## 5.10 ELASTICITY, PLANE STRESS, AND PLANE STRAIN

The previous section illustrated application of the finite element weak statement procedure to a fluid mechanics problem. We now introduce the analogous development for elementary problem classes in two-dimensional structural mechanics. Under the assumption of a body in *equilibrium*, an internal force (*stress*) distribution becomes induced by applied body forces (gravity) and surface forces (tractions) including pressure and restraint forces. This stress field results in an internal *strain* distribution which produces a displacement field, namely, the *deflected* state.

One desired output of a structural analysis for an elastic body is this displacement distribution, which is usually denoted as

$$\mathbf{u}(x_i) = u(x, y)\mathbf{i} + v(x, y)\mathbf{j} \tag{5.147}$$

In (5.147), $u$ and $v$ are the scalar components of a displacement field parallel to the global coordinate system $x_i$. Figure 5.11 illustrates (5.147) for a plate with an initially circular hole loaded by tensile traction forces with magnitude $T$ applied at each lateral edge. This section develops the finite element weak statement solution procedure for the elasticity statements of plane stress and plane strain. It may be omitted without loss of continuity in the following chapters.

It is of more than passing note that (5.147) and (5.134) present an *identical* appearing unknown. In truth, we are really dealing with finite element analysis in *computational continuum mechanics.* However, the preferred (and traditional) approach to finite element algorithm development in structural mechanics is to employ the *principle of virtual work* and the scalar *strain energy functional.* The classical extremization of this integral produces the statement of equilibrium, namely, Newton's law, and hence naturally contains all applied forces, thermal influences, and the effect of initial strains. The subsequent finite element solution procedure uses the basis function constitution to *approximately* evaluate the strain energy functional in terms of element nodal degrees of freedom $\{Q\}_e$. The resultant symmetric quadratic form is then "discretely extremized" with respect to all $\{Q\}_e$, producing the global matrix statement

$$S_e([K + H]_e\{Q\}_e - \{b\}_e) \equiv \{0\} \tag{5.2}$$

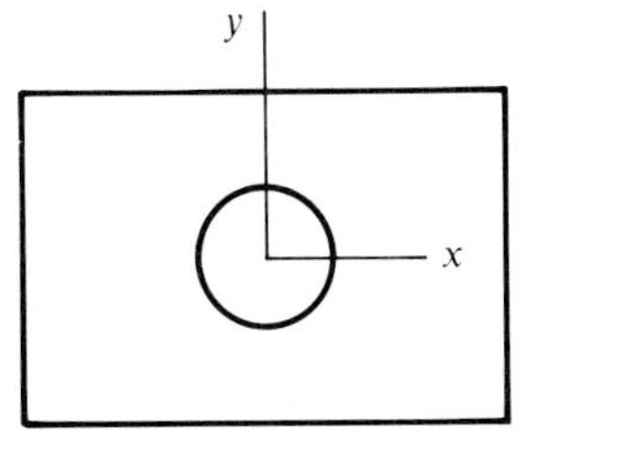


(*a*) Undeflected geometry

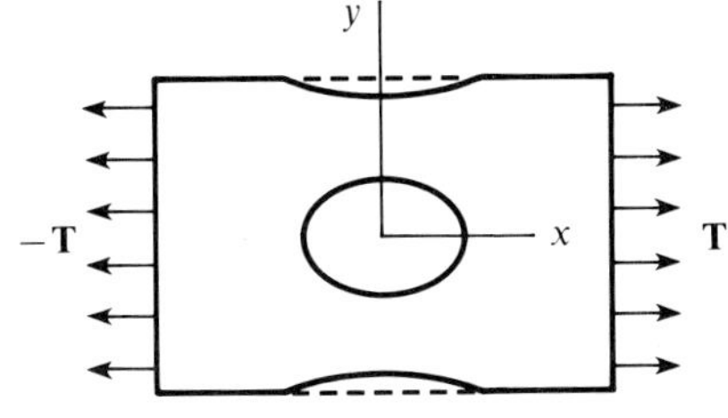


(*b*) Exaggerated deflected state

**FIGURE 5.11**
Deflection of a plate with a circular hole by tensile traction load of magnitude $T$.

In (5.2), $[K + H]_e$ is called the "element stiffness matrix" and $\{b\}_e$ is the equivalent nodal load matrix.

Forming the finite element discrete approximation to the weak statement written for the Newton's law statement of equilibrium is definitely more tedious to develop in full generality. However, for isothermal problems without initial strain, it can be formulated quite directly as illustrated in this section. The *LEARN.FE* code is not operational for this development, but it could be so extended based on the following material should you be so inclined.

**Newton's Law Statements**

The first step is to state Newton's law for two-dimensional linear elasticity for an ideal (homogeneous and isotropic) material. As the name implies, *plane stress* assumes all forces lie in the $x$–$y$ plane. The dependent variable is force per unit area, which is called *stress*. The planar scalar components are denoted $\sigma_x$, $\sigma_y$, and $\tau_{xy}$, and $\sigma_z \equiv 0 \equiv \tau_{yz} = \tau_{xz}$ by definition. Further, $\sigma_x$ and $\sigma_y$ are called the *normal stresses* while $\tau_{xy}$ is the *shear stress*. Denoting the scalar components of the volume-specific (gravity) body force as $B_x$ and $B_y$, the Newton's law statement of static equilibrium is

$$\frac{\partial \sigma_x}{\partial x} + \frac{\partial \tau_{xy}}{\partial y} + B_x = 0 \qquad (5.148a)$$

$$\frac{\partial \sigma_y}{\partial y} + \frac{\partial \tau_{xy}}{\partial x} + B_y = 0 \qquad (5.148b)$$

The *continuum mechanics* analysis relates the stress and body force distribution to the deflected state of the continua. The new dependent variable is *strain*, which is displacement per unit length with planar components $\varepsilon_x$, $\varepsilon_y$, and $\gamma_{xy}$. For the linear theory, $\varepsilon_z = -\nu(\varepsilon_x + \varepsilon_y)/(1 - \nu)$, $\varepsilon_x$ and $\varepsilon_y$ are the normal strains, $\gamma_{xy}$ is the shear strain, and $\nu$ is the (dimensionless) *Poisson ratio*. The *constitutive* relation between stress and strain is termed *Hooke's law*, which for a homogeneous isotropic material has the matrix form

$$\{\sigma\} = [C]\{\varepsilon\} \qquad (5.149)$$

In (5.149), $\{\sigma\} \equiv \{\sigma_x, \sigma_y, \tau_{xy}\}^T$, $\{\varepsilon\} \equiv \{\varepsilon_x, \varepsilon_y, \gamma_{xy}\}^T$, and the Hooke's law matrix $[C]$ is

$$[C] \equiv \frac{E}{1 - \nu^2}\begin{bmatrix} 1 & \nu & 0 \\ \nu & 1 & 0 \\ 0 & 0 & \frac{1}{2}(1 - \nu) \end{bmatrix} \qquad (5.150)$$

where $E$ is *Young's modulus* for the material with units of force per area.

The partial differential equation system governing the displacement vector (5.147) is then established by combining (5.148) to (5.150), along with the linear strain-displacement definitions $\varepsilon_x \equiv \partial u/\partial x$, $\varepsilon_y \equiv \partial v/\partial y$, and $\gamma_{xy} \equiv (\partial u/\partial y + \partial v/\partial x)/2$. A suggested exercise is to verify that this operation yields

$$L(u) = \nabla^2 u + \frac{1+\nu}{1-\nu}\frac{\partial}{\partial x}\left(\frac{\partial u}{\partial x} + \frac{\partial v}{\partial y}\right) + \frac{B_x}{G} = 0 \tag{5.151a}$$

$$L(v) = \nabla^2 v + \frac{1+\nu}{1-\nu}\frac{\partial}{\partial y}\left(\frac{\partial u}{\partial x} + \frac{\partial v}{\partial y}\right) + \frac{B_y}{G} = 0 \tag{5.151b}$$

where $G \equiv E/[2(1+\nu)]$ is termed the *elastic shear modulus* and is identical to the Lamé parameter $\mu$. Thus, for *plane stress*, each scalar component of the displacement vector $\mathbf{u}(x, y)$ satisfies a nonhomogeneous Laplace-type equation with a directional diffusion coefficient equal to $(1 + (1+\nu)/(1-\nu))$ in the coordinate direction parallel to the displacement component and unity in the orthogonal direction. Further, each equation is coupled to the other via the mixed derivative term $\partial^2(\cdot)/\partial x\,\partial y$, and the body force component is a source term.

The *plane strain* problem statement is distinguished by a modest modification to (5.151). As the name implies, the strain field is now $\{\varepsilon\} \equiv \{\varepsilon_x, \varepsilon_y, \gamma_{xy}\}^T$ and $\varepsilon_z \equiv 0 \equiv \gamma_{yz} = \gamma_{xz}$, and the resultant stress field is $\{\sigma\} = \{\sigma_x, \sigma_y, \tau_{xy}\}^T$ with $\sigma_z = \nu(\sigma_x + \sigma_y)$. Newton's law (5.148) is unchanged, as is Hooke's law (5.149), but the constitutive matrix $[C]$ is now

$$[C] \equiv \frac{E}{(1+\nu)(1-2\nu)}\begin{bmatrix} 1-\nu & \nu & 0 \\ \nu & 1-\nu & 0 \\ 0 & 0 & \frac{1}{2}(1-2\nu) \end{bmatrix} \tag{5.152}$$

Combining the linear strain-displacement definitions, which are also unchanged, with (5.148), (5.149) and (5.152) yields

$$L(u) = \nabla^2 u + \frac{1}{1-2\nu}\frac{\partial}{\partial x}\left(\frac{\partial u}{\partial x} + \frac{\partial v}{\partial y}\right) + \frac{B_x}{G} = 0 \tag{5.153a}$$

$$L(v) = \nabla^2 v + \frac{1}{1-2\nu}\frac{\partial}{\partial y}\left(\frac{\partial u}{\partial x} + \frac{\partial v}{\partial y}\right) + \frac{B_y}{G} = 0 \tag{5.153b}$$

We also suggest you complete the steps leading to (5.153), which is functionally identical to (5.151) for redefinition of the directional diffusion coefficient as $1 + 1/(1-2\nu)$ and unity.

Thus, Newton's law statements for plane stress and plane strain constitute the coupled partial differential equation (PDE) systems (5.151) and (5.153). Since the directional diffusion coefficients are uniformly positive, each PDE describes a linear elliptic boundary value problem. For a well-posed statement, we must thus define a combination of fixed and flux gradient (Dirichlet and Neumann) constraints all around the entire boundary $\partial\Omega$ of the domain $\Omega$ defining the structure. The Dirichlet constraint is a known value for either $u$ or $v$, (5.147), as results for a fixed support or a symmetry line, for example.

The Neumann constraint involves resolution of applied forces (surface tractions) as established from a free-body diagram of a boundary segment (see Fig. 5.12). The state of stress at the surface involves $\{\sigma\} = \{\sigma_x, \sigma_y, \tau_{xy}\}^T$, which can be expressed in local normal-tangent scalar components as $\{\sigma\} = \{\sigma_n, \tau_{ns}\}^T$. The externally applied traction (force/unit area) is $\mathbf{T}$, and from Fig. 5.12 the force balance for a boundary segment with unit normal vector $\mathbf{n}$ and tangent vector $\mathbf{s}$ is $\mathbf{T} = \sigma_{nn}\mathbf{n} + \tau_{ns}\mathbf{s}$.

The scalar components of $\mathbf{T}$ parallel to the global $(x, y)$ coordinate axes are $T_x$ and $T_y$, which are established using the vector dot product with the associated unit vectors $\mathbf{i}$ and $\mathbf{j}$:

$$\begin{aligned} T_x = \mathbf{T}\cdot\mathbf{i} &= (\sigma_{nn}\mathbf{n} + \tau_{ns}\mathbf{s})\cdot\mathbf{i} \\ &= \sigma_{nn}\mathbf{n}\cdot\mathbf{i} + \tau_{ns}\mathbf{s}\cdot\mathbf{i} \\ &= \sigma_{nn}\cos\theta + \tau_{ns}(-\sin\theta) \end{aligned} \tag{5.154a}$$

$$\begin{aligned} T_y = \mathbf{T}\cdot\mathbf{j} &= \sigma_{nn}\mathbf{n}\cdot\mathbf{j} + \tau_{ns}\mathbf{s}\cdot\mathbf{j} \\ &= \sigma_{nn}\sin\theta + \tau_{ns}\cos\theta \end{aligned} \tag{5.154b}$$

where $\theta$ is the included angle between $\mathbf{n}$ and the $x$ axis. The next step is to express $\sigma_{nn}$ and $\tau_{ns}$ in (5.154) in terms of $\{\sigma\} = \{\sigma_x, \sigma_y, \tau_{xy}\}$, and then invoke Hooke's law and the linear strain-displacement definitions to replace $\{\sigma\}$. This involves rather laborious detail which we elect to omit;

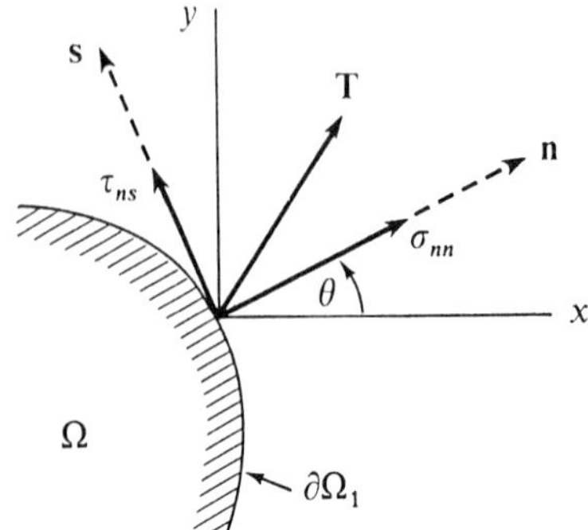


**FIGURE 5.12**
Free-body diagram for a boundary segment $\partial\Omega_1$ of $\Omega$ with traction $\mathbf{T}$.

the end result is (Huebner and Thornton, 1982, App. C)

$$T_x = \lambda \cos\theta\left(\frac{\partial u}{\partial x} + \frac{\partial v}{\partial y}\right) + \mu\left(2\cos\theta\frac{\partial u}{\partial x} + \sin\theta\left(\frac{\partial u}{\partial y} + \frac{\partial v}{\partial x}\right)\right) \qquad (5.155a)$$

$$T_y = \lambda \sin\theta\left(\frac{\partial u}{\partial x} + \frac{\partial v}{\partial y}\right) + \mu\left(2\sin\theta\frac{\partial v}{\partial y} + \cos\theta\left(\frac{\partial v}{\partial x} + \frac{\partial u}{\partial y}\right)\right) \qquad (5.155b)$$

In (5.155), $\mu$ remains the second Lamé parameter ($G$) while $\lambda \equiv E\nu/(1+\nu)(1-2\nu) = 2\nu G/(1-2\nu)$ is the first Lamé parameter. Equation (5.155) expresses the nonhomogeneous Neumann boundary condition for (5.151) and (5.153) for application of surface tractions $T_x$ and $T_y$ on boundary segments $\partial\Omega_1$. The homogeneous Neumann boundary conditions for (5.151) and (5.153) appropriate for symmetry planes and traction-free boundary segments $\partial\Omega$ are determined by inspection.

### Finite Element Algorithm

The next step is to form the finite element approximation to weak statements written for (5.151) and (5.153) with (5.155). We coalesce the dual development by defining the directional diffusion coefficient $k_\nu$

$$k_\nu \equiv \begin{cases} 1 + \dfrac{1+\nu}{1-\nu} & \text{for plane stress} \\ 1 + \dfrac{1}{1-2\nu} & \text{for plane strain} \end{cases} \qquad (5.156)$$

Then, both (5.151) and (5.153) can be written in expanded form as

$$\begin{aligned} L(u) &= k_\nu \frac{\partial^2 u}{\partial x^2} + \frac{\partial^2 u}{\partial y^2} + (k_\nu - 1)\frac{\partial^2 v}{\partial x\,\partial y} + b_x = 0 \\ &= k_\nu \frac{\partial}{\partial x}\left(\frac{\partial u}{\partial x} + \frac{\partial v}{\partial y}\right) + \frac{\partial}{\partial y}\left(\frac{\partial u}{\partial y} - \frac{\partial v}{\partial x}\right) + b_x = 0 \qquad (5.157a) \end{aligned}$$

$$\begin{aligned} L(v) &= k_\nu \frac{\partial^2 v}{\partial y^2} + \frac{\partial^2 v}{\partial x^2} + (k_\nu - 1)\frac{\partial^2 u}{\partial x\,\partial y} + b_y = 0 \\ &= k_\nu \frac{\partial}{\partial y}\left(\frac{\partial v}{\partial y} + \frac{\partial u}{\partial x}\right) + \frac{\partial}{\partial x}\left(\frac{\partial v}{\partial x} - \frac{\partial u}{\partial y}\right) + b_y = 0 \qquad (5.157b) \end{aligned}$$

where $(b_x, b_y)$ are the body force components $(B_x, B_y)$ normalized by $G$.

As always, the finite element algorithm involves a discrete approximate evaluation of a weak statement, now for (5.157). Letting $\{q\} \equiv \{u, v\}^T$ denote the dependent variable pair that is the solution to (5.157), then on the discretization $\Omega^h \bigcup \partial\Omega^h$ of $\Omega \bigcup \partial\Omega$, the finite element approximate solution definition is

$$\{q(x, y)\} \simeq \{q^h(x, y)\} = \bigcup_e \begin{Bmatrix} \{N_k\}^T\{U\}_e \\ \{N_k\}^T\{V\}_e \end{Bmatrix} \tag{5.158}$$

In (5.158), $\{U\}_e$, and $\{V\}_e$ denote the respective nodal displacements on any element $\Omega_e$ of $\Omega^h = \bigcup_e \Omega_e$. Substituting (5.158), the weak statement (5.23) for (5.157) becomes the pair of statements

$$\{WS^h\} = S_e \int_\Omega \begin{Bmatrix} -\{N_k\}L(u^h) \\ -\{N_k\}L(v^h) \end{Bmatrix} d\tau \equiv \{0\} \tag{5.159}$$

The negative signs are inserted for convenience, and both expressions in (5.159) must be formed and solved simultaneously to determine the global nodal discrete displacement fields $\{U\}$ and $\{V\}$.

Considering (5.157*a*) first, expand (5.159) as

$$\begin{aligned} WS^h(u^h) &\equiv -\int_\Omega \{N_k\}L(u^h)\, d\tau \\ &= -S_e\left(\int_{\Omega_e} \{N_k\}\left[k_v \frac{\partial}{\partial x}\left(\frac{\partial u^h}{\partial x} + \frac{\partial v^h}{\partial y}\right) + \frac{\partial}{\partial y}\left(\frac{\partial u^h}{\partial y} - \frac{\partial v^h}{\partial x}\right) + b_x^h\right] d\tau\right) \\ &= S_e\left(\int_{\Omega_e}\left[\frac{\partial\{N_k\}}{\partial x} k_v\left(\frac{\partial u^h}{\partial x} + \frac{\partial v^h}{\partial y}\right) + \frac{\partial\{N_k\}}{\partial y}\left(\frac{\partial u^h}{\partial y} - \frac{\partial v^h}{\partial x}\right) - \{N_k\}\, b_x^h\right] d\tau \right. \\ &\quad \left. - \int_{\partial\Omega} \{N_k\}\left[k_v\left(\frac{\partial u^h}{\partial x} + \frac{\partial v^h}{\partial y}\right)\cos\theta + \left(\frac{\partial u^h}{\partial y} - \frac{\partial v^h}{\partial x}\right)\sin\theta\right] d\sigma\right) \end{aligned} \tag{5.160}$$

where $\theta$ remains the angle between **n** of $\partial\Omega$ and the $x$ axis (Fig. 5.12). For boundary segments $\partial\Omega_e$ that coincide with $\partial\Omega_1$ where a traction is applied, the integrand in the surface integral in (5.160) is replaced using (5.155) which yields $G^{-1}T_x$. The integrand vanishes identically on boundary segments that are traction-free; hence the computational form of the weak statement for $L(u^h)$ is

$$WS^h(u^h) = S_e\left(\int_{\Omega_e}\left(k_v \frac{\partial\{N_k\}}{\partial x}\frac{\partial\{N_k\}^T}{\partial x} + \frac{\partial\{N_k\}}{\partial y}\frac{\partial\{N_k\}^T}{\partial y}\right)d\tau\{U\}_e\right.$$
$$+\int_{\Omega_e}\left(k_v \frac{\partial\{N_k\}}{\partial x}\frac{\partial\{N_k\}^T}{\partial y} - \frac{\partial\{N_k\}}{\partial y}\frac{\partial\{N_k\}^T}{\partial x}\right)d\tau\{V\}_e$$
$$-G^{-1}\int_{\Omega_e}\{N_k\}\{N_k\}^T d\tau\{BX\}_e$$
$$\left.-G^{-1}\int_{\partial\Omega_e\cap\partial\Omega_1}\{N_k\}\{N_k\}^T d\sigma\{TX\}_e\right) \equiv \{0\} \qquad (5.161a)$$

In (5.161$a$), $\{BX\}_e$ contains the nodal values of the (unnormalized) body force $B_x$, recall (5.151) and (5.153). Further, in the last term the notation $\partial\Omega_e \cap \partial\Omega_1$ is read, "where the element boundary segment $\partial\Omega_e$ coincides (intersects) with $\partial\Omega_1$."

Proceeding through the same detail for $L(v)$, (5.157$b$), inserting (5.155$b$) yields the weak statement finite element approximation for $L(v^h)$ as

$$WS^h(v^h) = S_e\left(\int_{\Omega_e}\left(k_v \frac{\partial\{N_k\}}{\partial y}\frac{\partial\{N_k\}^T}{\partial y} + \frac{\partial\{N_k\}}{\partial x}\frac{\partial\{N_k\}^T}{\partial x}\right)d\tau\{V\}_e\right.$$
$$+\int_{\Omega_e}\left(k_v \frac{\partial\{N_k\}}{\partial y}\frac{\partial\{N_k\}^T}{\partial x} - \frac{\partial\{N_k\}}{\partial x}\frac{\partial\{N_k\}^T}{\partial y}\right)d\tau\{U\}_e$$
$$-G^{-1}\int_{\Omega_e}\{N_k\}\{N_k\}^T d\tau\{BY\}_e$$
$$\left.-G^{-1}\int_{\partial\Omega_e\cap\partial\Omega_1}\{N_k\}\{N_k\}^T d\sigma\{TY\}_e\right) \equiv \{0\} \qquad (5.161b)$$

Thus, (5.161) expresses the finite element weak statement algorithm for the linear plane stress and plane strain structural mechanics problem classes for arbitrary trial basis selection $\{N_k\}$ and for $k_v$ as defined in (5.156).

### Linear Basis Stiffness Matrices

Now evaluate the element stiffness matrix structure for (5.161) for the linear basis selection. The first term is of laplacian type directionally weighted by the diffusion coefficient $k_v$ which is a uniform constant throughout $\Omega$. For element stiffness matrix $[K]_e$ formation, this fact requires (only) a summation index constraint in the product of the element metric data $ZETAIJ_e$ with the element master matrix $[B2IKL]$; recall (5.121). Viewing the diffusion matrix parent form (5.25), the modification to accept directional $k_v$ and invoking the chain rule in the first term of (5.161*a*) yields

$$
\begin{aligned}
[KU(U)]_e &\equiv \int_{\Omega_e} \left( k_v \frac{\partial\{N_1\}}{\partial x} \frac{\partial\{N_1\}^T}{\partial x} + \frac{\partial\{N_1\}}{\partial y} \frac{\partial\{N_1\}^T}{\partial y} \right) d\tau \\
&= \left( k_v \frac{\partial\{N_1\}}{\partial \zeta_i} \frac{\partial \zeta_i}{\partial x} \frac{\partial\{N_1\}^T}{\partial \zeta_k} \frac{\partial \zeta_k}{\partial x} + \frac{\partial\{N_1\}}{\partial \zeta_i} \frac{\partial \zeta_i}{\partial y} \frac{\partial\{N_1\}^T}{\partial \zeta_k} \frac{\partial \zeta_k}{\partial y} \right)_e \int_{\Omega_e} d\tau \\
&= A_e \left( \frac{\partial\{N_1\}}{\partial \zeta_i} \frac{\partial\{N_1\}^T}{\partial \zeta_k} \right) \left( k_v \frac{\partial \zeta_i}{\partial x} \frac{\partial \zeta_k}{\partial x} + \frac{\partial \zeta_i}{\partial y} \frac{\partial \zeta_k}{\partial y} \right)_e \\
&= (4A_e)^{-1} [B2IKL] (k_v ZETAI1_e ZETAK1_e \\
&\qquad + ZETAI2_e ZETAK2_e) \qquad \text{for } 1 \le (I, K) \le 3 \qquad (5.162a)
\end{aligned}
$$

Comparing (5.162*a*) and (5.121), the previous summation index $J$ with range $1 \le J \le n = 2$ has been replaced by the sum of metric products with coefficient $k_v$ and unity. In the similar manner, the corresponding term in (5.161*b*) becomes

$$
\begin{aligned}
[KV(V)]_e &= (4A_e)^{-1} [B2IKL] (ZETAI1_e ZETAK1_e \\
&\qquad + k_v ZETAI2_e ZETAK2_e) \qquad \text{for } 1 \le (I, K) \le 3 \qquad (5.162b)
\end{aligned}
$$

The second terms in (5.161) contain mixed derivatives involving the alternate displacement field component. For (5.161*a*), the element stiffness matrix contribution is

$$
\begin{aligned}
[KU(V)]_e &\equiv \int_{\Omega_e} \left( k_v \frac{\partial\{N_1\}}{\partial x} \frac{\partial\{N_1\}^T}{\partial y} - \frac{\partial\{N_1\}}{\partial y} \frac{\partial\{N_1\}^T}{\partial x} \right) d\tau \\
&= \left( k_v \frac{\partial\{N_1\}}{\partial \zeta_i} \frac{\partial \zeta_i}{\partial x} \frac{\partial\{N_1\}^T}{\partial \zeta_k} \frac{\partial \zeta_k}{\partial y} - \frac{\partial\{N_1\}}{\partial \zeta_i} \frac{\partial \zeta_i}{\partial y} \frac{\partial\{N_1\}^T}{\partial \zeta_k} \frac{\partial \zeta_k}{\partial x} \right)_e \int_{\Omega_e} d\tau \\
&= (4A_e)^{-1} [B2IKL] (k_v ZETAI1_e ZETAK2_e \\
&\qquad - ZETAI2_e ZETAK1_e) \qquad \text{for } 1 \le (I, K) \le 3 \qquad (5.163a)
\end{aligned}
$$

The corresponding contribution for (5.161*b*) is

$$
\begin{aligned}
[KV(U)]_e &= (4A_e)^{-1} [B2IKL] (k_v ZETAI2_e ZETAK1_e \\
&\qquad - ZETAI1_e ZETAK2_e) \qquad \text{for } 1 \le (I, K) \le 3 \qquad (5.163b)
\end{aligned}
$$

Finally, the body force term in either equation involves a product with the master matrix $A_e[B200L]$ while the surface traction term correspondingly uses $l_e[A200L]$.

Thus, the linear triangle basis matrix library for the finite element weak statement approximation (5.159) for the plane stress or plane strain problem class is now fully defined in Fortran nomenclature. In the form (5.2), the element-level statements are

$$S_e([KU(U)]_e\{U\}_e + [KU(V)]_e\{V\}_e - \{bx\}_e) = \{0\}$$
$$S_e([KV(V)]_e\{V\}_e + [KV(U)]_e\{U\}_e - \{by\}_e) = \{0\} \qquad (5.164)$$

These matrix statements must be solved simultaneously to handle the explicit nodal coupling. Therefore, the global matrix statement for computation is

$$S_e([K + H]_e\{Q\}_e - \{b\}_e)$$
$$= S_e\left(\begin{bmatrix}[KU(U)]_e & [KU(V)]_e \\ [KV(U)]_e & [KV(V)]_e\end{bmatrix}\begin{Bmatrix}\{U\}_e \\ \{V\}_e\end{Bmatrix} - \begin{Bmatrix}\{bx\}_e \\ \{by\}_e\end{Bmatrix}\right) = \{0\} \qquad (5.165)$$

The source (load) term definitions are

$$\{bx\}_e = A_e[B200L]\{BX/G\}_e + l_e[A200L]\{TX/G\}_e$$
$$\{by\}_e = A_e[B200L]\{BY/G\}_e + l_e[A200L]\{TY/G\}_e \qquad (5.166)$$

**Application Examples**

Equation (5.165) is the matrix statement of the plane stress–plane strain finite element algorithm with (5.162) and (5.163) providing the linear basis element stiffness matrix definitions. We now carry two informative examples to hand-completion for a prismatic bar loaded first by its own weight and then for an applied tractive force. Figure 5.13*a* illustrates the problem definitions, both of which possess a plane of symmetry; hence only a vertical half of the bar need be considered for analysis. Figure 5.13*b* shows the right-half computational domain $\Omega$ along with a candidate four-element discretization $\Omega^h$. For this problem statement in plane stress, the Dirichlet data are $U = 0$ at nodes on the symmetry line and $U = 0 = V$ at nodes on the rigid upper support, Fig. 5.13*c*. Further, both vertical boundary edges are traction-free and $\partial v^h/\partial x = 0$ is the homogeneous Neumann constraint along the symmetry line.

Only two finite element orientations are involved in the discretization, in Fig. 5.13*b*. Assume the discretization is uniform with each $\Omega_e$ of length $l$ and width $\omega$ as measured parallel to the $y$ and $x$ axes, respectively. The

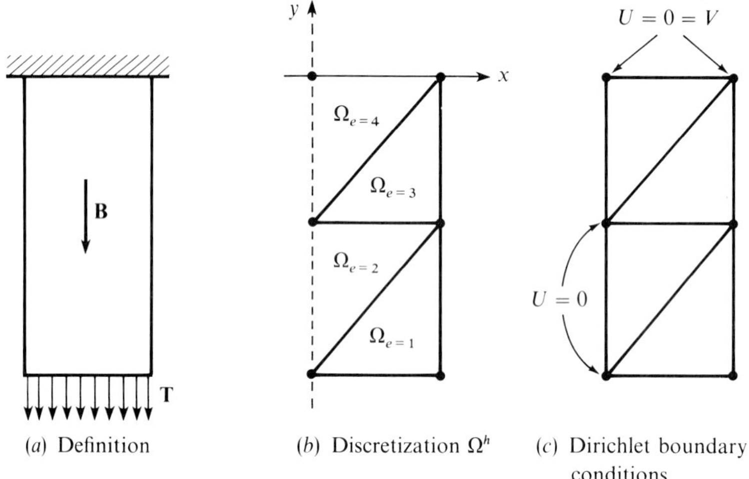


(a) Definition (b) Discretization $\Omega^h$ (c) Dirichlet boundary conditions

**FIGURE 5.13**
A prismatic bar acted on by gravity **B** and a traction **T**.

element metric data set $ZETAIJ_e$, recall (5.29) and (5.43), are the non-normalized (by $A_e$) entries in the element coordinate transformation. Hence

$$\left[\frac{\partial \zeta_i}{\partial x_j}\right]_{e=1,3} = \begin{bmatrix} Y_2 - Y_3 & X_3 - X_2 \\ Y_3 - Y_1 & X_1 - X_3 \\ Y_1 - Y_2 & X_2 - X_1 \end{bmatrix}_{e=1,3} = \begin{bmatrix} -l & 0 \\ l & -\omega \\ 0 & \omega \end{bmatrix}$$

$$\left[\frac{\partial \zeta_i}{\partial x_j}\right]_{e=2,4} = \begin{bmatrix} Y_2 - Y_3 & X_3 - X_2 \\ Y_3 - Y_1 & X_1 - X_3 \\ Y_1 - Y_2 & X_2 - X_1 \end{bmatrix}_{e=2,4} = \begin{bmatrix} 0 & -\omega \\ l & 0 \\ -l & \omega \end{bmatrix} \tag{5.167}$$

and the plane area of each finite element $\Omega_e$ is $A_e = 0.5l\omega$.

These data facilitate computing the element stiffness matrix set $[KU(U)]_e, \ldots, [KV(V)]_e$. For (5.162*a*), we obtain

$$\begin{aligned}
[KU(U)]_e &= (4A_e)^{-1}[B2IKL](k_v ZETAI1_e ZETAK1_e \\
&\quad + ZETAI2_e ZETAK2_e) \\
&= \frac{1}{4(l\omega)/2}([B211L](k_v ZETA11_e^2 + ZETA12_e^2) \\
&\quad + [B212L](k_v ZETA11_e ZETA21_e + ZETA12_e ZETA22_e) \\
&\quad + \cdots)
\end{aligned}$$

which is already fully expressed in (5.44*a*) except for inclusion of the $k_v$

multiplier. Recalling that $[B2IKL] = [DELTAIK]$ and using (5.167), you are encouraged to verify that

$$[KU(U)]_{e=1,3} = \frac{1}{2l\omega}\begin{bmatrix} k_v l^2 & -k_v l^2 & 0 \\ -k_v l^2 & k_v l^2 + \omega^2 & -\omega^2 \\ 0 & -\omega^2 & \omega^2 \end{bmatrix} \quad (5.168a)$$

$$[KU(U)]_{e=2,4} = \frac{1}{2l\omega}\begin{bmatrix} \omega^2 & 0 & -\omega^2 \\ 0 & k_v l^2 & -k_v l^2 \\ -\omega^2 & -k_v l^2 & k_v l^2 + \omega^2 \end{bmatrix} \quad (5.168b)$$

Comparing (5.162*a*) and (5.162*b*), the stiffness matrix element definitions are identical except for the location change for $k_v$. Thus, from (5.168*a*) it is easy to determine that

$$[KV(V)]_{e=1,3} = \frac{1}{2l\omega}\begin{bmatrix} l^2 & -l^2 & 0 \\ -l^2 & l^2 + k_v\omega^2 & -k_v\omega^2 \\ 0 & -k_v\omega^2 & k_v\omega^2 \end{bmatrix} \quad (5.168c)$$

$$[KV(V)]_{e=2,4} = \frac{1}{2l\omega}\begin{bmatrix} k_v\omega^2 & 0 & -k_v\omega^2 \\ 0 & l^2 & -l^2 \\ -k_v\omega^2 & -l^2 & l^2 + k_v\omega^2 \end{bmatrix} \quad (5.168d)$$

The operations yielding (5.168) are repeated, now for the cross-derivative stiffness matrices given in (5.163). Using the notation of (5.44), the expanded forms for the generic finite element $\Omega_e$ are

$$[KU(V)]_{e=1,3}$$
$$= (4A_e)^{-1}\begin{bmatrix} k_v\zeta_{11}\zeta_{12} - \zeta_{12}\zeta_{11} & k_v\zeta_{11}\zeta_{22} - \zeta_{12}\zeta_{21} & k_v\zeta_{11}\zeta_{32} - \zeta_{12}\zeta_{31} \\ k_v\zeta_{21}\zeta_{12} - \zeta_{22}\zeta_{11} & k_v\zeta_{21}\zeta_{22} - \zeta_{22}\zeta_{21} & k_v\zeta_{21}\zeta_{32} - \zeta_{22}\zeta_{31} \\ k_v\zeta_{31}\zeta_{12} - \zeta_{32}\zeta_{11} & k_v\zeta_{31}\zeta_{22} - \zeta_{32}\zeta_{21} & k_v\zeta_{31}\zeta_{32} - \zeta_{32}\zeta_{31} \end{bmatrix} \quad (5.169a)$$

$$[KV(V)]_{e=2,4}$$
$$= (4A_e)^{-1}\begin{bmatrix} k_v\zeta_{12}\zeta_{11} - \zeta_{11}\zeta_{12} & k_v\zeta_{12}\zeta_{21} - \zeta_{11}\zeta_{22} & k_v\zeta_{12}\zeta_{31} - \zeta_{11}\zeta_{32} \\ k_v\zeta_{22}\zeta_{11} - \zeta_{21}\zeta_{12} & k_v\zeta_{22}\zeta_{21} - \zeta_{21}\zeta_{22} & k_v\zeta_{22}\zeta_{31} - \zeta_{21}\zeta_{32} \\ k_v\zeta_{32}\zeta_{11} - \zeta_{31}\zeta_{12} & k_v\zeta_{32}\zeta_{21} - \zeta_{31}\zeta_{22} & k_v\zeta_{32}\zeta_{31} - \zeta_{31}\zeta_{32} \end{bmatrix} \quad (5.169b)$$

From (5.169), you may easily verify that $[KV(U)]_e = [KU(V)]_e^T$. Hence, only one stiffness matrix contribution is needed formed for each $\Omega_e$. Using the metric data (5.167), you may wish to verify that

$$[KU(V)]_{e=1,3} = \frac{1}{2l\omega}\begin{bmatrix} 0 & k_v l\omega & -k_v l\omega \\ -l\omega & -k_v l\omega + l\omega & k_v l\omega \\ l\omega & -l\omega & 0 \end{bmatrix}$$

$$= \frac{1}{2}\begin{bmatrix} 0 & k_v & -k_v \\ -1 & 1-k_v & k_v \\ 1 & -1 & 0 \end{bmatrix} \tag{5.170a}$$

$$[KU(V)]_{e=2,4} = \frac{1}{2}\begin{bmatrix} 0 & 1 & -1 \\ -k_v & 0 & k_v \\ k_v & -1 & 1-k_v \end{bmatrix} \tag{5.170b}$$

$$[KV(U)]_{e=1,3} = [KU(V)]^T_{e=1,3} = \frac{1}{2}\begin{bmatrix} 0 & -1 & 1 \\ k_v & 1-k_v & -1 \\ -k_v & k_v & 0 \end{bmatrix} \tag{5.170c}$$

$$[KV(U)]_{e=2,4} = [KU(V)]^T_{e=2,4} = \frac{1}{2}\begin{bmatrix} 0 & -k_v & k_v \\ 1 & 0 & -1 \\ -1 & k_v & 1-k_v \end{bmatrix} \tag{5.170d}$$

The last element matrix contribution to be formed is that due to either the traction or gravity body force (5.166). In two dimensions, the body force component $BY$ is a uniform constant equal to the unit thickness weight density. Thus, $A_e\{BY/G\}_e$ is equal to the weight $W_e$ of $\Omega_e$ divided by the shear modulus $G$, which can be written as $W_e G^{-1}\{1\}$. Recalling (5.125), (5.166) thus becomes

$$\{bx\}_e = \{0\}$$

$$\{by\}_e = A_e[B200L]\{BY/G\}_e = \frac{W_e G^{-1}}{12}\begin{bmatrix} 2 & 1 & 1 \\ 1 & 2 & 1 \\ 1 & 1 & 2 \end{bmatrix}\{1\}$$

$$= \frac{W_e}{G}\begin{Bmatrix} 1/3 \\ 1/3 \\ 1/3 \end{Bmatrix} \tag{5.171a}$$

Equation (5.171$a$) expresses the *equivalent nodal load* form of the gravity body force, which simply places one-third of the element weight, divided by $G$, at each node of $\Omega_e$. The development for a surface traction definition must also account for its distribution. For the orientation shown in Fig 5.13*a*, only *TY* exists and the load contribution becomes

$$\{bx\}_e = \{0\}$$

$$\{by\}_e = l_e[A200L]\{TY/G\}_e = \frac{l_e}{6G}\begin{bmatrix} 2 & 1 \\ 1 & 2 \end{bmatrix}\begin{Bmatrix} TY1 \\ TY2 \end{Bmatrix}_e$$

$$= \frac{l_e}{6G}\begin{Bmatrix} 2TY1 + TY2 \\ TY1 + 2TY2 \end{Bmatrix}_e \tag{5.171b}$$

The column matrix entries in (5.171*b*) express the *equivalent nodal load* associated with a distributed traction force parallel to the *y* axis.

This completes specification of all element matrix contributions necessary to form the global matrix statement (5.165) for the prismatic bar problems defined in Fig. 5.13. We must next focus on the element stiffness matrix $[K]_e$ completion for each generic domain $\Omega_e$, as shown in Fig. 5.14*a*. Figure 5.14*b* shows the local element node numbering and a convenient ordering for the associated global degrees of freedom (*UI*, *VI*), $1 \leq I \leq 4$.

The element stiffness matrix contributions to (5.165) are easily formed. For $\Omega_{e=1}$, hence using (5.168*a*), (5.168*c*), (5.170*a*), and (5.170*c*) and dividing out the $l$, $\omega$ factors in the first two matrices, you may verify that

$$[K]_e\{Q\}_{e=1} = \begin{bmatrix} [KU(U)]_e & [KU(V)]_e \\ [KV(U)]_e & [KV(V)]_e \end{bmatrix}\begin{Bmatrix} \{U\}_e \\ \{V\}_e \end{Bmatrix}_{e=1}$$

$$= \frac{1}{2}\begin{bmatrix} \begin{bmatrix} k_v l/\omega & -k_v l/\omega & 0 \\ -k_v l/\omega & k_v l/\omega + \omega/l & -\omega/l \\ 0 & -\omega/l & \omega/l \end{bmatrix} & \begin{bmatrix} 0 & k_v & -k_v \\ -1 & 1-k_v & k_v \\ 1 & -1 & 0 \end{bmatrix} \\ \begin{bmatrix} 0 & -1 & 1 \\ k_v & 1-k_v & -1 \\ -k_v & k_v & 0 \end{bmatrix} & \begin{bmatrix} l/\omega & -l/\omega & 0 \\ -l/\omega & l/\omega + k_v\omega/l & -k_v\omega/l \\ 0 & -k_v\omega/l & k_v\omega/l \end{bmatrix} \end{bmatrix}\begin{Bmatrix} U1 \\ U2 \\ U4 \\ V1 \\ V2 \\ V4 \end{Bmatrix}_{e=1} \tag{5.172a}$$

and the $\{Q\}_{e=1}$ entries are globally numbered. Note indeed that $[K]_e$ in (5.172*a*) is symmetric, as it will be for any element $\Omega_e$ in a discretization. The corresponding expression for $\Omega_{e=2}$ in Fig. 5.14*b* is

$$[K]_e\{Q\}_{e=2} = \frac{1}{2}\begin{bmatrix} \begin{bmatrix} \omega/l & 0 & -\omega/l \\ 0 & k_v l/\omega & -k_v l/\omega \\ -\omega/l & -k_v l/\omega & k_v l/\omega + \omega/l \end{bmatrix} & \begin{bmatrix} 0 & 1 & -1 \\ -k_v & 0 & k_v \\ k_v & -1 & 1-k_v \end{bmatrix} \\ \begin{bmatrix} 0 & -k_v & k_v \\ 1 & 0 & -1 \\ -1 & k_v & 1-k_v \end{bmatrix} & \begin{bmatrix} k_v\omega/l & 0 & -k_v\omega/l \\ 0 & l/\omega & -l/\omega \\ -k_v\omega/l & -l/\omega & l/\omega + k_v\omega/l \end{bmatrix} \end{bmatrix}\begin{Bmatrix} U1 \\ U4 \\ U3 \\ V1 \\ V4 \\ V3 \end{Bmatrix}_{e=2} \tag{5.172b}$$

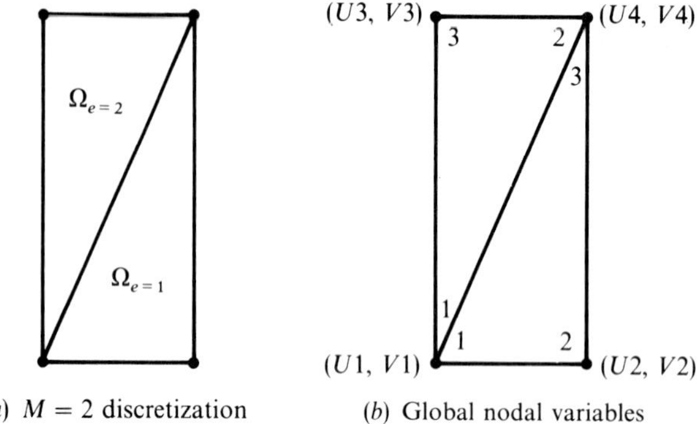


(a) $M = 2$ discretization

(b) Global nodal variables

**FIGURE 5.14**
An $M = 2$ finite element discretization segment of a prismatic bar.

**Example 5.3.** A hand computation can be completed for the $M = 2$ discretization shown in Fig. 5.14*a*. The needed step is to assemble (5.172*a*) and (5.172*b*) to form the global stiffness matrix as expressed by (5.165). As always, assembly amounts to row-wise addition of matrix terms over all nodal degrees of freedom shared by more than one finite element as keyed by the diagonal entry in $[K]_e$. In Fig. 5.14*b*, the only shared degree of freedom for assembly is $V1$, since the homogeneous Dirichlet constraint removes $U1$, $U4$, and $V4$ from further consideration. Thus, on clearing the multiplier in (5.172) the reduced global matrix statement for the $M = 2$ discretization is

$$S_e([K]_e\{Q\}_e - \{b\}_e) = \{0\}$$

$$= \begin{bmatrix} l/\omega + k_\nu \omega/l & -1 & -l/\omega \\ -1 & k_\nu l/\omega + \omega/l & 1 - k_\nu \\ -l/\omega & 1 - k_\nu & l/\omega + k_\nu \omega/l \end{bmatrix} \times \begin{Bmatrix} V1 \\ U2 \\ V2 \end{Bmatrix} + \frac{2W_e}{3G} \begin{Bmatrix} 1 + 1 \\ 0 \\ 1 \end{Bmatrix} \tag{5.173}$$

Note that the assembled global stiffness matrix is also symmetric. The last step is to solve (5.173) for $\{Q\} = \{V1, U2, V2\}^T$. The general solution statement is $\{Q\} = [K]^{-1}\{b\}$, and the inverse of the $3 \times 3$ global stiffness matrix is easy to form once the values of $l$, $\omega$, $k_\nu = 1 + (1 + \nu)/(1 - \nu) = 2/(1 - \nu)$ and the shear modulus $G = E/(2(1 + \nu))$ are given. For a mild steel, representative values are $\nu \simeq \frac{1}{4}$ and $E \simeq 3 \times 10^7$ lb/in$^2$; hence $k_\nu \simeq \frac{8}{3}$ and $G \simeq 10^7$ lb/in$^2$. The very coarse $M = 2$ discretization cannot handle a very large $l/\omega$; hence assume it to be equal to 3, which corresponds to a bar aspect ratio of 1.5 (recall that we have a symmetric half-domain). Assuming each element has weight $W_e \simeq 150$ lb, then (5.173) becomes

$$\begin{bmatrix} 4 & -1 & -3 \\ -3 & 8.3 & -2 \\ -3 & -2 & 4 \end{bmatrix} \begin{Bmatrix} V1 \\ U2 \\ V2 \end{Bmatrix} = -10^{-5} \begin{Bmatrix} 2 \\ 0 \\ 1 \end{Bmatrix} \tag{5.174}$$

and the solution is

$$\{Q\} = \begin{Bmatrix} V1 \\ U2 \\ V2 \end{Bmatrix} = 10^{-5} \begin{Bmatrix} -7.35 \\ -4.05 \\ -7.79 \end{Bmatrix} \text{in} \tag{5.175}$$

Thus, the bottom of the bar has become deflected downward and has also become narrower since $U2 < 0$. Figure 5.15*a* compares the original bar shape and the gravity-induced deflection (on an exaggerated scale).

**Example 5.4.** Repeat the previous analysis after replacement of the gravity body force with a uniform surface traction. The $M = 2$ discretization stiffness matrix given in (5.173) is unaltered. The only change is the load evaluation, the matrix entries for which occur only on the $\Omega_{e=1}$ lower boundary (Fig. 5.13*a*). Using (5.171*b*), the form is

$$\{by\}_{e=1} = l_e[A200L]\left\{\frac{TY}{G}\right\}_e$$

$$= \frac{T_e}{G}\begin{Bmatrix} 1/2 \\ 1/2 \end{Bmatrix}$$

where $T_e - l_e TY_e$ has units of force (since in two dimensions $TY_e$ is force/

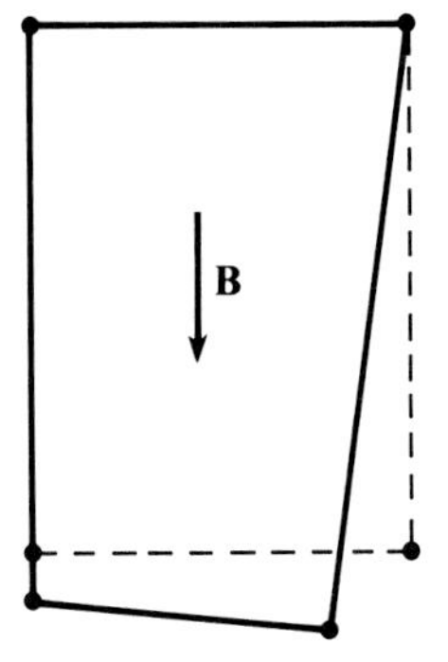


(*a*) Gravity body force **B**

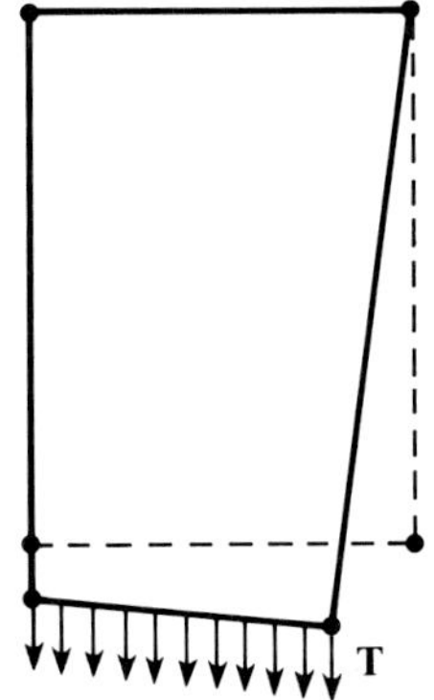


(*b*) Uniform surface traction **T**

**FIGURE 5.15**
Original and load-deflected shapes for a prismatic bar.

length). The resultant global matrix statement replacement for (5.173) is

$$\begin{bmatrix} l/\omega + k_v\omega/l & -1 & -l/\omega \\ -1 & k_v l/\omega + \omega/l & 1 - k_v \\ -l/\omega & 1 - k_v & l/\omega + k_v\omega/l \end{bmatrix} \begin{Bmatrix} V1 \\ U2 \\ V2 \end{Bmatrix} + \frac{T_e}{G} \begin{Bmatrix} 1 \\ 0 \\ 1 \end{Bmatrix} = \{0\} \tag{5.176}$$

For the same aspect ratio and material properties, and for $T_e \simeq 155$ lb (which makes the nodal displacement $U2$ identical to the previous case), (5.176) becomes

$$\begin{bmatrix} 4 & -1 & -3 \\ -3 & 8.3 & -2 \\ -3 & -2 & 4 \end{bmatrix} \begin{Bmatrix} V1 \\ U2 \\ V2 \end{Bmatrix} = -1.55 \times 10^{-5} \begin{Bmatrix} 1 \\ 0 \\ 1 \end{Bmatrix} \tag{5.177}$$

with solution

$$\{Q\} = \begin{Bmatrix} V1 \\ U2 \\ V2 \end{Bmatrix} = 10^{-5} \begin{Bmatrix} -7.29 \\ -4.05 \\ -7.87 \end{Bmatrix} \text{ in} \tag{5.178}$$

The resultant deflection is compared (on an exaggerated scale) to the original shape in Fig. 5.15*b*. The subtle distinction between the deflected state as caused by the body force and the traction force is barely evident for these very crude mesh approximate solutions. Nevertheless, the instructive goal has been fulfilled.

## 5.11 SUMMARY

This chapter has developed the weak statement finite element solution algorithm for two-dimensional boundary value problems as implemented using the triangular element domain and associated bases. A concomitant requirement was automated generation of suitable, nonuniform grids made up as the nonoverlapping sum ("union") of triangular elements. The macro domain algebraic methodology, the straightforward extension of the one-dimensional procedure, was developed and examined through several code exercises. The auxiliary requirement with grid generation is creation of the discretization database defining all node-element connections which is the MEL array in *LEARN.FE*. The reader hopefully has a good working knowledge of this macro grid-generation methodology and its implementation in a useful code environment.

Following grid generation, attention turned to the finite element basis definitions for the straight-sided triangular element $\Omega_e$. The linear and quadratic polynomial bases were identified as $\{N_k(\zeta_i)\}$, $k = 1, 2$ and $1 \le i \le n + 1 = 3$, which introduced the two-dimensional extension of the *natural coordinate system* $\zeta_i$ introduced in Chap. 4. The coordinate transformation, from rectangular cartesian to "natural," played the central role in determining specific element matrix forms in the algorithm statements for two-dimensional diffusion with boundary convection. All details were completed, for both $k = 1$ and $k = 2$ basis definitions, including a range of simple hand-executable example problems permitting full illustration of all details including *assembly*.

The third key development was a Fortran-compatible nomenclature for the *element master matrix library* structure. The added complexity stemming from variable parameters, and/or axisymmetric geometries, then clearly constituted rather modest extensions on specific basic statements. Each variation became explicitly stated using the defined hypermatrix nomenclature. The development capstone was a direct translation of the weak statement finite element approximation procedure for the general two-dimensional problem into the algorithm Fortran statement [recall (5.127)]. This also emphasized the utility of the one-dimensional master matrix library for boundary condition implementation for two-dimensional problems. The problem statement generalization to fluid mechanics and plane stress–plane strain then completed the technical development sections of this chapter.

The auxiliary topic of stationary iteration procedures for approximate solution of the weak statement global matrix statement rounds out the chapter. The Jacobi, Gauss-Seidel, and successive overrelaxation (SOR) methods were defined and correlated. Each eliminates the need to equation-solve using the system matrix $[K + H]$. The *LEARN.FE* code contains a Gauss-Seidel subroutine that can be used for approximately solving problems containing relatively larger numbers of elements in the discretization $\Omega^h$. Otherwise, the Gauss elimination direct solver can be used to advantage, including modest mesh convergence studies.

## 5.12 ADDITIONAL REFERENCES

Akin, J. E. (1982): *Application and Implementation of Finite Element Methods*, Academic Press, London.

Ames, W. F. (1977): *Numerical Methods for Partial Differential Equations*, Academic Press, New York.

Conte, S. D. (1975): *Elementary Numerical Analysis*, McGraw-Hill, New York.

Huebner, K. and E. A. Thornton (1982): *The Finite Element Method for Engineers, 2d ed.*, Wiley-Interscience, New York.

Hughes, T. R. J. (1987): *The Finite Element Method*, Prentice-Hall, Englewood Cliffs, N.J.

Varga, R. S. (1965): *Matrix Iterative Analysis*, Prentice-Hall, Englewood Cliffs, N.J.

# CHAPTER 6

# TWO-DIMENSIONAL QUADRILATERAL ELEMENTS

## 6.1 OVERVIEW

This chapter develops the finite element weak statement solution algorithm using the quadrilateral finite element and its associated basis functions for two-dimensional problems. The single key feature that distinguishes quadrilateral element methodology from triangular is the pervasive use of local coordinate transformations using the quadratic ($k = 2$) *serendipity* basis function. Specifically, the global $(x, y)$ coordinate system spanning the solution domain discretization $\Omega^h$ is interpolated on every finite element domain $\Omega_e$ using $\{N_2^+\}$ and the associated vertex and midside nodal coordinate arrays $\{X\}_e$ and $\{Y\}_e$. The word *isoparametric* characterizes this topic, which lends great geometric versatility to the quadrilateral element formulation. We develop this methodology in completeness in this chapter.

Of course, as amply verified in Chaps. 4 and 5, the end product of the finite element algorithm for any basis choice and any problem dimension is the global matrix statement

$$[K + H]\{Q\} = \{b\} \tag{6.1}$$

or equivalently

$$S_e([K + H]_e\{Q\}_e - \{b\}_e) = \{0\} \tag{6.2}$$

Therefore, our *sole* formulational requirement is to establish the master matrix library for $[K]_e$ and $[H]_e$ and those matrices involved in the element data matrix $\{b\}_e$ for the quadrilateral finite element basis family. As before, we first develop the pertinent methodology ingredients and then proceed to practical problem statements.

# A. METHODOLOGY

## 6.2 QUADRILATERAL FINITE ELEMENT MESH

The Chap. 5 development on triangular element mesh generation needs only modest extension for quadrilateral element discretizations. In truth, the quadrilateral *isoparametric coordinate transformation* (to be developed) is the procedure utilized under *GRID* to construct all two-dimensional discretizations by *LEARN.FE*.

Figure 6.1 illustrates the three basic distinctions for two-dimensional quadrilateral meshes. Any macro region $\Omega^H$ may be discretized, in a defined nonuniform manner, and the resultant discretization $\Omega^h$ cast onto (*a*) a

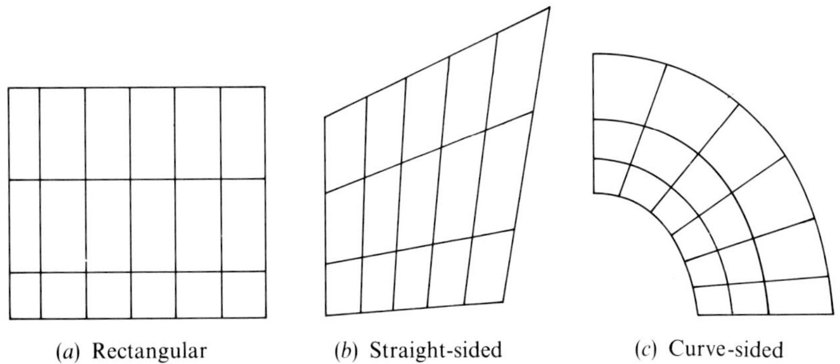


**FIGURE 6.1**
Quadrilateral region $\Omega$ subdivided into a mesh $\Omega^h$ of quadrilateral elements $\Omega_e$.

rectangular parallelepiped, (*b*) a general straight-sided quadrilateral, and/or (*c*) a curve-sided quadrilateral. Viewing Fig. 6.1, a quadrilateral finite element $\Omega_e$ always contains at least four vertex nodes. The local numbering convention is again counterclockwise as shown in Fig. 6.2. The $(n = 2)$ two-dimensional straight-sided quadrilateral element geometry is therefore defined by four coordinate pairs $[(X_i, Y_i)_e, 1 \leq i \leq 4$ (Fig. 6.2*a*)]. For the $(n = 3)$ three-dimensional problem statement, the straight-sided hexahedron element would thus be defined by the eight vertex node coordinate triples $(X_i, Y_i, Z_i)_e$, $1 \leq i \leq 8$.

In distinction to the quadratic triangle element (recall Fig. 5.4), the curve-sided $k \geq 2$ quadrilateral element requires at least four additional node coordinate pairs for definition. These are denoted by " × " in Fig. 6.2*b*, and each geometric node $(X_i, Y_i)_e$, $5 \leq i \leq 8$ is located within the middle third of the respective element boundary segment and usually ordered as illustrated. For the three-dimensional curve-sided hexahedron element, there are at least 24 node coordinate triples required to define $\Omega_e$. It is quite apparent, therefore, that isoparametric element methodology generates a large volume of geometric data.

Our purpose in this section is to introduce the fundamentals of isoparametric quadrilateral grid generation methodology for $n = 2$. The *LEARN.FE* code provides this capability under the *TYPE* and *GRID* commands and creates nonuniform macro region discretization unions using the one-dimensional progression ratio methodology described in Section 5.2; see (5.3) and (5.4).

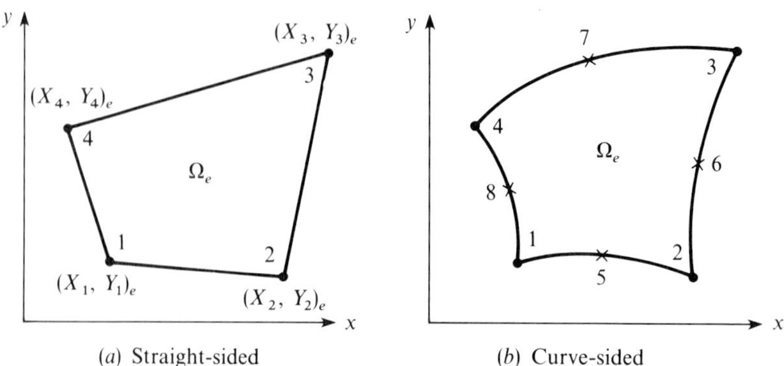

(*a*) Straight-sided (*b*) Curve-sided

**FIGURE 6.2**
Generic two-dimensional quadrilateral element $\Omega_e$.

## Rectangular Macro Domains

The following exercise illustrates creation of the basic quadrilateral mesh $\Omega^h$ as the union of rectangular elements $\Omega_e$. Under command name *TYPE*, the integer variable definitions N = 2, NNODEL = 4, and NTRAN = 0 key the construction. Refer to Code Exercise 5.1 for the triangle comparison to the following exercise.

**Code Exercise 6.1** Access the input file for this code exercise, and the following will appear on your monitor.

```
TITL
***** CODE EXERCISE 6.1,M=4, QUADRILATERALS *****
TYPE   [K         N      NNODEL    REFL  NPRN  NTRAN    NAXI]
        1         2        4        1.0   1 1    0
PRIN   [NBUG(*)]
        1         0        1          3
GRID   [FOR N(XL XR(*)PR(*)),NEM(PR*)]
                 2.0      3.0        1.0
        2
                 1.0      2.0        1.0
        2
STOP
```

Execute this file, and the following will scroll onto your screen.

```
TITL
***** EXERCISE 6.1, M=4, QUADRILATERALS *****
TYPE       K          N      NNODEL     REFL   NPRN    NTRAN   NAXI
           1          2        4         1.0   1 1       0       0
PRIN          NBUG(1)         (2)        (3)    (4)     (5)     (6)
                  1            0          1      3       0       0
GRID       NI       NPRI     NTRAN
           1         1         0
           XL       XR1       PR1       XR2      PR2      XR3      PR3
           2.0      3.0       1.0
           NEM1     NEM2      NEM3
           2
GRID       NI       NPRI      NTRAN
           2         1         0
           YL       YU1       PR1       YU2      PR2      YU3      PR3
           1.0      2.0       1.0
           NEM1   NEM2        NEM3
           2

        ELEMENT NODE CONNECTION ARRAY(MEL)
        ELEMENT NO.     NODE1     NODE2     NODE3     NODE4
                1       1         2         5         4
                2       2         3         6         5
                3       4         5         8         7
                4       5         6         9         8
```

```
        NODE COORDINATES (X REAL VARIABLE)
                2.00    2.50     3.00
                2.00    2.50     3.00
                2.00    2.50     3.00

        NODE COORDINATES (Y REAL VARIABLE)
                1.00    1.00     1.00
                1.50    1.50     1.50
                2.00    2.00     2.00

        NODE COORDINATES (X,INTEGER MAP) EXPONENT E:-2
                200     250     300
                200     250     300
                200     250     300

        NODE COORDINATES (Y,INTEGER MAP) EXPONENT E:-2
                200     200     200
                150     150     150
                100     100     100

STOP
```

Recall under *TYPE* that NPRN = 1, 1 is the specification for a mesh constructed from a single macro region $\Omega^H$. Further, NEM = (2, 2) produces an $M = M_1 \times M_2 = 2 \times 2$ uniform mesh for $p_i = (1, 1)$. Finally, note that the MEL array contains the element-node connection table for a union of quadrilateral domains $\Omega_e$ with four nodes each.

As with triangular meshes, NPRN keys the grid generator to creation of nonuniform quadrilateral meshes on macro region unions. The following code exercise illustrates the specification procedure.

**Code Exercise 6.2.** Access the input file for this code exercise, and the following will appear on your monitor.

```
TITL
***** CODE EXERCISE 6.2, M=16, NON-UNIFORM QUADS *****
TYPE    [K      N       NNODEL     REFL   NPRN   NTRAN   NAXI]
         1      2         4         1.0    2 2     0
PRIN    [NBUG(*)]
         1       0        0           2
GRID    [FOR N (XL XR(*) PR(*)), NEM (PR*)]
                2.0       3.0       0.8         4.0     1.25
         2      2
                0.0       1.0       0.8         2.0     1.25
         2      2
STOP
```

Note that NPRN = (2, 2) keys the expanded data read under *GRID*. Execute this file, and the following output will scroll onto your monitor.

```
TITL
***** CODE EXERCISE 6.2,M=16, NON-UNIFORM QUADS *****
TYPE    K       N       NNODEL    REFL   NPRN   NTRAN   NAXI
        1       2         4        1.0   2 2      0       0
PRIN      NBUG(1)        (2)      (3)    (4)     (5)     (6)
                1         0        0      2       0       0
GRID    NI      NPRI     NTRAN
        1       2         0
        XL      XR1       PR1      XR2       PR2      XR3      PR3
        2.0     3.0       0.8      4.0       1.25
        NEM1    NEM2      NEM3
        2       2
GRID    NI      NPRI     NTRAN
        2       2         0
        YL      YU1       PR1     YU2       PR2       YU3      PR3
        1.0     2.0       0.8     3.0       1.25
        NEM1    NEM2     NEM3
        2       2
        NODE COORDINATES(X,INTEGER MAP) EXPONENT E:-2
                200      255       300      344      400
                200      255       300      344      400
                200      255       300      344      400
                200      255       300      344      400
                200      255       300      344      400

        NODE COORDINATES(Y,INTEGER MAP) EXPONENT E:-2
                300      300       300      300      300
                244      244       244      244      244
                200      200       200      200      200
                155      155       155      155      155
                100      100       100      100      100
STOP
```

Under command *PRIN*, NBUG (4) = 2 outputs the node coordinate integer map only. Compare this output to that of Code Exercise 5.2; the node coordinate distributions for the discretization $\Omega^h$ are identical, with refinement located in the centroidal region. Note that this discretization is indeed the union of nonuniform rectangular quadrilateral finite elements $\Omega_e$. Now alter this input file with progression ratio $p_i$ definitions that will place the refined grid region around the periphery of $\Omega^h$, then reexecute and request a graphic. As you wish, experiment further with $M = M_1 \times M_2$ and NPRN to complete your familiarization with this *LEARN.FE* grid generation methodology.

## Straight-Sided Quadrilateral Macro Domains

Access the input file for **Code Exercise 6.3**, and the following will appear on your monitor.

```
TITL
** CODE EXERCISE 6.3, M=12, QUADRILATERAL MACRO **
TYPE      K         N      NNODEL   REFL     NPRN   NTRAN     NAXI
          1         2        4        1.0     1 1     1
PRIN    [NBUG(*)]
           1         0          0          2
GRID
                    1.0        2.5        1.0
           4
                    1.0        2.0        1.0
           3
                    1.0        2.5        3.0        1.0
                    1.0        0.9        2.7        2.0
STOP
```

As in Chap. 5, note that NTRAN = 1 keys the general quadrilateral vertex node data set definition under *GRID*. An $M = 4 \times 3$ uniform mesh is requested. Execute this file, and the following will appear on your screen.

```
TITL
*** CODE EXERCISE 6.3, M=12, QUADRILATERAL MACRO ***
TYPE      K        N        NNODEL     REFL   NPRN   NTRAN   NAXI
          1        2          4         1.0    1 1     1       0

PRINT     NBUG(1)        (2)        (3)        (4)    (5)    (6)
                1         0          0          2      0      0

GRID    NI       NPRI       NTRAN
        1        1          1
        XL       XR1        PR1        XR2        PR2    XR3    PR3
        1.0      2.5        1.0
        NEM1     NEM2       NEM3
        4
GRID    NI       NPRI       NTRAN
        2        1          1
        YL       YU1        PR1        YU2        PR2    YU3    PR3
        1.0      2.0        1.0
        NEM1     NEM2       NEM3
        3

        NODE COORDINATES(X,INTEGER MAP) EXPONENT E:-2
                 100        138        175        213        250
                 100        142        183        225        266
                 100        146        192        238        283
                 100        150        200        250        300

        NODE COORDINATES(Y,INTEGER MAP) EXPONENT E:-2
                 200        217        235        252        270
                 166        177        188        199        210
                 133        137        141        145        150
                 100         97         95         92         90
STOP
```

Setting NBUG(4) = 2 again restricts the node map to the integer form. Request a graphic, and verify that the $p_1 = 1 = p_2$ uniform quadrilateral element mesh comparable to Fig. 5.1 has been created. Now experiment with alternative grid generation parameters and NPRN to enhance your familiarity with the use of quadrilateral macros.

## Curve-Sided Quadrilateral Macro Domains

The *isoparametric* quadrilateral element transformation, developed in the next section, is employed to create curvilinear meshes on a curve-sided macro region domain $\Omega^H$. NTRAN = 2, 3 keys the definition of parabolic or circular boundary curves under command *TYPE*. Refer back to Code Exercise 5.4 for the triangle mesh comparison to the following quadrilateral mesh exercise.

**Code Exercise 6.4.** Access this input file, which will display the following on your monitor.

```
TITL
*** CODE EXERCISE 6.4, CURVE SIDED MACRO DOMAIN ***
TYPE     [K        N       NNODEL      REFL NPRN  NTRAN     NAXI]
          1        2         4          1.0  1 1    3

PRIN    [NBUG(*)]
          1         0          0              2

GRID    [FOR N (XL XR(*)PR(*)), NEM(PR*)]
                  1.0       2.0       1.0
         4
                  0.0       1.0       1.0
         4

                  1.0     2.0      0.0      0.0
                  0.0     0.0      2.0      1.0
                  1.5     1.4142   0.0      0.7071
                  0.0     1.4142   1.5      0.7071
STOP
```

Execute the file, and the following will scroll onto your screen.

```
TITL
*** CODE EXERCISE 6.4, CURVE SIDED MACRO DOMAIN ***
TYPE        K        N       NNODEL      REFL  NPRN  NTRAN     NAXI
            1        2         4          1.0   1 1    3         0

PRINT      NBUG(1)       (2)        (3)        (4)     (5)       (6)
                 1        0          0          2       0         0
```

```
GRID    NI      NPRI      NTRAN
        1       1         3
        XL      XR1       PR1        XR2       PR2    XR3     PR3
        1.0     2.0       1.0
        NEM1    NEM2      NEM3
        4
GRID    NI      NPRI      NTRAN
        2       1         3
        YL      YU1       PR1        YU2       PR2    YU3     PR3
        0.0     1.0       1.0
        NEM1    NEM2      NEM3
        4
        X1        X2          X3           X4
        1.0000    2.0000      0.0000       0.0000
        Y1        Y2          Y3           Y4
        0.0000    0.0000      2.0000       1.0000
        X5        X6          X7           X8
        1.5000    1.4142      0.0000       0.7071
        Y5        Y6          Y7           Y8
        0.0000    1.4142      1.5000       0.7071
        NODE COORDINATES(X,INTEGER MAP)EXPONENT E:-2
                0         0         0         0        0
               38        47        57        66       76
               70        88       106       123      141
               92       115       138       161      184
              100       125       150       175      200

        NODE COORDINATES(Y,INTEGER MAP)EXPONENT E:-2
              100       125       150       175      200
               92       115       138       161      184
               70        88       106       123      141
               38        47        57        66       76
                0         0         0         0        0
STOP
```

Verify that this command file has created a uniform $M = 4 \times 4$ quadrilateral discretization on the quarter-circle domain defined in Fig. 5.3. Experiment further with the grid definition commands including NPRN and NTRAN = 2 to create nonuniform meshes, and so on, for your familiarization.

This completes the discussion on quadrilateral mesh generation using *LEARN.FE*. We now proceed to development of the associated finite element basis family.

## 6.3 TWO-DIMENSIONAL QUADRILATERAL ELEMENT BASES

As amply illustrated, the finite element method constitutes a highly ordered procedure for constructing an approximate solution trial space $\phi_i(x_j)$,

$1 \le i \le N$ and $1 \le j \le n$, using a discretization $\Omega^h$ of a problem definition domain $\Omega$ on an $n$-dimensional region $\mathbb{R}^n$. At the risk of some repetition, *any* approximate solution for an $n = 2$ dimensional problem is

$$T^N(x, y) = \sum_{i=1}^{N} a_i \phi_i(x, y) \tag{6.3}$$

The finite element procedure produces (6.3) as an assembly of operations on the union of elements $\Omega_e$ forming the mesh $\Omega^h$. The quadrilateral finite element basis definition yields

$$T_e(x_j) \equiv \{N_k^+(\eta_i)\}^T\{Q\}_e \tag{6.4}$$

In (6.4), the superscript "plus" notation for the basis distinguishes the quadrilateral element from the triangle. The intrinsic coordinates spanning the quadrilateral $\Omega_e$ constitute the *tensor product* system, which is denoted $\eta_i$, rather than the triangle *natural coordinate* system $\zeta_i$.

Aside from these minor points, (6.4) is the very familiar statement for the element approximation. We thus continue to observe that finite element methodology is functionally and notationally invariant as the generalizations become established. This section develops the linear and quadratic degree finite element bases for the general quadrilateral. These correspond to $k = 1, 2$ in (6.4), and will be termed *bilinear* and *biquadratic* to keep the distinguishing aspects in mind.

### Bilinear Basis

Figure 6.1 illustrates the variety of available quadrilateral basis finite element shapes, ranging from the rectangular parallelepiped to the general curve-sided quadrilateral. As with the triangle element, the $k = 1$ basis $\{N_1^+\}$ must involve polynomials expressible in the global coordinate system $x_i$ with powers no higher than unity. The geometry of rectangular and straight-sided quadrilateral finite element domains $\Omega_e$ is completely defined by the coordinates of the four intersections (vertices) of the boundary segment generators. Hence, four expansion coefficients must be involved in the global coordinate equivalent of (6.4). The polynomial that involves $x$ and $y$ to powers no higher than unity is

$$T_e(x, y) = a_1 + a_2 x + a_3 y + a_4 xy \tag{6.5}$$

and the $a_i$, $1 \le i \le 4$, are the to-be-determined expansion coefficients. Comparing (6.5) and (5.6), the triangular element expression, the distinction in (6.5) is the fourth *bilinear* term, which is so called since $xy$ remains linear in both arguments.

The requirement is to equate (6.5) and (6.4), and hence establish the elements of $\{N_1^+(\eta_j)\}$, which is a column matrix of order four (to match the four entries that must occur in $\{Q\}_e$, which replaces the $a_i$). Figure 6.3*a* illustrates the geometry of the generic rectangular parallelepiped $\Omega_e$, which for convenience is drawn with sides parallel to the global coordinate system. By convention, the length and width of $\Omega_e$ are defined equal to $2a$ and $2b$, and the vertex nodes with coordinate pairs $(X_i, Y_i)_e$, $1 \leq i \leq 4$, are numbered counterclockwise starting at the lower left corner. Conversely, as shown in Fig. 6.3*b*, the coordinates of these nodes in the local $\eta_i$ coordinate system are normalized to unity for the origin defined at the element centroid. Obviously, then, the local intrinsic $\eta_i$ coordinate system is cartesian, and for the restriction in Fig. 6.3 is the following simple scaling of $x_j$

$$\eta_1 \equiv \frac{x - X_c}{a} \tag{6.6a}$$

$$\eta_2 \equiv \frac{y - Y_c}{b} \tag{6.6b}$$

In (6.6), $(X_c, Y_c)_e$ is the global coordinate pair of the centroid of $\Omega_e$, and the convenience of using $2a$ and $2b$ as the "measures" of $\Omega_e$ is apparent.

With these geometry definitions, the next step is to equate (6.4) and (6.5), to determine the elements of $\{N_1^+\}$. The direct assault is not really

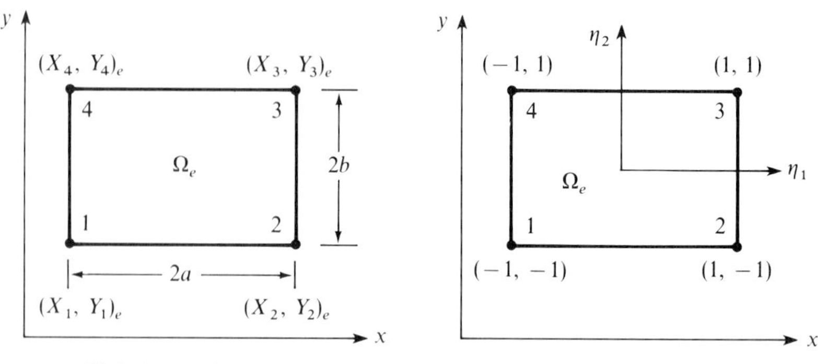


(*a*) Global coordinate system (*b*) Transformed local coordinate system

**FIGURE 6.3**
Generic rectangular parallelepiped element $\Omega_e$.

necessary, however, since with a little "insight" one can immediately write down the elements of $\{N_1^+\}$ by inspection of Fig. 6.3*b* as

$$\{N_1^+(\eta_i)\} = \frac{1}{4}\begin{Bmatrix}(1-\eta_1)(1-\eta_2)\\(1+\eta_1)(1-\eta_2)\\(1+\eta_1)(1+\eta_2)\\(1-\eta_1)(1+\eta_2)\end{Bmatrix} \tag{6.7}$$

That (6.7) is correct is easy to verify using the definition (6.4). For example,

$$\begin{aligned}T_e(x = X_1, y = Y_1) \equiv Q1 &= T_e(\eta_1 = -1, \eta_2 = -1)\\ &= \{N_1^+(\eta_1 = -1, \eta_2 = -1)\}^T\{Q\}_e\\ &= \tfrac{1}{4}\{4, 0, 0, 0\}\begin{Bmatrix}Q1\\Q2\\Q3\\Q4\end{Bmatrix}_e\\ &= Q1\end{aligned} \tag{6.8}$$

The reader is encouraged to verify that (6.7) indeed produces $QI_e$ at each node of $\Omega_e$ for all selections $(X_i, Y_i)_e$.

The ease with which (6.7) was established would certainly be a hollow accomplishment if the bilinear basis $\{N_1^+\}$ were restricted to a rectangular parallelpiped finite element $\Omega_e$. That this is not the case is readily verified. Figure 6.4 shows the global and local coordinate descriptions for a generic straight-sided quadrilateral element $\Omega_e$ with no particular orientation

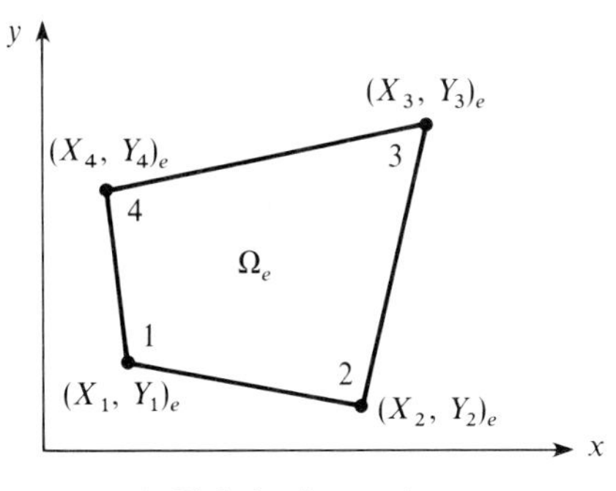


(*a*) Global reference frame

(*b*) Local reference frame

**FIGURE 6.4**
Generic quarilateral element $\Omega_e$.

and no sides parallel to each other or to a global coordinate axis. Comparing Figs. 6.4*b* and 6.3*b*, it is *obvious* that the global element geometry bears *no impact* whatsoever on the finite element basis description. Hence

$$T_e(x_j) = \{N_1^+(\eta_i)\}^T\{Q\}_e \tag{6.9}$$

and (6.7) accurately defines the distribution of $T_e$ on $\Omega_e$!

What has changed is emergence of a nontrivial coordinate transformation between $(x, y)$ and $(\eta_1, \eta_2)$ to replace (6.6). As amply illustrated in Chap. 5, the transformation from $x_j$ to the argument of $\{N_k(\cdot)\}$ is always needed in order to evaluate a finite element master matrix. Equation (6.6) meets this need only for the highly restrictive cartesian parallelogram geometry.

The resolution of this absolutely key issue is amazingly straightforward. The global $x$, $y$ coordinate system certainly varies linearly on the physical domain $\Omega_e$, Fig. 6.4*a*, as does the (assumed) dependence of $T_e$. Equation (6.9) is the associated finite element basis description for the dependent variable, and this form is certainly appropriate for interpolating *all* variables possessing bilinear dependence on $\Omega_e$. Hence

$$x_e = \{N_1^+(\eta_i)\}^T\{X\}_e \tag{6.10a}$$

$$y_e = \{N_1^+(\eta_i)\}^T\{Y\}_e \tag{6.10b}$$

states the *exact* interpolation of the global coordinate system $x_i$ on $\Omega_e$, which in turn defines the needed coordinate transformation

$$(x_i)_e = x_i(\eta_j)_e \tag{6.11}$$

In (6.10), the bilinear basis $\{N_1^+\}$ is as defined in (6.7), and the column matrices $\{X\}_e$ and $\{Y\}_e$ in (6.11) contain the element vertex node coordinate arrays $(X_i)_e$ and $(Y_i)_e$, $1 \le i \le 4$.

Equation (6.10) illustrates a truly *fundamental concept*: a (any) finite element basis serves to define the transformation of coordinates from the global to a local system. We skirted this issue in Chap. 5 to avoid introducing too much "finite element technology" before getting down to practical problem details. If you look back, (5.12) clearly states the (inverse) transformation $\zeta_i = \zeta_i(x_j)_e$, which is analogous to (6.6). The forward transformation for the triangular domain $\Omega_e$ is thus

$$x_e = \{N_1(\zeta_i)\}^T\{X\}_e \tag{6.12a}$$

$$y_e = \{N_1(\zeta_i)\}^T\{Y\}_e \tag{6.12b}$$

which is directly comparable to (6.10). We never had to deal with (6.12) in Chap. 5 since (5.12) was already available. However, since the quadrilateral element transformation is *quasi-linear*, (6.10) and its biquadratic generalization now play a central role.

### Biquadratic Bases

This subheading is plural since more than one biquadratic finite element basis can be defined. The distinction between the complete biquadratic and "serendipity" bases is that the latter utilizes only element surface geometric nodes. The complete biquadratic polynomial expressed in global coordinates contains nine terms:

$$\begin{aligned} T_e(x, y) &= a_1 + a_2 x + a_3 y + a_4 xy \\ &\quad + a_5 x^2 + a_6 y^2 + a_7 x^2 y + a_8 xy^2 + a_9 x^2 y^2 \\ &= \{1, x, y, \ldots, xy^2, x^2y^2\}\{a\}_e \end{aligned} \tag{6.13}$$

The "serendipity" approximation, a name associated with a fortuitous circumstance and referenced by Zienkiewicz (1978), amounts to omission of the last term in (6.13), i.e., $a_9 \equiv 0$.

For either basis, the number of geometric nodes for $\Omega_e$ equals the number of coefficients $a_i$ in (6.13). The *GRID* function in *LEARN.FE* utilizes the serendipity basis, with global nodal distribution and numbering convention as illustrated in Fig. 6.5*a*. The four vertex nodes remain numbered 1 to 4, while nodes 5 to 8 are sequentially located midside on the four boundary segments of $\Omega_e$. If an element boundary nodal coordinate set lies on a straight line, (6.13) exactly reproduces the associated bilinear

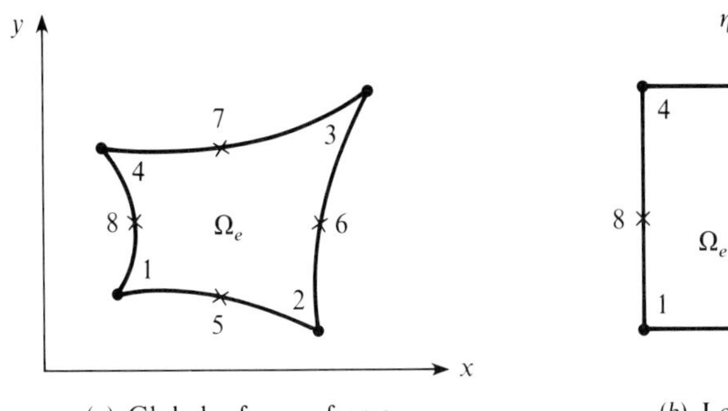


(*a*) Global reference frame (*b*) Local reference frame

**FIGURE 6.5**
Serendipity isoparametric quadrilateral element $\Omega_e$.

polynomial. Conversely, if the nonvertex node departs from a straight line, the resultant curved boundary segment will be a parabola. That this occurs can be "seen" on viewing Fig. 6.5*b*, the isoparametric element definition in the local transform space spanned by $\eta_i$. Since either $\eta_1$ or $\eta_2$ is a constant on every boundary segment $\partial\Omega_e$, the biquadratic polynomial reduces to a one-dimensional quadratic on the running variable, which defines a parabola. [Under *GRID*, NTRAN = 3 invokes a correction to the parabolic boundary geometry (NTRAN = 2) to move generated nodes of $\Omega^h$ to lie on circular arc segments.]

Returning to the key issue, the requirement is to establish the elements of the serendipity and/or complete biquadratic basis $\{N_2^+(\eta_i)\}$. The direct approach pursued in Chap. 5 with triangles is again replaced with *insight* and experience. In the local reference frame, (Fig. 6.5*b*), the nodal coordinates of $\Omega_e$ take only $(-1, 0, 1)$ as values. For the basis definition replacement of (6.13), that is,

$$T_e(x, y) = \{N_2^+(\eta_i)\}^T\{Q\}_e \tag{6.14}$$

the fundamental requirement is that every element of $\{N_2^+(\cdot)\}$ respectively vanish at all nodes of $\Omega_e$ except one, where it must take the value of unity. The ordering of nodal degrees of freedom in $\{Q\}_e$, and hence the node numbering scheme, determines where each element is unity. Starting with $\{N_1^+(\eta)\}$, as given in (6.7) wherein each element is (0, 1) at nodes 1 to 4 in Fig. 6.5*b*, the serendipity biquadratic finite element basis can be verified to be

$$\{N_2^+(\eta_i)\} = \frac{1}{4}\begin{Bmatrix} (1-\eta_1)(1-\eta_2)(-\eta_1-\eta_2-1) \\ (1+\eta_1)(1-\eta_2)(\eta_1-\eta_2-1) \\ (1+\eta_1)(1+\eta_2)(\eta_1+\eta_2-1) \\ (1-\eta_1)(1+\eta_2)(-\eta_1+\eta_2-1) \\ 2(1-\eta_1^2)(1-\eta_2) \\ 2(1+\eta_1)(1-\eta_2^2) \\ 2(1-\eta_1^2)(1+\eta_2) \\ 2(1-\eta_1)(1-\eta_2^2) \end{Bmatrix} \tag{6.15}$$

It is suggested you take a few moments to prove to yourself that each biquadratic polynomial in (6.15) appropriately reduces to (0, 1) at all eight nodes of $\Omega_e$, for the node ordering given in Fig. 6.5*b*.

For completeness, Fig. 6.6 shows the conventional finite element domain definitions for the full biquadratic basis. The distinction from Fig. 6.5 is the addition of node number 9 at the centroid in the transform

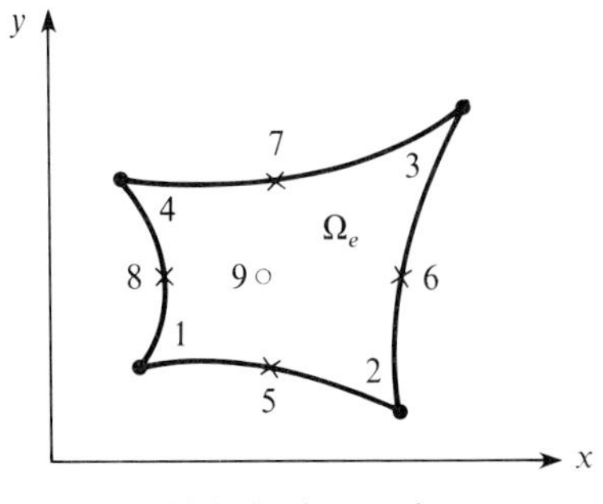

(a) Global reference frame

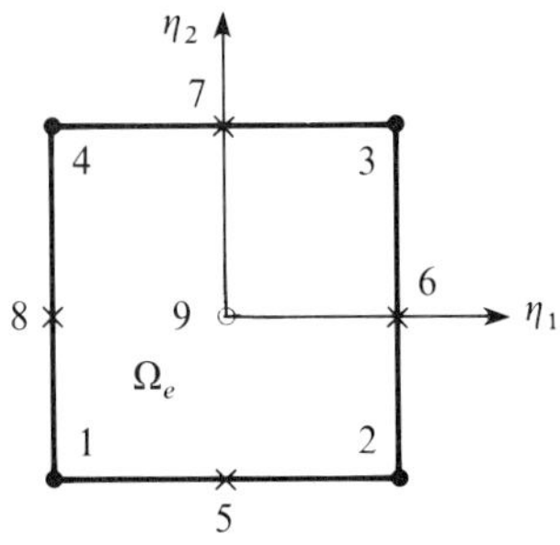

(b) Local reference frame

**FIGURE 6.6**
Full biquadratic isoparametric quadrilateral element $\Omega_e$.

domain (Fig. 6.6*b*), which maps to a nominal centroidal location in the physical domain (Fig. 6.6*a*). The associated finite element basis, denoted with superscript asterisk, can be directly verified to be

$$\{N_2^*(\eta_i)\} = \frac{1}{4}\left\{\begin{array}{l} (1-\eta_1)(1-\eta_2)(\eta_1\eta_2) \\ (1+\eta_1)(1-\eta_2)(-\eta_1\eta_2) \\ (1+\eta_1)(1+\eta_2)(\eta_1\eta_2) \\ (1-\eta_1)(1+\eta_2)(-\eta_1\eta_2) \\ 2(1-\eta_1^2)(1-\eta_2)(-\eta_2) \\ 2(1+\eta_1)(1-\eta_2^2)(\eta_1) \\ 2(1-\eta_1^2)(1+\eta_2)(\eta_2) \\ 2(1-\eta_1)(1-\eta_2^2)(-\eta_1) \\ 4(1-\eta_1^2)(1-\eta_2^2) \end{array}\right\} \tag{6.16}$$

The remaining topic is establishment of the coordinate transformation from the global to local system; recall (6.11). Either (6.15) or (6.16) provides the *isoparametric* transformation

$$x_e = \{N_2^+(\eta_i)\}^T\{X\}_e \tag{6.17a}$$

$$y_e = \{N_2^+(\eta_i)\}^T\{Y\}_e \tag{6.17b}$$

Equation (6.17) is no more than the $k = 2$ generalization of (6.12) involving the vertex and midside (and centroidal) global coordinates $(X_i, Y_i)_e$, $1 < i \leq 8$ or 9, of the quadrilateral finite element domain $\Omega_e$. Note that (6.17) explicitly defines the transformation as nonlinear, since the basis $\eta_i$ polynomials are biquadratic. Hence, $\eta_i$ coordinate lines in physical space are curved, to smoothly interpolate the $\partial\Omega_e$ boundary segment curvatures across $\Omega_e$.

This completes introduction of the family of bilincar and biquadratic finite element bases for use with quadrilateral domains $\Omega_e$. We now proceed to implementation aspects in the weak statement approximation.

## 6.4 STEADY-STATE DIFFUSION EQUATION

As in previous chapters, introductory development for the quadrilateral basis formulation is for a simple diffusion problem statement

$$L(T) = -\nabla \cdot k\,\nabla T - s = 0 \qquad \text{on } \Omega \subset \mathbb{R}^2 \tag{6.18}$$

$$l(T) = -k\,\nabla T \cdot \mathbf{n} - q_n = 0 \qquad \text{on } \partial\Omega_1 \tag{6.19}$$

$$T = T_b \qquad \text{on } \partial\Omega_2 \tag{6.20}$$

Recall that $\nabla \equiv \mathbf{i}\,\partial/\partial x + \mathbf{j}\,\partial/\partial y$ is the vector derivative operator (gradient). Further, $k = k(x_i)$ is the variable conductivity, $s$ is the internal source term, $q_n$ is the imposed heat flux normal to boundary segment $\partial\Omega_1$, and $T_b$ is the fixed temperature (distribution) on boundary segment $\partial\Omega_2$. Further, $\Omega \subset \mathbb{R}^2$ reads, "on the region $\Omega$ lying on the two-dimensional (euclidean) space $\mathbb{R}^2$." Finally, since (6.18) defines an elliptic boundary value problem, the nonoverlapping sum (union) of $\partial\Omega_1$ and $\partial\Omega_2$ must completely enclose $\Omega$.

The development of the weak statement for (6.18) to (6.20) and its finite element approximation follows the standard recipe; recall Section 5.4. Hence, for any approximation $T^N$ to the exact solution $T$ to (6.18) to (6.20), we have

$$T^N(x, y) \equiv \sum_{i=1}^{N} a_i \phi_i(x_j) \tag{6.21}$$

where $\phi_i(x_j)$ is the trial function set, and the weak statement is

$$\begin{aligned} WS &\equiv \int_\Omega \phi_i(x_j) L(T^N)\, d\tau \\ &= \int_\Omega \phi_i(x_j)(-\nabla \cdot k\,\nabla T^N - s)\, d\tau \\ &= \int_\Omega (\nabla\phi_i \cdot k\,\nabla T^N - \phi_i s)\, d\tau - \int_{\partial\Omega} \phi_i k\,\nabla T^N \cdot \mathbf{n}\, d\sigma \\ &= \int_\Omega (\nabla\phi_i \cdot k\,\nabla T^N - \phi_i s)\, d\tau + \int_{\partial\Omega_1} \phi_i q_n\, d\sigma - \int_{\partial\Omega_2} \phi_i k\,\nabla T^N \cdot \mathbf{n}\, d\sigma \\ &\equiv 0 \qquad \text{for } 1 \le i \le N \end{aligned} \tag{6.22}$$

A finite element implementation introduces the discretization $\Omega^h$ of $\Omega$ and replaces (6.21) with the *union* of finite element approximations cast in terms of the (quadrilateral) basis. Hence

$$T^h(x, y) \equiv \bigcup_e T_e(x_j) \tag{6.23}$$

$$T_e(x_j) \equiv \{N_k^+(\eta_i)\}^T\{Q\}_e \tag{6.24}$$

and the resultant finite element discrete approximation to the Galerkin symmetric weak statement (6.22) is

$$WS^h = S_e\Bigg(\int_{\Omega_e} \nabla\{N_k^+\}\cdot k_e\nabla\{N_k^+\}^T\,d\tau\{Q\}_e$$

$$-\int_{\Omega_e}\{N_k^+\}s_e\,d\tau + \int_{\partial\Omega_1}\{N_k^+\}q_{n_e}\,d\sigma - \int_{\partial\Omega_2}\{N_k^+\}k_e\nabla T_e\cdot\mathbf{n}\,d\sigma\Bigg) = \{0\} \tag{6.25}$$

As always, (6.25) requires the analyst to specify the basis degree $k$ and the distribution, hence interpolation, of the *problem data*, specifically, conductivity $k_e$, internal source $s_e$, normal flux $q_{n_e}$, and the Dirichlet data $T_b$. Note again that (6.25) will yield the estimate of normal heat flux on boundary segments $\partial\Omega_2$. Combining all data into the source column matrix $\{b\}_e$, and denoting the element conductivity matrix as $[K]_e$, (6.25) is the matrix statement

$$S_e([K]_e\{Q\}_e - \{b\}_e) \equiv \{0\} \tag{6.26}$$

as a specific case of (6.2). We now proceed to fill in the algebraic detail.

## Bilinear Basis

We now generate the element matrix $[K]_e$ and the matrix library involved in forming $\{b\}_e$. For $k = 1$ in (6.24), and assuming for simplicity that the conductivity $k_e$ is a constant on $\Omega_e$, the first term in (6.25) yields

$$[K]_e \equiv \bar{k}_e\int_{\Omega_e}\nabla\{N_1^+\}\cdot\nabla\{N_1^+\}^T\,d\tau \tag{6.27}$$

In sharp distinction to the $\{N_1(\zeta_i)\}$ triangular element form [recall (5.25)], we may not extract $\nabla\{N_1^+\}$ from the integrand in (6.27) since it is a function of $\eta_i$. Specifically

$$\begin{aligned}
\nabla\{N_1^+\} &\equiv \mathbf{i}\frac{\partial\{N_1^+\}}{\partial x} + \mathbf{j}\frac{\partial\{N_1^+\}}{\partial y} \\
&= \mathbf{i}\frac{\partial\{N_1^+\}}{\partial \eta_i}\left(\frac{\partial \eta_i}{\partial x}\right)_e + \mathbf{j}\frac{\partial\{N_1^+\}}{\partial \eta_i}\left(\frac{\partial \eta_i}{\partial y}\right)_e \\
&= \frac{\partial\{N_1^+\}}{\partial \eta_i}\left[\mathbf{i}\left(\frac{\partial \eta_i}{\partial x}\right)_e + \mathbf{j}\left(\frac{\partial \eta_i}{\partial y}\right)_e\right] \qquad 1 \le i \le 2
\end{aligned} \tag{6.28}$$

and from (6.7)

$$\begin{aligned}
\frac{\partial\{N_1^+\}}{\partial \eta_i} &= \frac{1}{4}\frac{\partial}{\partial \eta_i}\begin{Bmatrix}(1-\eta_1)(1-\eta_2)\\(1+\eta_1)(1-\eta_2)\\(1+\eta_1)(1+\eta_2)\\(1-\eta_1)(1+\eta_2)\end{Bmatrix} \\
&= \frac{1}{4}\begin{Bmatrix}-(1-\eta_2)\\(1-\eta_2)\\(1+\eta_2)\\-(1+\eta_2)\end{Bmatrix} \qquad \text{for } i = 1 \\
&= \frac{1}{4}\begin{Bmatrix}-(1-\eta_1)\\-(1+\eta_1)\\(1+\eta_1)\\(1-\eta_1)\end{Bmatrix} \qquad \text{for } i = 2
\end{aligned} \tag{6.29}$$

Equations (6.28) and (6.29) verify that $\nabla\{N_1^+\}$ is a function of $\eta_i$, and hence must remain under the integral in (6.27). Further, (6.28) confirms that the inverse coordinate transformation (from $\eta_i$ to $x_j$) on $\Omega_e$ is required to evaluate $[K]_e$, as anticipated. It is readily available by element operations on (6.10), the forward transformation. In general terms, (6.28) requires elements of the transformation *jacobian* matrix $[\partial\eta_i/\partial x_j]_e$ [recall (5.28)], which is the inverse of the forward transformation

$$[J]_e \equiv \left[\frac{\partial \eta_i}{\partial x_j}\right]_e = \left[\frac{\partial x_j}{\partial n_i}\right]_e^{-1} \tag{6.30}$$

The elements of $[J]_e$ are explicitly defined via the tensor index generalization of (6.10) as

$$\left[\frac{\partial x_j}{\partial \eta_i}\right]_e = \frac{\partial}{\partial \eta_i}\{N_1^+(\eta_i)\}^T\{XJ\}_e \qquad \text{for } 1 \le (i, j = J) \le 2 \tag{6.31}$$

Equation (6.31) is easy to evaluate using the derivatives established in (6.29). Thus, for example, for $i = 1$ and $j = 1$ (hence $\{XJ\}_e = \{X1\}_e = \{X\}_e$)

$$\left(\frac{\partial x_1}{\partial \eta_1}\right)_e = \frac{\partial}{\partial \eta_1}\{N_1^+(\eta_i)\}^T\{X1\}_e = \{X1\}_e^T \frac{\partial}{\partial \eta_1}\{N_1^+\}$$

$$= \tfrac{1}{4}\{X_1, X_2, X_3, X_4\}_e \begin{Bmatrix} -(1-\eta_2) \\ (1-\eta_2) \\ (1+\eta_2) \\ -(1+\eta_2) \end{Bmatrix}$$

$$= \tfrac{1}{4}[(1-\eta_2)(X_2 - X_1)_e + (1+\eta_2)(X_3 - X_4)_e] \qquad (6.32a)$$

which is a polynomial in $\eta_2$ and the $x$ coordinates $(X_i)_e$ of the vertex nodes of $\Omega_e$; see Fig. 6.4*a*. It is left as an exercise to verify that the other three entries in the $2 \times 2$ matrix $[J]_e$ are

$$\left(\frac{\partial x_1}{\partial \eta_2}\right)_e = \tfrac{1}{4}[(1-\eta_1)(X_4 - X_1)_e + (1+\eta_1)(X_3 - X_2)_e] \qquad (6.32b)$$

$$\left(\frac{\partial x_2}{\partial \eta_1}\right)_e = \tfrac{1}{4}[(1-\eta_2)(Y_2 - Y_1)_e + (1+\eta_2)(Y_3 - Y_4)_e] \qquad (6.32c)$$

$$\left(\frac{\partial x_2}{\partial \eta_2}\right)_e = \tfrac{1}{4}[(1-\eta_1)(Y_4 - Y_1)_e + (1+\eta_1)(Y_3 - Y_2)_e] \qquad (6.32d)$$

Therefore, each matrix entry in (6.31) is a polynomial in either $\eta_1$ or $\eta_2$ involving the vertex node coordinates $(X_i)_e$ or $(Y_i)_e$, $1 \le i \le 4$, on $\Omega_e$. Using elementary matrix algebra, the elements of (6.30) are then

$$[J]_e \equiv \left[\frac{\partial \eta_i}{\partial x_j}\right]_e = \left[\frac{\partial x_j}{\partial \eta_i}\right]_e^{-1} = \begin{bmatrix} \dfrac{\partial x_1}{\partial \eta_1} & \dfrac{\partial x_1}{\partial \eta_2} \\ \dfrac{\partial x_2}{\partial \eta_1} & \dfrac{\partial x_2}{\partial \eta_2} \end{bmatrix}_e^{-1}$$

$$= \frac{1}{\det [J]_e}\begin{bmatrix} \dfrac{\partial x_2}{\partial \eta_2} & -\dfrac{\partial x_1}{\partial \eta_2} \\ -\dfrac{\partial x_2}{\partial \eta_1} & \dfrac{\partial x_1}{\partial \eta_1} \end{bmatrix}_e$$

$$= \frac{1}{\det (6.31)}[\text{cof } (6.31)]_e^T \qquad (6.33)$$

Specifically, $[J]_e$ is constituted of the transformed cofactor matrix of (6.31), with entries (6.32), divided by the determinant of (6.31).

Hence, the element coordinate transformation (6.10) provides entirely for evaluating the differentiation contained in matrix $[K]_e$ in terms of rational polynomials in $\eta_i$ and $(XJ)_e$. Turn back to (5.29) in Chap. 5 and compare (6.33) to the element transformation expression on the triangle natural coordinate system $\zeta_i$, $1 \le i \le n+1 = 3$. Recalling that $A_e$ contained therein is evaluated as the determinant (5.11), the essential similarities are quite apparent. However, the significant distinction that $\nabla\{N_1^+\}$ is not a constant on $\Omega_e$ persists to complicate the present formulation.

Proceeding directly with the task at hand, to evaluate (6.27), the matrix product becomes

$$\nabla\{N_1^+\}\cdot\nabla\{N_1^+\}^T = \frac{\partial\{N_1^+\}}{\partial\eta_i}\left[\mathbf{i}\left(\frac{\partial\eta_i}{\partial x}\right)_e + \mathbf{j}\left(\frac{\partial\eta_i}{\partial y}\right)_e\right]\cdot\left[\mathbf{i}\left(\frac{\partial\eta_k}{\partial x}\right)_e + \mathbf{j}\left(\frac{\partial\eta_k}{\partial y}\right)_e\right]\frac{\partial\{N_1^+\}^T}{\partial\eta_k}$$

$$= \frac{\partial\{N_1^+\}}{\partial\eta_i}\frac{\partial\{N_1^+\}^T}{\partial\eta_k}\left(\frac{\partial\eta_i}{\partial x_j}\frac{\partial\eta_k}{\partial x_j}\right)_e \qquad 1 \le (i, j, k) \le 2 \qquad (6.34)$$

on introducing the dummy tensor indices $k$ and $j$ to allow the compact form [as in (5.30)]. The last step for (6.27) is to transform the differential element $d\tau \equiv dx\,dy$ to a function of $\eta_i$. The well-known equivalence yields

$$d\tau = dx\,dy = \det(6.31)\,d\eta_1\,d\eta_2 \qquad (6.35)$$

In other words, the coefficient of proportionality is the determinant of the forward transformation matrix (6.31). Hence, using (6.34) and (6.35), the conductivity matrix is

$$[K]_e \equiv \bar{k}_e \int_{\Omega_e} \nabla\{N_1^+\}\cdot\nabla\{N_1^+\}^T\,d\tau$$

$$= \bar{k}_e \int_{\Omega_e} \frac{\partial\{N_1^+\}}{\partial\eta_i}\frac{\partial\{N_1^+\}^T}{\partial\eta_k}\left(\frac{\partial\eta_i}{\partial x_j}\frac{\partial\eta_k}{\partial x_j}\right)_e \det_e d\eta_1\,d\eta_2$$

$$\text{for } 1 \le (i, j, k) \le n = 2 \qquad (6.36)$$

The quadrilateral element matrix $[K]_e$ in (6.36) certainly appears more complicated than its triangle counterpart, (5.25) to (5.39). This occurs since both the metric data and the basis derivatives contain $\eta_i$, the integrals of which cannot be evaluated in closed form for the general case. However, this does not represent an impasse since formulas for the numerical approximation of integrals, termed *numerical quadrature*, are well established. The reader has most likely encountered this methodology with a *Newton-Cotes* or *Simpson's rule* procedure introduced in a first course in computer science.

The numerical integration method of choice for finite element matrix evaluation is *Gaussian quadrature*, a variable accuracy procedure that exactly integrates polynomials on regular regions. Specifically, a $P$th-order gaussian quadrature rule integrates exactly a polynomial of degree $2P - 1$ on $\Omega_e$ in transform space, that is, the unit square. Letting $k_{\alpha\beta}$ denote the integral of the matrix element $\kappa_{\alpha\beta}$ in (6.36), the two-dimensional gaussian quadrature rule is

$$\begin{aligned} k_{\alpha\beta} &\equiv \int_{-1}^{+1}\int_{-1}^{+1} \kappa_{\alpha\beta}(\eta_1, \eta_2)\, d\eta_1\, d\eta_2 \\ &= \sum_{p=1}^{P}\sum_{q=1}^{Q} H_p H_q \kappa_{\alpha\beta}(\eta_1 = \eta_1^p, \eta_2 = \eta_2^q) \end{aligned} \tag{6.37}$$

In (6.37), $P$ is the order of the quadrature rule, $Q = P$ for a symmetric evaluation, $H_p$ and $H_q$ are the gaussian rule coefficients (called "weights"), and $\eta_i^p$ and $\eta_i^q$ are specific coordinates on $\Omega_e$. Figure 6.7 illustrates these "Gauss points" for $1 \leq (P = Q) \leq 3$, and Table 6.1 lists the coordinates and the corresponding weights.

With this development, we return again to evaluate $[K]_e$, (6.36). One must first determine the order $P$ of the gaussian quadrature rule that will accurately evaluate the integrals. Recalling (6.29), the first two terms in

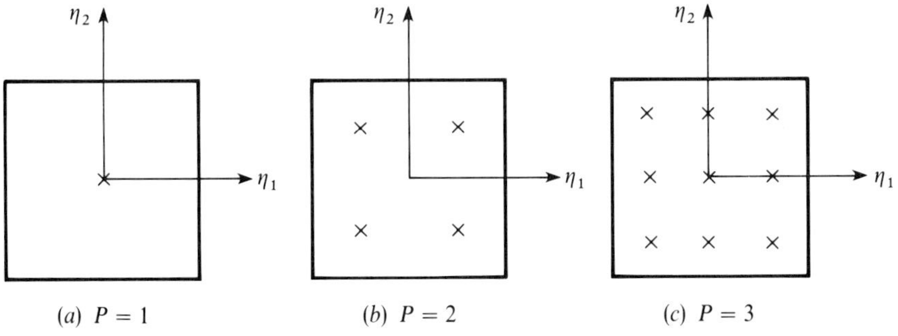


**FIGURE 6.7**
Quadrature coordinates $\eta_i^p$ for symmetric gaussian integration of order $P = Q$, equation (6.37), on $\Omega_e$.

TABLE 6.1
**Gaussian quadrature rule coordinates and weights**

| | Coordinates | | Weights | |
|---|---|---|---|---|
| Order ($P$) | $\eta_1^p$ | $\eta_2^q$ | $H_p$ | $H_q$ |
| 1 | 0.0 | 0.0 | 1.0 | 1.0 |
| 2 | $\pm 1/\sqrt{3}$ | $\pm 1/\sqrt{3}$ | 1.0 | 1.0 |
| 3 | 0.0 | 0.0 | 8/9 | 8/9 |
| | $\pm\sqrt{0.6}$ | $\pm\sqrt{0.6}$ | 5/9 | 5/9 |

(6.36) are linear monomials, hence the product is at most quadratic in the $\eta_i$. Each metric term $(\partial\eta_i/\partial x_j)_e$ [see (6.33)], is a linear polynomial in $\eta_i$ divided by the jacobian determinant (6.31), which is a biquadratic polynomial, recall (6.32). In the product of metric terms, one $\det_e$ in the denominator will cancel with $\det_e$ from the $d\tau$ transformation (6.35), thus the metric product is a rational function of a quadratic and a biquadratic polynomial. Hence, the product of all terms forming each element $\kappa_{\alpha\beta}(\eta_i)$ of $[K]_e$ can be no more than cubic in $\eta_i$. Thus, the $P = 2 = Q$ quadrature rule is appropriate since $2P - 1 = 2(2) - 1 = 3$. The following example proceeds through the algebraic detail for illustration.

**Example 6.1.** Form the $\alpha = 1 = \beta$ element $k_{11}$ of the thermal conductivity matrix $[K]_e$ for the generic quadrilateral element $\Omega_e$ (Fig. 6.4*a*). The primitive geometric data are $\{X_i\}_e$ and $\{Y_i\}_e$, $1 \le i \le 4$, and the element thermal conductivity is $\bar{k}_e$. The (1, 1) element in $[K]_e^T$ results for the product of the first entries in the derivative of $\{N_1^+\}$ and $\{N_1^+\}^T$ in (6.36). This coefficient is obtained from (6.29); hence

$$k_{\alpha=1,\beta=1} = \bar{k}_e \int_{\Omega_e} \left(\frac{1}{4}\right)^2 \frac{\partial(1-\eta_1)(1-\eta_2)}{\partial\eta_i} \frac{\partial(1-\eta_1)(1-\eta_2)}{\partial\eta_k} \times \left(\frac{\partial\eta_i}{\partial x_j}\frac{\partial\eta_k}{\partial x_j}\right)_e \det_e d\eta_1\, d\eta_2 \qquad \text{for } 1 \le (i, j, k) \le n$$

$$= \frac{\bar{k}_e}{16} \sum_{i=1}^{2} \sum_{j=1}^{2} \sum_{k=1}^{2} \int_{-1}^{+1}\int_{-1}^{+1} \frac{\partial(1-\eta_1)(1-\eta_2)}{\partial\eta_i} \frac{\partial(1-\eta_1)(1-\eta_2)}{\partial\eta_k} \times \left(\frac{\partial\eta_i}{\partial x_j}\frac{\partial\eta_k}{\partial x_j}\right)_e \det_e d\eta_1\, d\eta_2 \qquad (6.38)$$

By terms in (6.38), we have from (6.29)

$$\frac{\partial}{\partial \eta_1}(1-\eta_1)(1-\eta_2) = -(1-\eta_2)$$

$$\frac{\partial}{\partial \eta_2}(1-\eta_1)(1-\eta_2) = -(1-\eta_1) \tag{6.39}$$

The components of the jacobian $(\partial \eta_i/\partial x_j)_e$, from the definition (6.33) and using (6.32) are

$$\left(\frac{\partial \eta_1}{\partial x_1}\right)_e \equiv \frac{1}{\det_e}\left(\frac{\partial x_2}{\partial \eta_2}\right)_e = \frac{1}{4\det_e}[(1-\eta_1)(Y_4-Y_1)_e + (1+\eta_1)(Y_3-Y_2)_e]$$

$$\left(\frac{\partial \eta_1}{\partial x_2}\right)_e \equiv \frac{1}{\det_e}\left(-\frac{\partial x_1}{\partial \eta_2}\right)_e = \frac{-1}{4\det_e}[(1-\eta_1)(X_4-X_1)_e + (1+\eta_1)(X_3-X_2)_e]$$

$$\left(\frac{\partial \eta_2}{\partial x_1}\right)_e \equiv \frac{1}{\det_e}\left(-\frac{\partial x_2}{\partial \eta_1}\right)_e = \frac{-1}{4\det_e}[(1-\eta_2)(Y_2-Y_1)_e + (1+\eta_2)(Y_3-Y_4)_e]$$

$$\left(\frac{\partial \eta_2}{\partial x_2}\right)_e \equiv \frac{1}{\det_e}\left(\frac{\partial x_1}{\partial \eta_1}\right)_e = \frac{1}{4\det_e}[(1-\eta_2)(X_2-X_1)_e + (1+\eta_2)(X_3-X_4)_e] \tag{6.40}$$

Finally, $\det_e$ is the determinant of (6.31); hence

$$\det_e = \det\begin{bmatrix} \dfrac{\partial x_1}{\partial \eta_1} & \dfrac{\partial x_1}{\partial \eta_2} \\ \dfrac{\partial x_2}{\partial \eta_1} & \dfrac{\partial x_2}{\partial \eta_2} \end{bmatrix}_e = \left(\frac{\partial x_1}{\partial \eta_1}\right)_e\left(\frac{\partial x_2}{\partial \eta_2}\right)_e \quad \left(\frac{\partial x_1}{\partial \eta_2}\right)_e\left(\frac{\partial x_2}{\partial \eta_1}\right)_e$$

$$\begin{aligned} &= \tfrac{1}{16}[((1-\eta_2)(X_2-X_1)_e + (1+\eta_2)(X_3-X_4)_e) \\ &\quad \times ((1-\eta_1)(Y_4-Y_1)_e + (1+\eta_1)(Y_3-Y_2)_e) \\ &\quad - ((1-\eta_1)(X_4-X_1)_e + (1+\eta_1)(X_3-X_2)_e) \\ &\quad \times ((1-\eta_2)(Y_2-Y_1)_e + (1+\eta_2)(Y_3-Y_4)_e)] \end{aligned} \tag{6.41}$$

Equations (6.39) to (6.41) define all terms needed in (6.38), with summation over the three repeated indices $i$, $j$, $k$. However, in practice the integral is replaced by the $P = 2 = Q$ quadrature rule; hence (6.38) is really evaluated as

$$k_{11} = \frac{\bar{k}_e}{16}\sum_{i=1}^{2}\sum_{j=1}^{2}\sum_{k=1}^{2}\sum_{p=1}^{2}\sum_{q=1}^{2} H_p H_q[\text{integrand}_{11} \text{ in } (6.38)]_{i,j,k,p,q} \tag{6.42}$$

Thus, there are $2^5$ multiplication-addition operations required to form one element of $[K]_e$ for $\{N_1^+\}$. Since the matrix is symmetric, (6.42) need be repeated only $4 + 3 + 2 + 1 = 10$ times to complete evaluation of $[K]_e$ for any $\Omega_e$. The computer is particularly good at handling this tedium, so it merely constitutes the algebraic detail associated with the bilinear quadrilateral basis.

To complete the illustration, for all indices set equal to 1 in (6.42), whereupon $\eta_1^1 = -1/\sqrt{3} = -0.5777\cdots = \eta_2^1$, and $H_1 = 1 = H_2$, the corresponding term in $k_{11}$ is

$$k_{11} = \frac{\frac{\bar{k}_e}{16}(1+0.577)(1+0.577)[(1+0.577)(Y_4 - Y_1)_e + (1-0.577)(Y_3 - Y_2)_e]^2}{[(1+0.577)(X_4 - X_1)_e + (1-0.577)(X_3 - X_4)_e](\cdots)} + \cdots \tag{6.43}$$

as the factors of $\frac{1}{16}$ cancel in the metric data operations.

The reader is probably either overwhelmed [or uninterested (!)] at this point with the algebraic detail associated with the bilinear quadrilateral basis. Certainly, as stated, a significant amount of metric data handling is required for this basis choice. The following simplification cuts cleanly through this "fog" and represents a viable alternative for bilinear basis master matrix generation in the PC environment.

## Bilinear Basis Simplified

Equation (6.36) is the analytical statement of the element conductivity master matrix. A significant simplification would result if the metric data could be extracted from the integrand, as occurred (naturally) for the triangle element basis. A first thought might be to use the $P = 1 = Q$ quadrature rule, which would eliminate all $\eta_i$ dependence throughout the integrand, recall Table 6.1. The resulting numerical error in evaluating $[K]_e$ is generally unacceptable, however, which rules out the use of such a reduced-order quadrature rule.

However, the idea of underinterpolating (averaging) the metric data does have merit provided one has a quality discretization $\Omega^h$, that is, one with no elements having included angles near 0° or 180°. (The analyst should avoid such meshes anyway, since solution accuracy—even without interpolation error—is always adversely affected by a poor discretization.) Equation (6.40) clearly states each metric component is simply the linear interpolation of differences among the nodal coordinate set $\{X_i\}_e$ and $\{Y_i\}_e$ divided by $\det_e$. Further, as illustrated, the product of these metric coefficients, divided by $\det_e$, is the ratio of quadratic polynomials. The element-average value of the linear interpolate occurs for setting $\eta_1 \equiv 0 \equiv \eta_2$, whereupon we are led to define the element-average metric set $ETAIJ_e$ (pronounced "eta-i-j") with components

$$
\begin{aligned}
ETA11_e &\equiv \det_e \overline{\left(\frac{\partial \eta_1}{\partial x_1}\right)_e} = \tfrac{1}{4}(Y_4 - Y_1 + Y_3 - Y_2)_e \\
ETA12_e &\equiv \det_e \overline{\left(\frac{\partial \eta_1}{\partial x_2}\right)_e} = \tfrac{1}{4}(X_4 - X_1 + X_3 - X_2)_e \\
ETA21_e &\equiv \det_e \overline{\left(\frac{\partial \eta_2}{\partial x_1}\right)_e} = \tfrac{1}{4}(Y_2 - Y_1 + Y_3 - Y_4)_e \\
ETA22_e &\equiv \det_e \overline{\left(\frac{\partial \eta_2}{\partial x_2}\right)_e} = \tfrac{1}{4}(X_2 - X_1 + X_3 - X_4)_e
\end{aligned}
\tag{6.44}
$$

Then, since $\det_e$ is the determinant of (6.31), which contains the entries in $(\partial\eta_i/\partial x_j)_e$ transposed, the element average value of $\det_e$ is computed by using (6.44) as

$$DET_e \equiv \overline{\det_e} = ETA11_e ETA22_e - ETA12_e ETA21_e \tag{6.45}$$

[recall (6.41)].

**Example 6.2.** Evaluate the element-average metric data in (6.44) to (6.45) and compare them to the exact value for the finite elements shown in Figs. 6.3 and 6.4. For the rectangular parallelepiped element in Fig. 6.3*a*, one directly verifies that

$$
\begin{aligned}
ETA11_e &= \tfrac{1}{4}(Y_4 - Y_1 + Y_3 - Y_2)_e = \tfrac{1}{4}(2b + 2b) = b \\
ETA12_e &= 0 = ETA21_e \\
ETA22_e &= \tfrac{1}{4}(X_2 - X_1 + X_3 - X_4)_e = \tfrac{1}{4}(2a + 2a) = a \\
DET_e &= ETA11_e ETA22_e - 0 = ab
\end{aligned}
\tag{6.46}
$$

Obviously, (6.46) is identical with the exact expressions given in (6.40) and (6.41), since the multipliers of $(1 - \eta_i)$ and $(1 + \eta_i)$ are coincident throughout. Further, we see that $\det_e$, the "measure" of $\Omega_e$, is one-fourth its plane area.

The element shape given in Fig. 6.4 is hard to analyze in a completely general way; hence, focus on the two special cases shown in Fig. 6.8. For the nonrectangular parallelepiped (Fig. 6.8*a*), the included angles are $\alpha$ and $\beta = \pi - \alpha$. For any angle $\alpha$, you may easily verify that (6.46) for the average

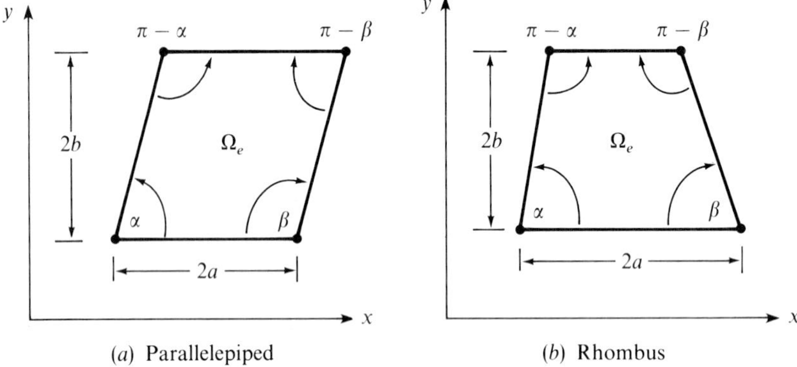


**FIGURE 6.8**
Special cases of bilinear quadratic element in Fig. 6.4.

metric data is identical with the analytical expressions (6.40) and (6.41). The measure $ab$ remains equal to one-fourth the plane area. Further, these observations hold for arbitrary orientation of the element in the global reference frame.

The case of the rhombus (Fig. 6.8$b$) can be handled analytically provided two sides are parallel to a global coordinate axis and $\alpha = \beta$. The included angles are complements as shown. For the given orientation, $ETA11_e$ and $ETA21_e = 0$ are again exact. Further, we have using (6.44)

$$ETA21_e = \tfrac{1}{4}(\Delta Y_{\text{left}} + \Delta Y_{\text{right}}) = 0$$

since $Y_2 - Y_1 = -(Y_3 - Y_4)$, and it is exact. Finally

$$ETA22_e = \tfrac{1}{4}(\Delta X_{\text{bottom}} + \Delta X_{\text{top}}) = \tfrac{1}{2}(\Delta X)_{\text{avg}}$$

which equals the analytical value only at $\eta_1 = 0 = \eta_2$.

The exact expression (6.41) for the measure reduces to

$$\det_e = \tfrac{1}{16}\{[(1 - \eta_2)\Delta X_{\text{bottom}} + (1 - \eta_2)\Delta X_{\text{top}}]_e(2\,\Delta Y) - 0\}$$

From (6.45) we have

$$DET_e = \tfrac{1}{4}(2\,\Delta Y)(\tfrac{1}{2}\,\Delta X_{\text{avg}}) - 0 = \tfrac{1}{4}\,\Delta Y\,\Delta X_{\text{avg}}$$

which is equal to $\det_e$ evaluated at the element centroid ($\eta_2 = 0$).

This example verifies that averaging the metric data is a reasonable assumption for a range of quadrilateral element geometric shapes typically encountered in a quality discretization; see Fig. 6.1. The attendant simplification to creation of the bilinear basis master matrix library is substantial.

Specifically, the element conductivity matrix (6.36) becomes

$$[K]_e = \bar{k}_e \int_{\Omega_e} \nabla\{N_1^+\} \cdot \nabla\{N_1^+\}^T \, d\tau$$

$$= \bar{k}_e \int_{\Omega_e} \frac{\partial\{N_1^+\}}{\partial\eta_i} \frac{\partial\{N_1^+\}^T}{\partial\eta_k} \left(\frac{\partial\eta_i}{\partial x_j}\frac{\partial\eta_k}{\partial x_j}\right)_e \det_e d\eta_1 \, d\eta_2$$

$$= \bar{k}_e \overline{\det_e \left(\frac{\partial\eta_i}{\partial x_j}\frac{\partial\eta_k}{\partial x_j}\right)_e} \int_{\Omega_e} \frac{\partial\{N_1^+\}}{\partial\eta_i} \frac{\partial\{N_1^+\}^T}{\partial\eta_k} d\eta_1 \, d\eta_2$$

$$= \bar{k}_e DET_e^{-1} ETAIJ_e ETAKJ_e \int_{\Omega_e} \frac{\partial\{N_1^+\}}{\partial\eta_i} \frac{\partial\{N_1^+\}^T}{\partial\eta_k} d\eta_1 \, d\eta_2$$

$$= \bar{k}_e DET_e^{-1} ETAIJ_e ETAKJ_e [B2IKB] \quad \text{for } 1 \le (I, J, K) \le 2 \qquad (6.47)$$

Recalling the Fortran nomenclature (5.100), $[B2IKB]$ is a set of two-dimensional matrices ($a = B$), involving two bases ($b = 2$) differentiated on $\eta_i$ and $\eta_k$ ($c = I, c = K$), and the suffix ($d = B$) denotes *B*ilinear basis (in distinction to $d = L$ for the *L*inear triangle basis). The form (6.47) is operationally identical to the triangular element formulation; recall (5.121). Finally, the integrals remaining in (6.47) are analytically evaluable, hence the master matrix library contains the following data.

$$[B211B] = \frac{1}{6}\begin{bmatrix} 2 & -2 & -1 & 1 \\ & 2 & 1 & -1 \\ & & 2 & -2 \\ (\text{sym}) & & & 2 \end{bmatrix}$$

$$[B222B] = \frac{1}{6}\begin{bmatrix} 2 & 1 & -1 & -2 \\ & 2 & -2 & -1 \\ & & 2 & 1 \\ (\text{sym}) & & & 2 \end{bmatrix}$$

$$[B212B] = \frac{1}{4}\begin{bmatrix} 1 & 0 & -1 & 0 \\ 0 & -1 & 0 & 1 \\ -1 & 0 & 1 & 0 \\ 0 & 1 & 0 & -1 \end{bmatrix} \qquad (6.48)$$

$$[B221B] = \frac{1}{4}\begin{bmatrix} 1 & 0 & -1 & 0 \\ 0 & -1 & 0 & 1 \\ -1 & 0 & 1 & 0 \\ 0 & 1 & 0 & -1 \end{bmatrix}$$

Figure 6.9 contains a listing excerpt from *LEARN.FE* subroutine FORMIT illustrating the bilinear basis organizational structure as developed using (6.44) to (6.48). The comparison to Fig. 5.8, the linear basis structure, confirms the modest distinctions that accrue with use of element-average metric data in the present instance.

The remaining completion steps involve the matrix evaluations for the source and boundary condition terms in (6.25). Using the element-average metric procedure, the first is

$$\begin{aligned}\int_{\Omega_e} \{N_1^+\} s_e \, d\tau &= \int_{\Omega_e} \{N_1^+\}\{N_1^+\}^T \{S\}_e \det_e d\eta_1 \, d\eta_2 \\ &\simeq DET_e \int_{\Omega_e} \{N_1^+\}\{N_1^+\}^T d\eta_1 \, d\eta_2 \{S\}_e \\ &= DET_e [B200B]\{S\}_e \end{aligned} \tag{6.49}$$

```
      ******************************************************************
C     SUBROUTINE FORMIT(IENERG)
C
C     FORM THE ALGORITHM MATRIX STATEMENT
C     MODIFY SMAT FOR DIRICHLET AND NEUMANN BOUNDARY DATA
C     ******************************************************************
        .
        .
        .
C     ******************************************************************
C     Q U A D R I L A T E R A L S
C     ******************************************************************
      ELSE IF (NDOFE.GE.4)THEN

C     3.3) BILINEAR QUADRILATERALS

C     DETERMINANT OF JACOBIAN
      EVALUATED AT ELEMENT CENTROID
            ETA11=((YG(ML4)-YG(ML1))+(YG(ML3)-YG(ML2)))/4.E0
            ETA21=((XG(ML1)-XG(ML4))+(XG(ML2)-XG(ML3)))/4.E0
            ETA12=((YG(ML1)-YG(ML2))-(YG(ML4)-YG(ML3)))/4.E0
            ETA22=((XG(ML2)-XG(ML1))-(XG(ML3)-XG(ML4)))/4.E0
              DET=4.*(ETA11*ETA22-ETA12*ETA21)
              COND=(AK(ML1)+AK(ML2)+AK(ML3)-AK(ML4))/4.
C             ELEMENT CONDUCTIVITY MATRIX
            DO 175I=1, NDOFE
               DO 176J=1, NDOFE
                  EMAT(I,J)=B211B(I,J)*(ETA11**2+ETA21**2)
     *                      +(B212B(I,J)+B221B(I,J))
     *                         *(ETA11*ETA12)
     *                      +(B221B(I,J)+B212B(I,J))
     *                         *(ETA21*ETA22)
     *                      +(B222B(I,J)*(ETA22**2+ETA12**2))
                  EMAT(I,J)=EMAT(I,J)*COND/DET
176             CONTINUE
175          CONTINUE
```

**FIGURE 6.9**
Excerpt from *LEARN.FE* subroutine FORMIT for bilinear quadrilateral $[K]_e$ formation with average metric data.

The integrals in the second line of (6.49) are analytically evaluated. The matrix library entry is

$$[B200B] = \frac{1}{9}\begin{bmatrix} 4 & 2 & 1 & 2 \\ & 4 & 2 & 1 \\ & & 4 & 2 \\ \text{(sym)} & & & 4 \end{bmatrix} \tag{6.50}$$

The flux boundary condition term evaluation is absolutely unchanged from the development in Chap. 5 [recall (5.41) and (5.42)]. Specifically,

$$\begin{aligned} \int_{\partial\Omega_1} \{N_1^+\} q_{n_e}\, d\sigma &= \int_{l_e} \{N_1\}\{N_1\}^T\, d\bar{x}\{QN\}_e \\ &= l_e[A200L]\{QN\}_e \end{aligned} \tag{6.51}$$

where $l_e$ is the length of the element side on $\partial\Omega_1$ and $[A200L]$ is as defined in (5.126).

## Biquadratic Basis

The averaged metric data procedure is not appropriate for use with either the serendipity or complete biquadratic basis formulations on elements with curved sides. Hence, the matrix element evaluations must be completed using gaussian quadrature. The thermal conduction matrix for the serendipity basis is

$$\begin{aligned} [K]_e &= \bar{k}_e \int_{\Omega_e} \nabla\{N_2^+\} \cdot \nabla\{N_2^+\}^T\, d\tau \\ &= \bar{k}_e \int_{\Omega_e} \frac{\partial\{N_2^+\}}{\partial\eta_i} \frac{\partial\{N_2^+\}^T}{\partial\eta_k} \left(\frac{\partial\eta_i}{\partial x_j}\frac{\partial\eta_k}{\partial x_j}\right)_e \det_e d\eta_1\, d\eta_2 \qquad \text{for } 1 \le (i, j, k) \le 2 \end{aligned} \tag{6.52}$$

with (6.15) to (6.17) defining the basic data. The serendipity basis $\{N_2^+\}$ contains eight entries; hence (6.52) is an $8 \times 8$ symmetric matrix. The individual terms constituting each expression in (6.52) are considerably more lengthy than those for $\{N_1^+\}$; hence have not been expressed in expanded form. The quadrature rule evaluation for the generic entry in (6.52) is

$$k_{\alpha\beta} = \frac{\bar{k}_e}{16} \sum_{i=1}^{2} \sum_{j=1}^{2} \sum_{k=1}^{2} \sum_{p=1}^{P} \sum_{q=1}^{P} H_p H_q [\text{integrand } \kappa_{\alpha\beta} \text{ in (6.52)}]_{i,j,k,p,q} \tag{6.53}$$

and $P = 3 = Q$ is appropriate. Conversely, $P = 4 = Q$ is required for the full biquadratic because of the last entry in (6.16). *LEARN.FE* does not contain the organizational structure for these two basis formulations; however, FORMIT is keyed for the entry point should you elect to implement either.

## B. PRACTICAL PROBLEMS

### 6.5 STEADY HEAT CONDUCTION WITH BOUNDARY CONVECTION

We now implement the averaged-metric bilinear quadrilateral element basis for the generic problem class. Recalling Section 5.6, the partial differential equation system is

$$L(T) = -\nabla \cdot k(x_i)\nabla T - s(x_i) = 0 \qquad \text{on } \Omega \subset \mathbb{R}^2 \tag{5.82}$$

$$l(T) = k\nabla T \cdot \mathbf{n} + h(T - T_r) = 0 \qquad \text{on } \partial\Omega_1 \tag{5.83}$$

$$l(T) = k\nabla T \cdot \mathbf{n} + q_n = 0 \qquad \text{on } \partial\Omega_2 \tag{5.84}$$

$$T(\mathbf{x}_b) = T_b \qquad \text{on } \partial\Omega_3 \tag{5.85}$$

where conductivity $k$ and source $s$ are assumed variable and the solution domain boundary $\partial\Omega$ is the union of $\partial\Omega_1$, $\partial\Omega_2$ and $\partial\Omega_3$. While we seek to establish the bilinear basis algorithm, the statement of finite element approximation to the weak statement (5.20) for (5.82) to (5.85) is *absolutely* unchanged from (5.86):

$$\begin{aligned}
WS^h = S_e\Bigg(&\int_{\Omega_e} \nabla\{N_k\} \cdot k_e \nabla\{N_k\}_e^T \, d\tau \{Q\}_e \\
&- \int_{\Omega_e} \{N_k\} s_e \, d\tau + \int_{\partial\Omega_1} \{N_k\} h_e \{N_k\}^T \{Q - TR\}_e \, d\sigma \\
&+ \int_{\partial\Omega_2} \{N_k\} q_{n_e} \, d\sigma - \int_{\partial\Omega_3} \{N_k\} k_e \nabla T_e \cdot \mathbf{n} \, d\sigma\Bigg) \\
&= S_e([K + H]_e\{Q\}_e - \{SC\}_e - [H]_e\{TR\}_e + \{QF\}_e - \{FF\}_e) \\
&= S_e([K + H]_e\{Q\}_e - \{b\}_e) = \{0\}
\end{aligned} \tag{5.86}$$

Recall that in (5.86), for any $\Omega_e$ of the discretization $\Omega^h$, $[K + H]_e$ is the conductivity matrix, $\{SC\}_e$ results from the distributed source, $[H]_e\{TR\}_e$

is the boundary thermal convection source, $\{QF\}_e$ is the boundary fixed flux source, and the boundary flux term $\{FF\}_e$ on $\partial\Omega_3$ does not enter the global matrix statement due to Dirichlet data imposition.

All the ingredients of the master matrix definition structure are at hand. For the averaged-metric bilinear basis, the Fortran master matrix statement for (5.86) is

$$\begin{aligned} WS^h = S_e(&\bar{k}_e DET_e^{-1} ETAIJ_e ETAKJ_e[B2IKB]\{Q\}_e \\ &+ l_e\{H\}_e^T[A3000L]\{Q\}_e - DET_e[B200B]\{S\}_e \\ &- l_e\{H\}_e^T[A3000L]\{TR\}_e - l_e[A200L]\{QN\}_e) \\ = \{0\} &\qquad \text{for } 1 \le (I, J, K) \le n = 2 \end{aligned} \tag{6.54}$$

Comparing (6.54) with the triangle basis master matrix statement (5.127), one is again struck by the uniformity the finite element algorithm presents as an approximation procedure. The distinction on basis choice, hence element shape, is contained exclusively in the $B$ matrix and metric data arrays, the element measure ($A_e$ vs. $DET_e$) and the summation limit on the $(I, K)$ index set. Note in particular that all master matrices for boundary data imposition are identical in each basis statement, as predicted by the theory. The measure $l_e$ is the appropriate side length $\partial\Omega_e$ and $[A3000L]$ and $[A200L]$ are defined in (5.118) and (5.126).

Equation (6.54) is the global matrix statement of the bilinear basis algorithm for steady-state conduction with boundary convection. We suggest that you continue with Study Problem 3 and compare the bilinear and triangle basis solutions. The precursor to this involves the following.

**Code Exercise 6.5.** Generate a $k = 1$ bilinear basis solution for a thick-walled pipe of internal radius $r_1 = 1$ ft and external radius $r_2 = 2$ ft using an $M = 4 \times 4$ uniform discretization of a (symmetric) quarter-circle region of the pipe. The discretization is shown in Fig. 6.10, which is identical to Fig. 5.3 with removal of the diagonal lines used to form the triangular element mesh.

The creation file for this discretization was given in Code Exercise 6.4. As before, it must be expanded to define the specific problem statement using the *MATL*, *ROBN*, *DIRI*, *FORM*, and *SOLV*, or *GAUS* commands. Thus, the input file you should create for this code exercise is

```
TITL
***** EXERCISE 6.5, M=16, PIPE WITH TB FIXED *****
TYPE     [K          N    NNODEL       REFL    NPRN      NTRAN   NAXI]
          1          2       4          1.0     1 1         3
PRIN     [NBUG(*)]
          1          0       0            2
```

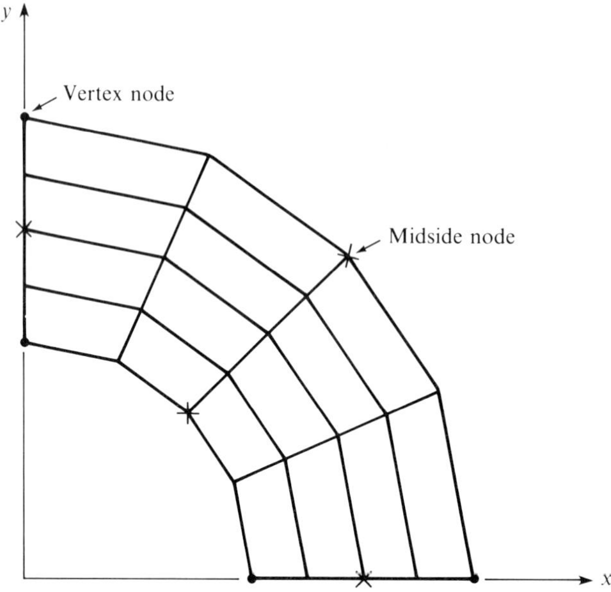


**FIGURE 6.10**
Isoparametric quadrilateral macro domain discretization.

```
GRID      [FOR N(XL XR(*)PR(*)), NEM(PR*)]
                   1.0       2.0       1.0
            4
                   0.0       1.0       1.0
            4
                    1.0   2.0        0.0      0.0
                    0.0   0.0        2.0      1.0
                    1.5   1.4142     0.0      0.7071
                    0.0   1.4142     1.5      0.7071
MATL      [FOR MACRO(*),K(*)]
            1      10.0
ROBN      [NROB,NODE(*),(H(*),TR(*))]
          5
          1    6    11      16       21
                20.00                  1500.00
                20.00                  1500.00
                20.00                  1500.00
                20.00                  1500.00
                20.00                  1500.00

DIRI       [(NDIR,REPEAT),NODE(*),(NRPT,QB(*))]
          5   1   0
          5          10             15             20            25
          5       306.8528

FORM
SOLV
STOP
```

Comparing this to the execution file for Code Exercise 5.5, note that the sole modification is under *TYPE* wherein NNODEL = 4 (rather than 3) indicating the quadrilateral element.

Now execute this data file and the following will scroll onto your monitor.

```
TITL
***** EXERCISE 6.5, M=16, PIPE WITH TB FIXED *****
TYPE     K       N   NNODEL       REFL    NPRN    NTRAN   NAXI
         1       2     4           1.0     1 1      3       0
PRINT    NBUG(1)      (2)        (3)       (4)     (5)     (6)
               1       0          0         2       0       0
GRID     NI       NPRI     NTRAN
         1        1        3
         XL       XR1      PR1      XR2      PR2      XR3      PR3
         1.0      2.0      1.0
         NEM1     NEM2     NEM3
         4
GRID     NI       NPRI     NTRAN
         2        1        3
         YL       YU1      PR1      YU2      PR2      YU3      PR3
         0.0      1.0      1.0
         NEM1     NEM2     NEM3
         4
         X1        X2        X3        X4
         1.0000    2.0000    0.0000    0.0000
         Y1        Y2        Y3        Y4
         0.0000    0.0000    2.0000    1.0000
         X5        X6        X7        X8
         1.5000    1.4142    0.0000    0.7071
         Y5        Y6        Y7        Y8
         0.0000    1.4142    1.5000    0.7071

         NODE COORDINATES(X INTEGER MAP)EXPONENT E:-2
                0         0         0         0      0
               38        47        57        66     76
               70        88       106       123    141
               92       115       138       161    184
              100       125       150       175    200

         NODE COORDINATES(Y INTEGER MAP)EXPONENT E:-2
              100       125       150       175    200
               92       115       138       161    184
               70        88       106       123    141
               38        47        57        66     76
                0         0         0         0      0

MATL     MACRO CONDUCTIVITIES
          1       10.0

ROBN     NROB:      5
         JROB:      1      6        11        16        21
                    H              TREF
                   20.0           1500.0
                   20.0           1500.0
                   20.0           1500.0
                   20.0           1500.0
                   20.0           1500.0
```

```
DIRI      NFIX:    5
          JFIX:    5       10      15      20      25
     306.8528   306.8528     306.8528     306.8528      306.8528

FORM
SOLV     NODAL SOLUTION (INTEGER MAP)EXPONENT E:0
                993       772       592       439      306
                993       772       592       439      306
                993       772       592       439      306
                993       772       592       439      306
                993       772       592       439      306
STOP
```

Listed following *SOLV* is the integer map of the bilinear basis nodal temperature solution. Comparing these data to Code Exercise 5.5, the nodal distributions are indeed identical. This should not surprise you, since the problem has variation in only one (curvilinear) coordinate direction and both bases are linear. You will find this to be the case for Study Problem 3 in its entirety, which you should verify by repeating the computational experiment sequence detailed in Chap. 5.

Any axisymmetric problem exhibits this coordinate degeneracy. However, for problems that are fully two-dimensional via data specifications, distinct nodal temperature distributions will generally result as a consequence of the added bilinear term in the quadrilateral basis. A verification will result for the following Code Exercise.

**Code Exercise 6.6.** Determine the temperature distribution in the thick axisymmetric cylinder for the azimuthally varying convection and fixed-temperature distributions defined by the following file.

```
TITL
*** CODE EXERCISE 6.6, M=6x6 CYLINDER, VARIABLE H AND TREF ***
TYPE    [K         N        NNODEL     REFL   NPRN    NTRAN     NAXI]
         1         2          4         1.0    1 1      3

PRIN    [NBUG(*)]
         1         0          0          2

GRID    [FOR N (XL XR(*)PR(*)), NEM(PR*)]
                   1.0       2.0       1.0
         6
                   0.0       1.0       1.0
         6

                   1.0      2.0       0.0       0.0
                   0.0      0.0       2.0       1.0
                   1.5      1.4142    0.0       0.7071
                   0.0      1.4142    1.5       0.7071

MATL     [FOR MACRO(*),K(*)]
     1        10.0
```

```
ROBN     [NROB,NODE(*),(H(*)TR(*))]
        7      0      1
        1      2      3      4      5      6      7
                    40.0                  750.0
                    40.0                  750.0
                    35.0                  625.0
                    30.0                  500.0
                    25.0                  375.0
                    20.0                  250.0
                    20.0                  250.0
DIRI     [NDIR, NODE(*),TB(*)]
        7
       43     44     45     46     47     48     49
     300.0         300.0         400.0         500.0         600.0
     700.0         700.0
FORM
SOLV
STOP
```

Prepare and execute this data file; the bilinear basis nodal solution distribution you should obtain is

```
SOLV: NODAL SOLUTION (INTEGER MAP) EXPONENT E:0
        300     300     400     500     600     700     700
        381     400     442     502     564     606     624
        445     453     476     506     536     559     567
        491     493     500     509     519     526     529
        531     528     521     512     504     498     496
        577     568     544     513     485     467     460
        641     627     572     510     457     422     412
```

Now modify the file to execute the linear triangle basis algorithm, then compare the resultant nodal temperature solution to the above. Proceed with data and geometry variations that you deem appropriate to complete the familiarization.

## 6.6 AXISYMMETRIC CONDUCTION WITH BOUNDARY CONVECTION

The generalized axisymmetric diffusion problem statement was given in Chap. 5 as

$$L(q) = -\nabla \cdot k(x_i)\nabla q - s(x_i)$$

$$= -\frac{1}{r}\frac{\partial}{\partial r}\left(kr\frac{\partial q}{\partial r}\right) - \frac{\partial}{\partial z}\left(k\frac{\partial q}{\partial z}\right) - s(r, z) = 0 \qquad \text{on } \Omega \subset \mathbb{R}^2 \quad (5.88)$$

$$l(q) = k\nabla q \cdot \mathbf{n} + aq - b = 0 \qquad \text{on } \partial\Omega_1 \quad (5.89)$$

$$q(\mathbf{x}_b) = q_b(\mathbf{x}_b) \qquad \text{on } \partial\Omega_2 \quad (5.90)$$

where $\nabla$ in the $r$, $z$ coordinate system is expanded for clarity, and $a$ and $b$ in (5.89) unify the applicable boundary conditions. The finite element approximation to the associated weak statement remains (5.92):

$$WS^h = S_e\left(\int_{\Omega_e} \nabla\{N_k\}\cdot k_e\nabla\{N_k\}^T\,d\tau\{Q\}_e - \int_{\Omega_e}\{N_k\}s_e\,d\tau\right.$$
$$\left. + \int_{\partial\Omega_1}\{N_k\}(a_eq_e - b_e)\,d\sigma - \int_{\partial\Omega_2}\{N_k\}k_e\,\nabla q_e\cdot\mathbf{n}\,d\sigma\right) = \{0\} \quad (5.92)$$

The element domain differential element for (5.92) is $d\tau = r\,dr\,dz$; hence the conduction matrix for the bilinear basis selection becomes

$$[K]_e = \int_{\Omega_e}\nabla\{N_1^+\}\cdot k_e\nabla\{N_1^+\}^T r_e\,dr\,dz$$
$$= \{K\}_e^T\int_{\Omega_e}\{N_1^+\}\nabla\{N_1^+\}\cdot\nabla\{N_1^+\}^T\{N_1^+\}^T\,\det{}_e\,d\eta_1\,d\eta_2\{R\}_e \quad (6.55)$$

on interpolation of $k_e$ and $r_e$ using $\{N_1^+\}$ and the appropriate nodal arrays. The integrand in (6.55) is now rather more complicated, especially so for the residual $\eta_i$ dependence in the transformation of $\nabla$ to the local coordinate system. The highest degree for the resulting polynomials in (6.55) is biquadratic; hence the $P = 3 = Q$ gaussian quadrature rule (6.37) is now required. Equation (6.42) expresses the associated tightly nested DO loop for the representative element in $[K]_e$, on replacement of the $(p, q)$ summation limit of 2 by 3 and removal of $\bar{k}_e/16$ as the premultiplier.

For any reasonable mesh refinement, this gaussian quadrature DO loop is quite demanding in the PC environment. One practical resolution remains *commission* of interpolation error, as an extension on that introduced in (6.47). Electing to use element centroidal values for both conduction and element radius, (6.55) is approximated as

$$[K]_e \simeq \bar{k}_e\bar{r}_e DET_e^{-1}ETAIJ_e ETAKJ_e[B2IKB] \quad (6.56a)$$

Conversely, if centroidal evaluation of either variable is considered too much of an approximation, an improved-accuracy approximate form would be

$$[K]_e \simeq \bar{k}_e DET_e^{-1}ETAIJ_e ETAKJ_e\{R\}_e^T[B30IKB] \quad (6.56b)$$

The nodal distribution of element radius (or conductivity) is thus utilized with the bilinear basis, degree-1 hypermatrix $[B30IKB]$, a set of $4 \times 4$ square matrices with elements that are $4 \times 1$ column matrices. The respective element library additions are contained in *LEARN.FE* for $1 \leq (I, K) \leq 2$; as an example,

$$[B3012B] = \frac{1}{36}\begin{bmatrix} \begin{Bmatrix}4\\2\\1\\2\end{Bmatrix} & \begin{Bmatrix}2\\4\\2\\1\end{Bmatrix} & \begin{Bmatrix}-2\\-4\\-2\\-1\end{Bmatrix} & \begin{Bmatrix}-4\\-2\\-1\\-2\end{Bmatrix} \\ \begin{Bmatrix}-4\\-2\\-1\\-2\end{Bmatrix} & \begin{Bmatrix}-2\\-4\\-2\\-1\end{Bmatrix} & \begin{Bmatrix}2\\4\\2\\1\end{Bmatrix} & \begin{Bmatrix}4\\2\\1\\2\end{Bmatrix} \\ \begin{Bmatrix}-2\\-1\\-2\\-4\end{Bmatrix} & \begin{Bmatrix}-1\\-2\\-4\\-2\end{Bmatrix} & \begin{Bmatrix}1\\2\\4\\2\end{Bmatrix} & \begin{Bmatrix}2\\1\\2\\4\end{Bmatrix} \\ \begin{Bmatrix}2\\1\\2\\4\end{Bmatrix} & \begin{Bmatrix}1\\2\\4\\2\end{Bmatrix} & \begin{Bmatrix}-1\\-2\\-4\\-2\end{Bmatrix} & \begin{Bmatrix}-2\\-1\\-2\\-4\end{Bmatrix} \end{bmatrix} \tag{6.57}$$

Obviously, the role of $\bar{k}_e$ and $\{R\}_e$ can be interchanged in (6.56*b*), i.e., $\bar{r}_e$ and $\{K\}_e$.

The remaining domain integral to implement (5.92) is the source term. The form with minor interpolation error is

$$\begin{aligned}\int_{\Omega_e} \{N_1^+\} s_e \, d\tau &= \int_{\Omega_e} \{N_1^+\}\{N_1^+\}^T \{S\}_e r_e \det_e d\eta_1 \, d\eta_2 \\ &\simeq DET_e \{R\}_e^T \int_{\Omega_e} \{N_1^+\}\{N_1^+\}\{N_1^+\}^T d\eta_1 \, d\eta_2 \{S\}_e \\ &\equiv DET_e \{R\}_e^T [B3000B]\{S\}_e \end{aligned} \tag{6.58}$$

The corresponding master matrix library addition is

$$
[B3000B] = \frac{1}{36}\begin{bmatrix} \begin{Bmatrix} 9 \\ 3 \\ 1 \\ 3 \end{Bmatrix} & \begin{Bmatrix} 3 \\ 3 \\ 1 \\ 1 \end{Bmatrix} & \begin{Bmatrix} 1 \\ 1 \\ 1 \\ 1 \end{Bmatrix} & \begin{Bmatrix} 3 \\ 1 \\ 1 \\ 3 \end{Bmatrix} \\ \begin{Bmatrix} 3 \\ 3 \\ 1 \\ 1 \end{Bmatrix} & \begin{Bmatrix} 3 \\ 9 \\ 3 \\ 1 \end{Bmatrix} & \begin{Bmatrix} 1 \\ 3 \\ 3 \\ 1 \end{Bmatrix} & \begin{Bmatrix} 1 \\ 1 \\ 1 \\ 1 \end{Bmatrix} \\ \begin{Bmatrix} 1 \\ 1 \\ 1 \\ 1 \end{Bmatrix} & \begin{Bmatrix} 1 \\ 3 \\ 3 \\ 1 \end{Bmatrix} & \begin{Bmatrix} 1 \\ 3 \\ 9 \\ 3 \end{Bmatrix} & \begin{Bmatrix} 1 \\ 1 \\ 3 \\ 3 \end{Bmatrix} \\ \begin{Bmatrix} 3 \\ 1 \\ 1 \\ 3 \end{Bmatrix} & \begin{Bmatrix} 1 \\ 1 \\ 1 \\ 1 \end{Bmatrix} & \begin{Bmatrix} 1 \\ 1 \\ 3 \\ 3 \end{Bmatrix} & \begin{Bmatrix} 3 \\ 1 \\ 3 \\ 9 \end{Bmatrix} \end{bmatrix} \tag{6.59}
$$

which is also a degree-1 $4 \times 4$ hypermatrix. It is of interest to compare (6.59) with the triangle linear basis expression (5.94) and hence note the structural uniformity.

The axisymmetric algorithm formulation steps that remain involve various convection boundary conditions. You realize that this implementation is independent of the triangle-versus-quadrilateral basis choice; hence, the Chap. 5 material [(5.95) to (5.98), (5.132) and (5.133)] provides all needed definitions. For completeness, then, the metric-simplified bilinear basis algorithm statement (5.92) in Fortran library notation is

$$
\begin{aligned}
WS^h &= S_e([K + H]_e\{Q\}_e - \{b\}_e) \\
&= S_e(\bar{k}_e DET_e^{-1} ETAIJ_e ETAKJ_e \{R\}_e^T [B30IKB]\{Q\}_e \\
&\quad - DET_e\{R\}_e^T[B3000B]\{S\}_e \\
&\quad + l_e(\{A\}_e^T[A40000L]\{R\}_e)\{Q\}_e - l_e\{B\}_e^T[A3000L]\{R\}_e) \\
&= \{0\}
\end{aligned} \tag{6.60}
$$

The data arrays $\{A\}_e$ and $\{B\}_e$ contain nodal values of the interpolation of boundary coefficients $a(x_i)$ and $b(x_i)$ in (5.89). The implementation of (6.60) in *LEARN.FE* is a straightforward procedure left to the reader.

## 6.7 SUMMARY

This chapter has extended the developments of Chap. 5 for use of the generalized quadrilateral-shaped finite element domain $\Omega_e$ and its associated family of finite element bases $\{N_k^+(\eta_j)\}$. This involved few new ideas and concepts except for formalization of the local coordinate transformation $(x_i)_e = x_i(\eta_j)$, and introduction of numerical quadrature methods for evaluation of the integrals forming element matrices. The *isoparametric* element and the associated transformation embeds considerable geometrical versatility to the quadrilateral element discretization. This is utilized directly in *LEARN.FE* to establish the automated grid-generation capability.

CHAPTER
# 7

# THREE-DIMENSIONAL ELEMENT FAMILIES

## 7.1 OVERVIEW

Chapters 5 and 6 developed implementation of finite element bases, hence finite element domains $\Omega_e$, for two-dimensional problem statements. The high regularity of the methodology is quite evident, yielding a library of element matrices for use in *any* problem definition, including boundary conditions. This recipe for construction of a finite element method for elliptic problems is certainly extendible to three dimensions. Furthermore, the end product must be the global matrix statement

$$[K + H]\{Q\} = \{b\} \tag{7.1}$$

and equivalently,

$$S_e([K + H]_e\{Q\}_e - \{b\}_e) = \{0\} \tag{7.2}$$

where $S_e$ is the assembly operation for three-dimensional element matrices, and hence two-dimensional assembly on the boundaries.

Establishing the finite element weak statement approximation for three-dimensional problems is thus direct. It is completed in this brief chapter for the $n = 3$ extension of the $n = 2$ triangle and quadrilateral element families. The subsequent implementation into coding is not so elementary, mainly as the result of a substantial increase in database handling, and hence input/output facilities. Furthermore, and of greater current pertinence, the global matrix $[K + H]$ in (7.1) becomes very large for a useful discretization $\Omega^h$ for $n = 3$, hence, present-day PC capabilities are quite inadequate for meaningful problem statements. Granted, matrix iterative techniques exist that never form or handle $[K + H]$. However, when compared with a current PC clock cycle, the elapsed time needed to converge to the solution makes such procedures quite ponderous.

You, and the authors, appreciate these comments are dated, since the incredible advances in VLSI technology will put a PC/mini-CRAY on your desktop within a few years. Deferring to this eventuality, we herein develop the theory and library matrix nomenclature in completeness for the $n = 3$ problem statement. However, the current *LEARN.FE* code is devoid of the implementation, to avoid an unnecessary cluttering.

## 7.2 FINITE ELEMENT DISCRETIZATIONS FOR $n = 3$

The preceding chapters start with a Methodology section, wherein finite element discretization techniques are discussed, including code exercises. In the absence of such exercises, we seek only to illustrate suitable meshes as created by unions (the nonoverlapping sum) of suitable finite element domain shapes. The latitude for useful element geometries increases for $n = 3$, in comparison to the two available (triangle and quadrilateral) for $n = 2$. However, the generally useful $\Omega_e$ are

1. The tetrahedron, as the $n = 3$ extension of the triangle.
2. The hexahedron, the $n = 3$ extension of the quadrilateral.
3. The pentahedron, an $n = 3$ triangular cross-section prism.

Any of these elements may possess a mixture of planar and nonplanar bounding surface segments, with the latter created using direct extension of the isoparametric coordinate transformation methodology developed in Chap. 6.

Figure 7.1*a* illustrates a basic three-dimensional domain $\Omega$ discretized into a union of hexahedral elements $\Omega_e$, any of which may be curve-faced to some extent. In turn, each (any) hexahedron $\Omega_e$ can be resolved into a union of five (5) tetrahedron elements as illustrated in Fig. 7.1*b*. The removal of any one of these, to match a sharply sloping local boundary segment, for example, then yields a truncated hexahedron that has five boundary surface segments. Many other variations exist, as you may well imagine; the illustrations in Fig. 7.1 serve our purpose.

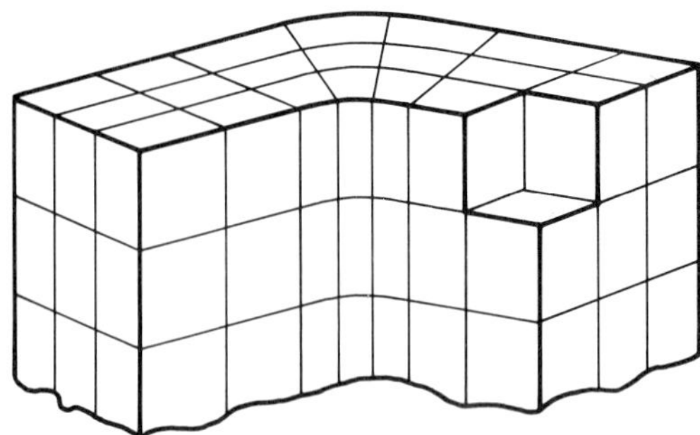

($a$) $\Omega^h$ on $\mathbb{R}^3$ formed using hexahedra

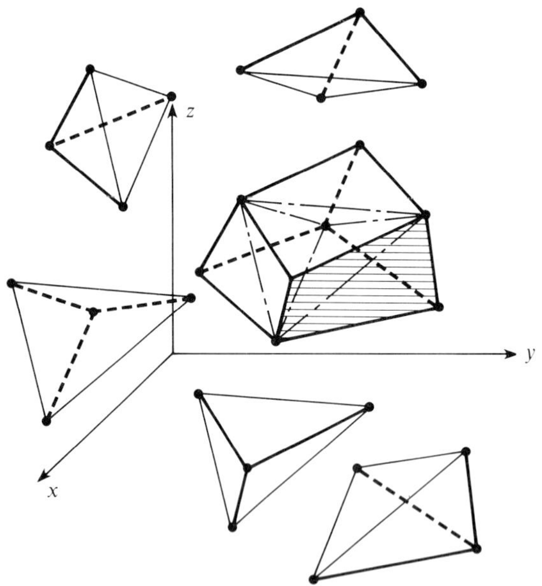


($b$) Hexahedron resolution into five tetrahedra

**FIGURE 7.1**
Illustration of candidate discretizations and finite element shapes in three dimensions. [*From Zienkiewicz and Taylor (1989, pp. 95–96).*]

## 7.3 THREE-DIMENSIONAL TETRAHEDRON BASES

The choice of approximate solution *trial space* $\phi_i(x_j)$, $1 \le i \le N$ and $1 \le j \le n$, for all $1 \le n \le 3$, remains *fundamental*, and it is always the code designer's choice. The $n$-dimensional generalization for *any* approximation is

$$T^N(x_j) \equiv \sum_{i=1}^{N} a_i \phi_i(x_j) \qquad \text{for } 1 \le j \le n \tag{7.3}$$

which contains the previous statements (6.3), (5.5), and (3.6). The finite element procedure "assembles" (7.3) as the union over the elements $\Omega_e$ of the discretization $\Omega^h$ of the finite element approximation

$$T_e(x_j) \equiv \{N_k(\cdot)\}^T\{Q\}_e \tag{7.4}$$

As before, when a mesh is introduced, the approximation $T^N$ is denoted $T^h$, where $h$ signifies a *measure* of $\Omega^h$. The finite element approximate solution definition is then

$$T(x_j) \simeq T^h(x_j) \equiv \bigcup_e T_e(x_j) \tag{7.5}$$

and $\bigcup$ denotes union, the nonoverlapping sum.

As illustrated numerous times, the key step to implementation is to establish the functional form of the finite element basis $\{N_k(\cdot)\}$ in (7.4), which is constituted of $k$th degree polynomials written on the element intrinsic coordinate system. This step is now completed for the generally useful three-dimensional finite element domain geometries.

## Linear Basis

The tetrahedron with four vertex nodes is the three-dimensional extension of the triangle families developed in Chap. 5. The linearly dependent *natural coordinate* system $\zeta_i$ generalizes directly; hence (7.4) for $k = 1$ becomes

$$T_e(x_j) \equiv \{N_1(\zeta_i)\}^T\{Q\}_e \tag{7.6}$$

and,

$$\{N_1(\zeta_i)\} \equiv \begin{Bmatrix} \zeta_1 \\ \zeta_2 \\ \zeta_3 \\ \zeta_4 \end{Bmatrix} \tag{7.7}$$

is the (obvious) extension to (4.6) and (5.8).

Establishing the natural coordinate system $\zeta_i$, $1 \le i \le n + 1 = 4$, is the direct expansion on the procedure presented in Section 5.3, Linear Basis, involving the element vertex node coordinates $\{X_i, Y_i, Z_i\}_e$. Figure 7.2 illustrates the generic tetrahedron, and the following equations define

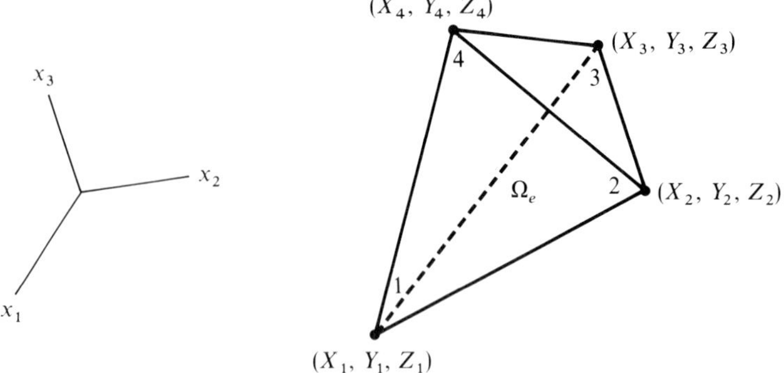


**FIGURE 7.2**
Generic linear basis tetrahedron element $\Omega_e$.

the elements of the linear basis (7.7)

$$\{N_1(\zeta_i)\} \equiv \begin{Bmatrix} \zeta_1 \\ \zeta_2 \\ \zeta_3 \\ \zeta_4 \end{Bmatrix} \equiv \frac{1}{6V_e} \begin{Bmatrix} a_1 + b_1 x + c_1 y + d_1 z \\ -(a_2 + b_2 x + c_2 y + d_2 z) \\ a_3 + b_3 x + c_3 y + d_3 z \\ -(a_4 + b_4 x + c_4 y + d_4 z) \end{Bmatrix} \tag{7.8}$$

The coefficient sets $a_i$, $b_i$, $c_i$, and $d_i$, for $1 \leq i \leq 4$, and the volume $V_e$ of the tetrahedron are determined as

$$V_e \equiv \frac{1}{6} \det \begin{bmatrix} 1 & X_1 & Y_1 & Z_1 \\ 1 & X_2 & Y_2 & Z_2 \\ 1 & X_3 & Y_3 & Z_3 \\ 1 & X_4 & Y_4 & Z_4 \end{bmatrix}_e \tag{7.9}$$

$$a_i \equiv \det \begin{bmatrix} X_j & Y_j & Z_j \\ X_m & Y_m & Z_m \\ X_p & Y_p & Z_p \end{bmatrix}_e \tag{7.10a}$$

$$b_i \equiv -\det \begin{bmatrix} 1 & Y_j & Z_j \\ 1 & Y_m & Z_m \\ 1 & Y_p & Z_p \end{bmatrix}_e \tag{7.10b}$$

$$c_i \equiv \det \begin{bmatrix} X_j & 1 & Z_j \\ X_m & 1 & Z_m \\ X_p & 1 & Z_p \end{bmatrix}_e \tag{7.10c}$$

$$d_i \equiv -\det \begin{bmatrix} X_j & Y_j & 1 \\ X_m & Y_m & 1 \\ X_p & Y_p & 1 \end{bmatrix}_e \tag{7.10d}$$

In definitions (7.10), the node coordinate index set $(i, j, m, p)$ permutes as (1, 2, 3, 4), and hence establishes all coefficients needed in (7.8), while det denotes the determinant operation. Thus, construction of the $\zeta_i$ coordinate system for $n = 3$ is straightforward as a nested DO loop with the element-node coordinate data base. You should compare (7.8) to (7.10) with (5.9) to (5.13) to verify the hierarchical structure as a function of $n$.

### Quadratic Basis

Having established the tetrahedron $\zeta_i$ coordinate system, construction of the $k = 2$ basis is the direct extension on the one- and two-dimensional forms. Recalling (4.14) and (5.15), the $n = 3$ quadratic finite element basis is

$$\{N_2(\zeta_i)\} = \begin{Bmatrix} \zeta_1(2\zeta_1 - 1) \\ \zeta_2(2\zeta_2 - 1) \\ \zeta_3(2\zeta_3 - 1) \\ \zeta_4(2\zeta_4 - 1) \\ 4\zeta_1\zeta_2 \\ 4\zeta_2\zeta_3 \\ 4\zeta_3\zeta_1 \\ 4\zeta_1\zeta_4 \\ 4\zeta_2\zeta_4 \\ 4\zeta_3\zeta_4 \end{Bmatrix} \tag{7.11}$$

Thereby, the planar-faced quadratic tetrahedron contains 4 vertex nodes and 6 mid-edge nodes which are conventionally ordered as shown in Fig. 7.3.

## 7.4 THREE-DIMENSIONAL HEXAHEDRON BASES

Following the notation of Chap. 6, finite element bases spanning the family of three-dimensional hexahedron element domains are denoted $\{N_k^+(\eta_i)\}$. The element approximation is

$$T_e(x_j) = \{N_k^+(\eta_i)\}^T\{Q\}_e \tag{7.12}$$

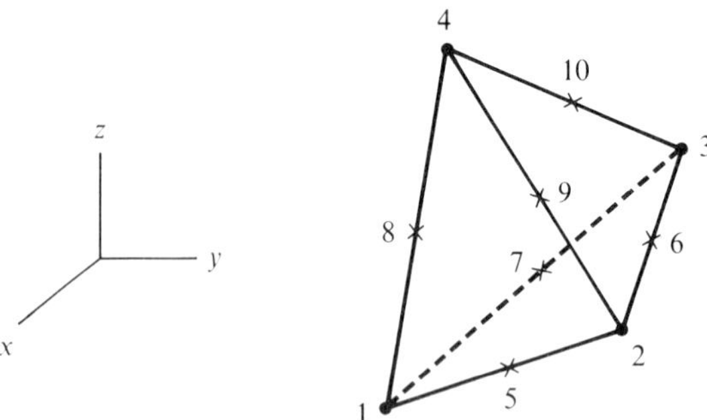


**FIGURE 7.3**
Quadratic basis planar-faced tetrahedron $\Omega_e$.

The requirement is to determine the entries in $\{N_k^+(\eta_i)\}$ as functions of the tensor product coordinate system $\eta_i$, $1 \le i \le n$.

### Trilinear Basis

The specification $k = 1$ in (7.12) denotes the trilinear basis. Figure 7.4 illustrates the appropriate planar-faced element domains. By inspection of (6.7), the corresponding $n = 3$ basis must be

$$\{N_1^+(\eta_i)\} = \frac{1}{8}\begin{Bmatrix} (1-\eta_1)(1-\eta_2)(1-\eta_3) \\ (1+\eta_1)(1-\eta_2)(1-\eta_3) \\ (1+\eta_1)(1+\eta_2)(1-\eta_3) \\ (1-\eta_1)(1+\eta_2)(1-\eta_3) \\ (1-\eta_1)(1-\eta_2)(1+\eta_3) \\ (1+\eta_1)(1-\eta_2)(1+\eta_3) \\ (1+\eta_1)(1+\eta_2)(1+\eta_3) \\ (1-\eta_1)(1+\eta_2)(1+\eta_3) \end{Bmatrix} \tag{7.13}$$

for the given node numbering sequence. Specifically, the nodal plane (1, 2, 3, 4) is spanned by $(\eta_1, \eta_2)$ and $\eta_3 = -1$, and the $\eta_i$ origin is at the geometric centroid of $\Omega_e$.

If $\Omega_e$ in Fig. 7.4*a* were rotated such that face boundary normals were all parallel to $x_i$, then the $\eta_i$ system would constitute a translation of origin only. Conversely, for $\Omega_e$ oriented arbitrarily, the coordinate transformation $\eta_i = \eta_i(x_j)$ is nontrivial and must be determined. Recalling the pertinent discussion in Section 6.3, Bilinear Basis, the inverse coordinate transformation $x_j = x_j(\eta_i)$ on $\Omega_e$ is

$$\begin{aligned} x_e &= \{N_1^+(\eta_i)\}^T\{X\}_e \\ y_e &= \{N_1^+(\eta_i)\}^T\{Y\}_e \\ z_e &= \{N_1^+(\eta_i)\}^T\{Z\}_e \end{aligned} \tag{7.14}$$

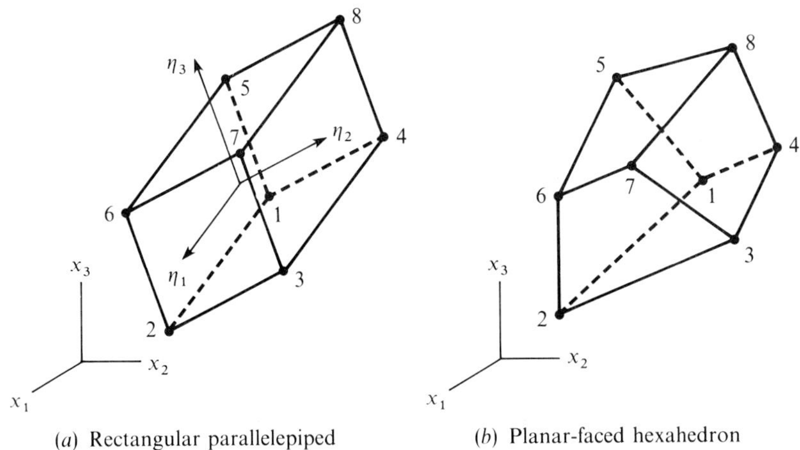


(*a*) Rectangular parallelepiped (*b*) Planar-faced hexahedron

**FIGURE 7.4**
Illustration of planar-faced trilinear basis element $\Omega_e$.

i.e.,

$$(x_j)_e = \{N_1^+(\eta_i)\}^T\{XJ\}_e \qquad \text{for } 1 \le (i, j = J) \le n \tag{7.15}$$

The element data arrays in (7.14) and (7.15) contain the global coordinate triples $(X, Y, Z)_e$ of the vertex nodes of $\Omega_e$, and (7.13) defines the basis for $n = 3$ while (6.7) is the definition for $n = 2$. Hence, all the element metric data handling procedures derived in Chap. 6 extend immediately to $n = 3$.

## Triquadratic Bases

As occurred for $n = 2$, there are several hexahedron element bases for $k = 2$ in (7.12), and all are appropriate for elements $\Omega_e$ with nonplanar boundary surfaces. Figure 7.5 illustrates the generic element domain wherein vertex nodes are denoted by (●), nonvertex edge nodes as (×), surface nodes as (○), and the centroidal node (△).

One serendipity $n = 3$ element is defined by omitting the six surface nodes (○) and the centroidal node (△) at the $\eta_i$ origin. The corresponding basis $\{N_2^+(\eta_i)\}$ has 20 entries for 8 vertex nodes and 12 nonvertex edge nodes. A second serendipity basis excludes only the centroidal node, and hence has 26 entries, including the 6 midsurface nodes. The complete (Lagrange) $k = 2$ basis $\{N_2^*(\eta_i)\}$ has 27 entries on addition of the centroidal node. The definitions for these bases are direct extensions of the $n = 2$ formulations given in (6.15) and (6.16), and are available in the literature (Zienkiewicz and Taylor, 1989).

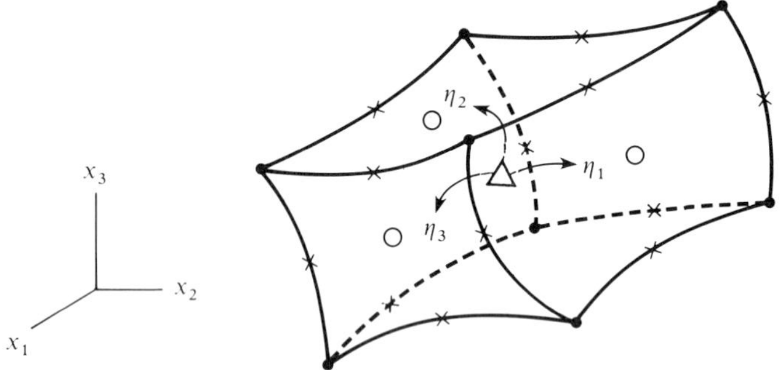


**FIGURE 7.5**
Illustration of $n = 3$ triquadratic basis element $\Omega_e$.

## 7.5 STEADY HEAT CONDUCTION WITH BOUNDARY CONVECTION

The candidate problem statement definition is the following modest extension of (5.82) to (5.85).

$$L(T) = -\nabla \cdot k(x_i)\nabla T - s(x_i) = 0 \qquad \text{on } \Omega \subset \mathbb{R}^3 \qquad (7.16)$$

$$l(T) = k\nabla T \cdot \mathbf{n} + h(T - T_r) = 0 \qquad \text{on } \partial\Omega_1 \qquad (7.17)$$

$$l(T) = k\nabla T \cdot \mathbf{n} + q_n = 0 \qquad \text{on } \partial\Omega_2 \qquad (7.18)$$

$$T(\mathbf{x}_b) = T_b \qquad \text{on } \partial\Omega_3 \qquad (7.19)$$

The finite element approximation to the weak statement (5.20) for (7.16) to (7.19) remains absolutely unchanged from (5.86). Specifically

$$\begin{aligned} WS^h = S_e\Bigg( & \int_{\Omega_e} \nabla\{N_k\} \cdot k_e \nabla\{N_k\}_e^T \, d\tau \{Q\}_e \\ & - \int_{\Omega_e} \{N_k\} s_e \, d\tau + \int_{\partial\Omega_1} \{N_k\} h_e \{N_k\}^T \{Q - TR\}_e \, d\sigma \\ & + \int_{\partial\Omega_2} \{N_k\} q_{n,e} \, d\sigma - \int_{\partial\Omega_3} \{N_k\} k_e \nabla T_e \cdot \mathbf{n} \, d\sigma \Bigg) \\ = {} & S_e([K + H]_e \{Q\}_e - \{SC\}_e - [H]_e \{TR\}_e + \{QF\}_e - \{FF\}_e) \\ = {} & S_e([K + H]_e \{Q\}_e - \{b\}_e) = \{0\} \end{aligned} \qquad (7.20)$$

In (7.20), recall that for any $\Omega_e$ of the discretization $\Omega^h$, $[K + H]_e$ is the element conduction plus boundary convection matrix, $\{SC\}_e$ results for the distributed source, $[H]_e\{TR\}_e$ is the boundary convection source,

$\{QF\}_e$ is the boundary fixed flux source, and the boundary flux term $\{FF\}_e$ on $\partial\Omega_3$ does not enter the global matrix statement due to Dirichlet data imposition.

All ingredients of the developed matrix terminology are extendible to the $n = 3$ weak statement approximation (7.20). For the trilinear basis, wherein use of element-averaged metric (and thermal) data is appropriate, the resultant Fortran master matrix statement is

$$\begin{aligned} WS^h = S_e(\bar{k}_e DET_e^{-1} ETAIJ_e ETAKJ_e[C2IKL]\{Q\}_e \\ + \bar{h}_e DET2_e[B200B]\{Q\}_e - DET_e[C200L]\{S\}_e \\ - \bar{h}_e DET2_e[B200B]\{TR\}_e - DET2_e[B200B]\{QN\}_e) \\ = \{0\} \qquad \text{for } 1 \leq (I, J, K) \leq n = 3 \end{aligned} \tag{7.21}$$

In (7.21), the master matrix prefix $C$ denotes the $n = 3$ dimensional form, and all boundary condition applications now involve the $n = 2$ bilinear basis forms, as signified by the $B$ prefix. Further, to distinguish between the various element average measures, the notation $DET2_e$ refers to the boundary condition (surface) determination [recall the appearance of $l_e$ in the $n = 2$ boundary statement (6.54)]. Finally, the computation of the metric data arrays $ETAIJ_e$ and $DET_e$ is the direct extension of the $n = 2$ definitions given in (6.44) and (6.45).

The tetrahedron basis form for (7.21) simply replaces $ETAIJ_e$ with $ZETAIJ_e$ and $DET_e$ and $DET2_e$ with constants times $V_e$ and $A_e$, respectively.

## 7.6 SUMMARY

This chapter has briefly presented the essential ingredients of the finite element weak statement algorithm extension to the three-dimensional problem class. Developing the theoretical issues required only a modest effort. The practical issue of implementation into a viable code in the PC environment is a much larger challenge being worked on.

## 7.7 ADDITIONAL REFERENCE

Zienkiewicz, O. C. and R. L. Taylor (1989), *The Finite Element Method*, Vol. 1, 4th ed. McGraw-Hill, London.

# CHAPTER 8

# UNSTEADY TRANSPORT WITH FLUID MOTION

## 8.1 OVERVIEW

Attention has been focused on developing the finite element weak statement approximation for the linear partial differential equation governing steady-state diffusion with various flux and fixed boundary condition combinations. In engineering and mathematical physics, differential statements of this type have become termed a *conservation law*. Our focus has been on conservation of thermal energy, specifically the first law of thermodynamics with boundary conditions.

The developed finite element weak statement solution procedure is directly extendible to full forms of conservation law systems, such as the nonlinear Navier-Stokes equations governing unsteady flow of a viscous and heat-conducting fluid. To bring this text to a logical completion, and to introduce the reader to the next level of pertinent issues, we now extend the weak statement concept, and develop its finite element approximation, to the conservation law statement governing time-dependent transport generalized for fluid convection (motion) and diffusion processes operating simultaneously.

## A. ONE-DIMENSIONAL PROBLEMS

### 8.2 ALGORITHM STATEMENT FOR CONVECTION-DIFFUSION

We first extend the problem statement to the unsteady energy equation including an imposed (one-dimensional) fluid velocity field $\mathbf{u} = u\mathbf{i}$. The solution domain remains the portion of the $x$ axis $0 < x < L$. For a fluid with uniform density $\rho$ and constant specific heat $c_p$, and dividing through by their product, the one-dimensional unsteady energy *conservation law* is

$$L(T) = \frac{\partial T}{\partial t} + u\frac{\partial T}{\partial x} - \frac{\partial}{\partial x}\left(\frac{k}{\rho c_p}\frac{\partial T}{\partial x}\right) - \frac{s}{\rho c_p} = 0 \qquad \text{on } \Omega \tag{8.1}$$

$$l(T) = k\frac{dT}{dn} + h(T - T_r) = 0 \qquad \text{on } \partial\Omega_1 \tag{8.2}$$

$$T(x_b) = T_b \qquad \text{on } \partial\Omega_2 \tag{8.3}$$

Note that (8.1) contains partial derivatives since $T = T(x, t)$ is a function of both space and time. In addition, a first-order spatial derivative term now appears which is multiplied by the fluid velocity component $u = u(x, t)$, which is assumed specified data. Finally, since (8.1) defines an initial-value problem in time, an initial condition specification is needed to close the statement. Hence, assume that

$$T(x, t = 0) \equiv T_o(x) \tag{8.4}$$

is given and defined on both the domain $\Omega$ and the boundaries $\partial\Omega_1 \bigcup \partial\Omega_2 = \partial\Omega$. As a final comment, for conductivity $k$ in (8.1) a uniform constant, then $k/\rho c_p \equiv \alpha$ defines the *thermal diffusivity*.

Development of the finite element approximation to a weak statement for (8.1) to (8.3) follows exactly the developed recipe. The statement of a (any) approximation is

$$T^N(x, t) \equiv \sum_{i=1}^{N} a_i(t)\phi_i(x) \tag{8.5}$$

where the expansion coefficient set $a_i$, $1 \le i \le N$, is assumed to carry the required time dependence. This choice ensures that all finite element bases developed thus far are immediately applicable to the unsteady problem, and the specific selection remains a fundamental choice. The symmetric Galerkin weak statement for (8.1) to (8.3) is

$$
\begin{aligned}
WS &\equiv \int_\Omega \phi_i L(T^N)\, d\tau = 0 \qquad \text{for } 1 \le i \le N \\
&= \int_\Omega \phi_i \left[ \frac{\partial T^N}{\partial t} + u \frac{\partial T^N}{\partial x} - \frac{\partial}{\partial x}\left( \frac{k}{\rho c_p} \frac{\partial T^N}{\partial x} \right) - \frac{s}{\rho c_p} \right] d\tau \\
&= \int_\Omega \left[ \phi_i \left( \frac{\partial T^N}{\partial t} + u \frac{\partial T^N}{\partial x} \right) + \frac{d\phi_i}{dx} \frac{k}{\rho c_p} \frac{\partial T^N}{\partial x} - \phi_i \frac{s}{\rho c_p} \right] d\tau \\
&\quad - \int_{\partial\Omega} \phi_i \frac{k}{\rho c_p} \frac{\partial T^N}{\partial n}\, d\sigma \\
&= \int_\Omega \left[ \phi_i \left( \frac{\partial T^N}{\partial t} + u \frac{\partial T^N}{\partial x} - \frac{s}{\rho c_p} \right) + \frac{d\phi_i}{dx} \frac{k}{\rho c_p} \frac{\partial T^N}{\partial x} \right] d\tau \\
&\quad + \int_{\partial\Omega_1} \phi_i \frac{h}{\rho c_p} (T^N - T_r)\, d\sigma - \int_{\partial\Omega_2} \phi_i \frac{k}{\rho c_p} \frac{\partial T^N}{\partial n}\, d\sigma = 0 \qquad (8.6)
\end{aligned}
$$

Observe in (8.6) that the thermal convection boundary condition substitution requires $h/\rho c_p$ as the multiplier but is otherwise unaltered from the steady-state definition. As in Chap. 4, the differential elements $d\tau$ and $d\sigma$ remain variables to become assigned as appropriate for one-dimensional or axisymmetric problem statements.

The finite element approximate evaluation of (8.6) employs a discretization $\Omega^h$ of $\Omega$ and a $k$th-degree basis on $\Omega_e$. The nodal values of the approximate solution thus become the *time-dependent* expansion coefficients. Hence, the unsteady generalization defined on the union of the $\Omega_e$ constituting $\Omega^h$ is

$$T^N(x, t) \to T^h(x, t) \equiv \bigcup_e T_e(x, t) \qquad (8.7)$$

and

$$T_e(x, t) \equiv \{N_k(\zeta_i)\}^T \{Q(t)\}_e \qquad (8.8)$$

The first term in (8.6) requires the partial derivative on time. From

(8.7) and (8.8) we obtain

$$\frac{\partial T^N}{\partial t} \to \frac{\partial T_e}{\partial t} = \frac{\partial}{\partial t}\{N_k(\zeta_i)\}^T\{Q(t)\}_e$$

$$= \{N_k(\zeta_i)\}^T \frac{d}{dt}\{Q\}_e \tag{8.9}$$

since the elements of $\{N_k\}$ are time-independent. Thus, the time partial derivative becomes the ordinary derivative on the entries in $\{Q\}_e$ which indeed are a function only of time. Denoting $d\{Q\}_e/dt$ as $\{Q\}'_e$, the common notation for an ordinary derivative, the finite element discrete approximation to (8.6) can be precisely written as

$$\begin{aligned} WS^h = S_e\Bigg(&\int_{\Omega_e} \{N_k\}\{N_k\}^T\, d\tau\{Q\}'_e + \int_{\Omega_e} \{N_k\}u_e \frac{d\{N_k\}^T}{dx}\, d\tau\{Q\}_e \\ &+ \int_{\Omega_e} \frac{d\{N_k\}}{dx}\left(\frac{k}{\rho c_p}\right)_e \frac{d\{N_k\}^T}{dx}\, d\tau\{Q\}_e - \int_{\Omega_e} \{N_k\}\{N_k\}^T\, d\tau\{S\}_e \\ &+ \int_{\partial\Omega_e \cap \partial\Omega_1} \{N_k\}\frac{h}{\rho c_p}\{N_k\}^T\, d\tau\{Q - T_r\}_e \\ &- \int_{\partial\Omega_e \cap \partial\Omega_2} \{N_k\}\left(\frac{k}{\rho c_p}\right)_e \frac{\partial T^h}{\partial n}\, d\sigma\Bigg) = \{0\} \end{aligned} \tag{8.10}$$

where $\{S\}_e$ contains the nodal values of $s/\rho c_p$.

The notation in (8.10) is quite familiar. Certainly, no problems are to be encountered in evaluating the element master matrices for the two new terms. However, in another regard (8.10) is a truly substantial departure from the steady-state form. To now, the finite element algorithm has always yielded the global matrix statement

$$S_e([K + H]_e\{Q\}_e - \{b\}_e) = \{0\} \tag{8.11}$$

However, (8.10) is not an algebraic matrix statement, but instead is a system of ordinary differential equations.

To verify and expand on this, the element matrix for the first term in (8.10) typically has the definition

$$[M]_e \equiv \int_{\Omega_e} \{N_k\}\{N_k\}^T\, d\tau \tag{8.12}$$

and for the new second term we could write

$$[U]_e \equiv \int_{\Omega_e} \{N_k\} u_e \frac{d\{N_k\}^T}{dx} \, d\tau \tag{8.13}$$

The remaining terms are all familiar and involve $[K]_e$, $[H]_e$, and $\{b\}_e$. Hence, (8.10) written in assembled element matrix form is

$$\begin{aligned} WS^h &= S_e([M]_e\{Q\}'_e + ([U]_e + [K]_e + [H]_e)\{Q\}_e - \{b\}_e) \\ &= [M]\frac{d\{Q(t)\}}{dt} + [U + K + H]\{Q(t)\} - \{b\} = \{0\} \end{aligned} \tag{8.14}$$

Thus, the global statement (8.10), that is (8.14), couples the time derivative of $\{Q\}$ to the balance of fluid convection, thermal conduction, boundary heat convection, and source term effects, all evaluated at any given time $t$.

Specifically, $WS^h$ in (8.14) allows one to evaluate the rate at which $\{Q(t)\}$ changes with time knowing its value at (say) time $t^n$. Hence, (8.14) can be used to evaluate a time Taylor series that will advance the solution over the interval $\Delta t \equiv t^{n+1} - t^n$. The traditional Taylor series uses derivative data at time $t^n$ yielding

$$\{Q(t^{n+1})\} = \{Q(t^n + \Delta t)\} = \{Q(t^n)\} + \Delta t \frac{d\{Q(t^n)\}}{dt} + O(\Delta t^2) \tag{8.15}$$

In (8.15), the notation "$+O(\Delta t^2)$" is read "plus the order of $\Delta t$-squared," which is the order of the *truncation error* associated with the expressed single term *explicit* Taylor series. This error can be made less significant by averaging the derivative evaluation half-way between $t^n$ and $t^{n+1}$. In this case, (8.15) is replaced by

$$\{Q(t^{n+1})\} = \{Q(t^n)\} + \frac{\Delta t}{2}\left(\frac{d\{Q(t^{n+1})\}}{dt} + \frac{d\{Q(t^n)\}}{dt}\right) + O(\Delta t^3) \tag{8.16}$$

For those familiar with ordinary differential equation (ODE) nomenclature, (8.15) is called the (explicit) *forward Euler* algorithm while (8.16) is termed the *trapezoidal rule*. A third algorithm which utilizes only $d\{Q(t^{n+1})\}/dt$ for the Taylor series is termed *backward Euler* and has a truncation error term of $O(\Delta t^2)$. The family of these *single-step* ODE *algorithms* can be compactly expressed as

$$\{Q\}^{n+1} = \{Q\}^n + \Delta t\left(\theta \frac{d\{Q\}^{n+1}}{dt} + (1-\theta)\frac{d\{Q\}^n}{dt}\right) + O(\Delta t^2, \Delta t^3)$$

$$\text{for } 0 \le \theta \le 1 \tag{8.17}$$

where $\{Q\}^{n+1} \equiv \{Q(t^{n+1})\}$, etc. The user-selected parameter $\theta$ defines the specific Taylor series, i.e., explicit Euler ($\theta = 0$), trapezoidal rule ($\theta = \frac{1}{2}$), or backward Euler ($\theta = 1$).

With this modest introduction, the weak statement (8.14) permits evaluating any derivative needed in (8.17), since (8.14) can be equation-solved at any time to produce

$$\frac{d\{Q\}}{dt} = [M]^{-1}(-[U + K + H]\{Q\} + \{b\}) \tag{8.18}$$

Substituting (8.18) into (8.17) yields

$$\begin{aligned}\{Q\}^{n+1} = \{Q\}^n &+ \Delta t\theta[M]^{-1}(-[U + K + H]\{Q\}^{n+1} + \{b\}^{n+1}) \\ &+ \Delta t(1 - \theta)[M]^{-1}(-[U + K + H]\{Q\}^n + \{b\}^n) \\ &\text{for } 0 \leq \theta \leq 1\end{aligned} \tag{8.19}$$

If (8.19) is then multiplied through by $[M]$, and noting that the matrix product $[M][M]^{-1}$ produces the identity matrix $[I]$, one obtains

$$\begin{aligned}[M](\{Q\}^{n+1} - \{Q\}^n) = \Delta t(\theta)(-[U + K + H]\{Q\}^{n+1} + \{b\}^{n+1}) \\ + \Delta t(1 - \theta)(-[U + K + H]\{Q\}^n + \{b\}^n)\end{aligned} \tag{8.20}$$

Now carrying the right side $\{Q\}$ terms multiplied by $\theta$ to the left side, and defining $\{\Delta Q\}^{n+1} \equiv \{Q\}^{n+1} - \{Q\}^n$ as the change in $\{Q\}$ over the time interval $\Delta t$, (8.20) becomes the global matrix statement

$$\begin{aligned}([M] + \theta\Delta t[U + K + H])\{\Delta Q\}^{n+1} = -\Delta t[U + K + H]\{Q\}^n \\ + \Delta t(\theta\{b\}^{n+1} + (1 - \theta)\{b\}^n)\end{aligned} \tag{8.21}$$

Finally, viewing the assembly form of the steady-state algorithm (8.11), and assuming for simplicity that the data in $\{b\}$ are not a function of time, (8.21) can be expressed in element matrix form as

$$\begin{aligned}S_e(([M]_e + \theta\Delta t[U + K + H]_e)\{\Delta Q\}_e \\ -\Delta t(\{b\}^n - [U + K + H]_e\{Q\}_e^n)) = \{0\}\end{aligned} \tag{8.22}$$

Comparing (8.22) to the steady-state algorithm (8.11) then yields the following associations for a code implementation. The steady-state element nodal unknown $\{Q\}_e$ is replaced by the change $\{\Delta Q\}_e$ in the nodal values over time interval $\Delta t$. The solution of (8.22) thus yields

$$\{Q\}_e^{n+1} = \{Q(t^n + \Delta t\}_e = \{Q\}_e^n + \{\Delta Q\}_e \tag{8.23}$$

The system matrix assembly of master matrices $[K]_e + [H]_e$ for the steady-state problem now includes constructions for the time and fluid convection additions to the problem statement. Finally, the problem "data" contained in $\{b\}_e$ now includes contributions due to the action of the solution at the old time station $t^n$. Note that $\{Q\}_e^n$ is indeed *always known information*, and hence correctly identified as *data*. These data start with the mapping of the *initial conditions* (8.4) onto the nodes of $\Omega^h$.

Hence, the generalization of the problem statement is a quite natural and uncomplicated extension on the finite element constructions already developed. We now proceed to specific forms and examples for one-dimensional problems.

## 8.3 MASTER MATRIX LIBRARY

Completion of the master matrix library for (8.22), for the unsteady convection-diffusion problem statement (8.1) to (8.4), is straightforward. [Note that (8.1) to (8.4) is sometimes termed the "advection-diffusion" problem, wherein advection implies the velocity field is given.] Using the Fortran nomenclature introduced in Chap. 5, the one-dimensional algorithm statement is

$$S_e((l_e[A200d] + \theta\Delta t(\bar{u}_e[A201d] - l_e^{-1}\bar{k}_e[A211d] + \bar{h}[A0d]))\{\Delta Q\}_e$$
$$-\Delta t(\{b\}_e - (\bar{u}_e[A201d] + l_e^{-1}\bar{k}_e[A211d] + \bar{h}_e[A0d])\{Q\}_e^n))$$
$$= \{0\} \quad (8.24)$$

for the simplest case where metric, thermal, and fluid velocity data need not be interpolated. These distribution complications are minor and become stated in the following discussions as a function of the matrix name parameter $d$ in (8.24), which takes the label $L$, $Q$, etc., as a function of basis degree $k$ in (8.8).

### Linear Basis

For the problem statement with no additional interpolation requirements, the sole new master matrix in (8.24) is $[A201L]$. From the definition in (8.10), see also (8.13), the evaluation yields

$$\int_{\Omega_e} \{N_1\}u_e \frac{d\{N_1\}^T}{dx}\,d\tau = \bar{u}_e \int_{\Omega_e} \begin{Bmatrix}\zeta_1\\ \zeta_2\end{Bmatrix} \frac{1}{l_e}\{-1, 1\}\,dx$$
$$= \frac{\bar{u}_e l_e}{2}\begin{Bmatrix}1\\ 1\end{Bmatrix}\frac{1}{l_e}\{-1, 1\}$$
$$= \frac{\bar{u}_e}{2}\begin{bmatrix}-1, 1\\ -1, 1\end{bmatrix} \quad (8.25)$$

Hence, the linear basis master matrix library for the algorithm statement (8.22) to (8.24) is

$$[M]_e \equiv l_e[A200L] \quad = \frac{l_e}{6}\begin{bmatrix} 2 & 1 \\ 1 & 2 \end{bmatrix} \tag{8.26a}$$

$$[U]_e \equiv \bar{u}_e[A201L] \quad = \frac{\bar{u}_e}{2}\begin{bmatrix} -1 & 1 \\ -1 & 1 \end{bmatrix} \tag{8.26b}$$

$$[K]_e = \bar{k}_e l_e^{-1}[A211L] = \bar{k}_e l_e^{-1}\begin{bmatrix} 1 & -1 \\ -1 & 1 \end{bmatrix} \tag{8.26c}$$

$$[H]_e = \bar{h}_e[A0L] \quad = \bar{h}_e\begin{bmatrix} 1 & 0 \\ 0 & 0 \end{bmatrix}\delta_{e1} \quad \text{or} \quad \bar{h}_e\begin{bmatrix} 0 & 0 \\ 0 & 1 \end{bmatrix}\delta_{eM} \tag{8.26d}$$

$$\{b\}_e = l_e[A200L]\{S\}_e \quad \text{and/or} \quad \bar{h}_e[A0L]\{TR\}_e \tag{8.26e}$$

Recall in (8.26) that $\bar{k}_e \equiv k_e/\rho c_p$ and $\bar{h}_e \equiv h_e/\rho c_p$ are the unsteady generalizations for uniform element thermal conductivity.

The needed additions to (8.26) for element metric data, fluid convection velocity, and/or thermal conductivity are readily obtained. For the convection term, for $u = u(x, t)$ (8.25) becomes

$$\begin{aligned}
\int_{\Omega_e} \{N_1\}u_e \frac{d\{N_1\}^T}{dx}\, d\tau &= \int_{\Omega_e} \{U(t)\}_e^T\{N_1\}\{N_1\}\frac{d\{N_1\}}{dx}\, dx \\
&= \{U(t)\}_e^T \int_{\Omega_e} \{N_1\}\{N_1\}\frac{1}{l_e}\begin{Bmatrix} -1 \\ 1 \end{Bmatrix} dx \\
&= \frac{\{U(t)\}_e^T}{6}\begin{bmatrix} -\begin{Bmatrix} 2 \\ 1 \end{Bmatrix} & \begin{Bmatrix} 2 \\ 1 \end{Bmatrix} \\ -\begin{Bmatrix} 1 \\ 2 \end{Bmatrix} & \begin{Bmatrix} 1 \\ 2 \end{Bmatrix} \end{bmatrix}
\end{aligned} \tag{8.27}$$

using the analytical integration statement (4.40). Hence, (8.27) defines the new master matrix $[A3001L]$, and we have

$$[U]_e \equiv \{U\}_e^T[A3001L] \tag{8.28b}$$

A suggested exercise is to verify that, for a uniform convection velocity $u_e$, insertion of $\{U\}_e^T = \bar{u}_e\{1\}^T$ into (8.28*b*) reproduces (8.26*b*). For variable element thermal conductivity, the generalization in (8.26) is

$$[K]_e \equiv l_e^{-1}\{K\}_e^T[A3011L] \tag{8.28c}$$

where the element array $\{K\}_e$ contains nodal values of $k/\rho c_p$ and $[A3011L] = \{A10\}[A211L]$ expressed as a degree-1 hypermatrix. Finally, for an axisymmetric one-dimensional geometry,

$$[M]_e = \int_{\Omega_e} \{N_1\}\{N_1\}^T r\,dr = \int_{\Omega_e} \{R\}_e^T\{N_1\}\{N_1\}\{N_1\}^T\,dx$$

$$= l_e\{R\}_e^T[A3000L] \tag{8.28a}$$

and $[A3000L]$ is given in (5.118).

## Quadratic Basis

For $k = 2$ in (8.8), the new addition to the master matrix library is

$$\int_{\Omega_e} \{N_2\}u_e \frac{d\{N_2\}^T}{dx}\,d\tau = \bar{u}_e \int_{\Omega_e} \{N_2\}\frac{d\{N_2\}^T}{dx}\,dx$$

$$= \frac{\bar{u}_e}{l_e}\int_{\Omega_e} \begin{Bmatrix} \zeta_1(2\zeta_1 - 1) \\ 4(\zeta_1\zeta_2) \\ \zeta_2(2\zeta_2 - 1) \end{Bmatrix} \{\zeta_2 - 3\zeta_1, 4(\zeta_1 - \zeta_2), 3\zeta_2 - \zeta_1\}\,dx$$

$$= \frac{\bar{u}_e}{6}\begin{bmatrix} -3 & 4 & -1 \\ -4 & 0 & 4 \\ 1 & -4 & 3 \end{bmatrix} \tag{8.29}$$

Thus, the quadratic basis master matrix library for the finite element algorithm (8.22) to (8.24) is

$$[M]_e \equiv l_e[A200Q] = \frac{l_e}{30}\begin{bmatrix} 4 & 2 & -1 \\ 2 & 16 & 2 \\ -1 & 2 & 4 \end{bmatrix} \tag{8.30a}$$

$$[U]_e \equiv \bar{u}_e[A201Q] = \frac{\bar{u}_e}{6}\begin{bmatrix} -3 & 4 & 1 \\ -4 & 0 & 4 \\ 1 & -4 & 3 \end{bmatrix} \tag{8.30b}$$

$$[K]_e \equiv \bar{k}_e l_e^{-1}[A211Q] = \frac{\bar{k}_e}{3l_e}\begin{bmatrix} 7 & -8 & 1 \\ -8 & 16 & -8 \\ 1 & -8 & 7 \end{bmatrix} \tag{8.30c}$$

$$[H]_e = \bar{h}_e[A0Q] = \bar{h}_e\begin{bmatrix} 1 & 0 & 0 \\ 0 & 0 & 0 \\ 0 & 0 & 0 \end{bmatrix}\delta_{e1} \quad \text{or} \quad \bar{h}_e\begin{bmatrix} 0 & 0 & 0 \\ 0 & 0 & 0 \\ 0 & 0 & 1 \end{bmatrix}\delta_{eM} \tag{8.30d}$$

$$\{b\}_e = l_e[A200Q]\{S\}_e \quad \text{and/or} \quad \bar{h}_e[A0Q]\{TR\}_e \tag{8.30e}$$

As in (8.26), $\bar{k}_e \equiv k_e/\rho c_p$ and $\bar{h}_e \equiv h_e/\rho c_p$ in (8.30). The generalizations needed for full quadratic interpolation of metric, fluid convection, and/or thermal conductivity data yield the following definitions.

$$[M]_e = l_e\{R\}_e^T[A3000Q] = \frac{l_e\{R\}_e^T}{420}\begin{bmatrix} \begin{Bmatrix} 39 \\ 20 \\ -3 \end{Bmatrix} & \begin{Bmatrix} 20 \\ 16 \\ -8 \end{Bmatrix} & \begin{Bmatrix} -3 \\ -8 \\ -3 \end{Bmatrix} \\ \begin{Bmatrix} 20 \\ 16 \\ -8 \end{Bmatrix} & \begin{Bmatrix} 16 \\ 192 \\ 16 \end{Bmatrix} & \begin{Bmatrix} -8 \\ 16 \\ 20 \end{Bmatrix} \\ \begin{Bmatrix} -3 \\ -8 \\ -3 \end{Bmatrix} & \begin{Bmatrix} -8 \\ 16 \\ 20 \end{Bmatrix} & \begin{Bmatrix} -3 \\ 20 \\ 39 \end{Bmatrix} \end{bmatrix} \tag{8.31a}$$

$$[U]_e = \{U\}_e^T[A3001Q] = \frac{\{U\}_e^T}{90}\begin{bmatrix} \begin{Bmatrix} -30 \\ -18 \\ 3 \end{Bmatrix} & \begin{Bmatrix} 36 \\ 24 \\ 0 \end{Bmatrix} & \begin{Bmatrix} -6 \\ -6 \\ -3 \end{Bmatrix} \\ \begin{Bmatrix} -18 \\ -48 \\ 6 \end{Bmatrix} & \begin{Bmatrix} 24 \\ 0 \\ 24 \end{Bmatrix} & \begin{Bmatrix} -6 \\ 48 \\ 18 \end{Bmatrix} \\ \begin{Bmatrix} 3 \\ 6 \\ 6 \end{Bmatrix} & \begin{Bmatrix} 0 \\ 24 \\ -36 \end{Bmatrix} & \begin{Bmatrix} -3 \\ 18 \\ 30 \end{Bmatrix} \end{bmatrix} \tag{8.31b}$$

$$[K]_e = l_e^{-1}\{K\}_e^T[A3011Q] = \frac{\{K\}_e^T}{30l_e}\begin{bmatrix} \begin{Bmatrix} 37 \\ 36 \\ -3 \end{Bmatrix} & \begin{Bmatrix} -44 \\ -32 \\ -4 \end{Bmatrix} & \begin{Bmatrix} 7 \\ -4 \\ 7 \end{Bmatrix} \\ \begin{Bmatrix} -44 \\ -32 \\ -4 \end{Bmatrix} & \begin{Bmatrix} 48 \\ 64 \\ 48 \end{Bmatrix} & \begin{Bmatrix} -4 \\ -32 \\ -44 \end{Bmatrix} \\ \begin{Bmatrix} 7 \\ -4 \\ 7 \end{Bmatrix} & \begin{Bmatrix} -4 \\ -32 \\ -44 \end{Bmatrix} & \begin{Bmatrix} -3 \\ 36 \\ 37 \end{Bmatrix} \end{bmatrix} \tag{8.31c}$$

You may well appreciate the ample use of (4.40) involved in evaluating the master matrix library additions (8.31)!

## 8.4 STEADY-STATE DIFFUSION WITH FLUID CONVECTION

This model problem is selected to forcefully illustrate the dominance of fluid convection over diffusion for small thermal diffusivity $\alpha = k/\rho c_p$. The important parameter that emerges on nondimensionalization of (8.1) is the *Peclet number* $\mathrm{Pe} \equiv \mathrm{RePr}$, where $\mathrm{Re} = \rho UL/\mu$ is the flow *Reynolds number*, $\mathrm{Pr} = \mu c_p/k$ is the *Prandtl number*, and $U$ and $L$ are reference velocity and length scales. Letting a superscript asterisk denote a dimensional variable, the nondimensionalization of (8.1) involves these definitions:

$$
\begin{aligned}
q &= \frac{T^* - T^*_{\min}}{T^*_{\max} - T^*_{\min}} & k &= \frac{k^*}{k^*_{\mathrm{ref}}} \\
u &= \frac{u^*}{U^*} & \rho c_p &= \frac{\rho^* c^*_p}{(\rho c_p)^*_{\mathrm{ref}}} \equiv 1 \\
\frac{\partial}{\partial x} &= L^* \frac{\partial}{\partial x^*} & h &= \frac{h^*}{k^*_{\mathrm{ref}} L^{-1}} \equiv \mathrm{Nu} \\
\frac{\partial}{\partial t} &= \frac{L^*}{U^*} \frac{\partial}{\partial t^*} & s^*/s^*_{\mathrm{ref}} &= \frac{s^* L^*}{U^*(T^*_{\max} - T^*_{\min})}
\end{aligned}
\tag{8.32}
$$

It is a suggested exercise to substitute (8.32) into (8.1), clear the common multiplier, and hence verify that the nondimensional form of (8.1) and (8.2) is

$$
L(q) = \frac{\partial q}{\partial t} + u\frac{\partial q}{\partial x} - \frac{1}{\mathrm{Pe}}\frac{\partial}{\partial x}\left(k\frac{\partial q}{\partial x}\right) - s = 0 \tag{8.33}
$$

$$
l(q) = k\frac{dq}{dn} + \mathrm{Nu}(q - T_r) = 0 \tag{8.34}
$$

The nondimensional appearance of (8.1) and (8.2) is thus a modest change, but recall that $k$ is now nondimensional, that is, equal to unity for a uniform constant thermal conductivity. Further, $s$ and $T_r$ are nondimensional, and the *Nusselt number* $\mathrm{Nu} = h^*L/k^*$ has replaced the thermal convection coefficient in (8.2).

The next code exercise sequence seeks to determine the steady-state solution of (8.33) for the Dirichlet (fixed) boundary conditions $q(x = 0) = 0$ and $q(x = 1) = 1$. The imposed fluid convection velocity and the conductivity are uniform constants, hence $u(x) = 1$ and $k = 1$ in (8.33), and no source is present, therefore $s = 0$. For these data, you should verify that the analytical solution to (8.33) is

$$q(x) = \frac{1 - e^{\mathrm{Pe}(x)}}{1 - e^{\mathrm{Pe}}} \tag{8.35}$$

Recalling that $x$ is nondimensional, $q(x)$ differs from $q(x = 0) = 0$ only in the immediate region $x \leq 1$ wherein the product Pe($x$) is close to Pe. Figure 8.1 graphs the solution (8.35) for $0.1 \leq \mathrm{Pe} \leq 100$. The collapse of the exponential character into the region $x \approx 1$ for increasing Pe is quite evident.

For reference, the Fortran statement you will be instructing *LEARN.FE* to solve is (8.22), i.e., (8.24), for the $k = 1$ and/or $k = 2$ master matrix libraries (8.26) or (8.30). The $k = 1$ algorithm matrix statement is, specifically

$$
\begin{aligned}
&([M] + \theta\,\Delta t[U + K])\{\Delta Q\} - \Delta t(\{b\} - [U + K]\{Q\}^n) \\
&\quad = S_e\Bigg((l_e[A200d] + \theta\,\Delta t(\bar{u}_e[A201d] + \frac{\bar{k}_e}{l_e\mathrm{Pe}}[A211d]))\{\Delta Q\}_e \\
&\qquad - \Delta t(\{b\} - \bar{u}_e[A201d] - \frac{\bar{k}_e}{l_e\mathrm{Pe}}[A211d])\{Q\}_e^n\Bigg) \\
&\quad = S_e\Bigg(\Bigg(l_e\begin{bmatrix} 2 & 1 \\ 1 & 2 \end{bmatrix} + \theta\,\Delta t\Bigg(\frac{1}{2}\begin{bmatrix} -1 & 1 \\ -1 & 1 \end{bmatrix} + \frac{l_e^{-1}}{\mathrm{Pe}}\begin{bmatrix} 1 & -1 \\ -1 & 1 \end{bmatrix}\Bigg)\Bigg)\{\Delta Q\}_e \\
&\qquad + \Delta t\Bigg(\frac{1}{2}\begin{bmatrix} -1 & 1 \\ -1 & 1 \end{bmatrix} + \frac{l_e^{\,1}}{\mathrm{Pe}}\begin{bmatrix} 1 & -1 \\ -1 & 1 \end{bmatrix}\Bigg)\{Q\}_e^n\Bigg) = \{0\}
\end{aligned} \tag{8.36}
$$

Equation (8.36) confirms that the needed additional data are integration time-step size $\Delta t$, implicitness factor $\theta$, the known data $\{Q\}^n$ at any time $t^n$, and the Peclet number Pe for any specific solution.

We now proceed with the code exercise sequence.

**Code Exercise 8.1.** Determine the finite element steady-state solution to (8.33) for the data specification leading to (8.35), using the linear basis algorithm on a uniform $M = 4$ element discretization of the domain $0 \leq x \leq 1$. Conduct the computational experiment for $\mathrm{Pe} = 1.0$.

In proceeding directly to steady state, any initial condition $\{Q\}^{n=0}$ is adequate (the steady state does not depend on the initial-condition), define $\theta = 1$, and any time-step $\Delta t$ is permissible provided the final time $t_f = t_o + \eta\Delta t$ is large enough. Conversely, you may set $\Delta t$ to a very large value and solve (8.36) directly for the steady-state solution. *LEARN.FE* provides

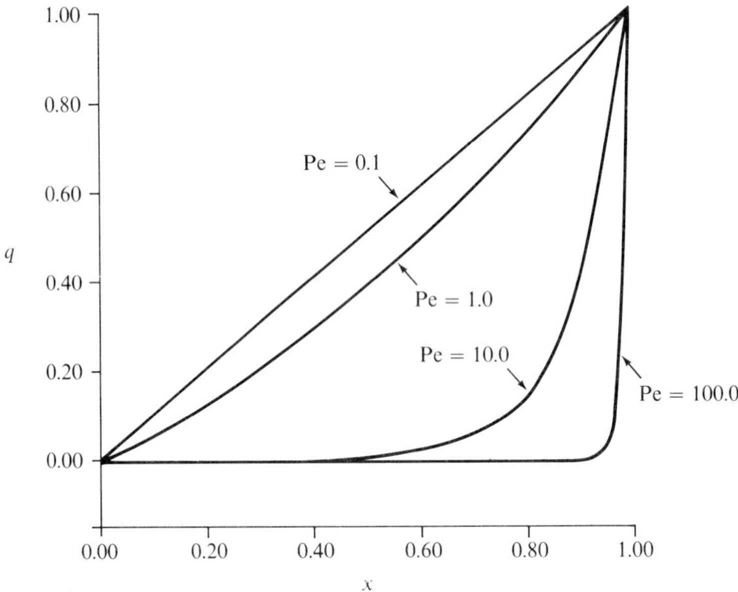


**FIGURE 8.1**
Analytical solutions for one-dimensional model problem.

this option under command *SOLV*, while the unsteady evolution is keyed via command *INTE*, which stands for "*INTE*grate." Access the input file for this code exercise and the following will scroll onto your monitor.

```
TITL
***** CODE EXERCISE 8.1:STEADY-STATE,M=4, PE=1 *****

TYPE    [ K  N  NNODEL      REFL NPR   NTRAN  NAXI ]
     1    1   2   1.       1

PRIN    [ NBUG(*) ]
     1    1   1   1        1

GRID    [ FOR N (XL XR(*)PR(*) ) NEM(PR*) ]
          1.00    2.00     1.00
     4

MATL [FOR NPR (KL(*),KR(*))]
     1    1.0       1.0

INIT [NINIT, NODE(*),QZERO(*)]
   5  0  0
   1    2      3    4      5
   0.0         0.25        0.50       0.75      1.0

DIRI [NDIRI,NODE(*),QB(*)]
     2  0  0
     1      5
    0.0       1.0
```

```
UVEL [OPTION,MAG]
  1    1.0

FORM
SOLV
  INTE [TZERO TSTOP TSTEP THETA DQMIN DQMAX TSTEPI  NTSMAX]
         0.0   50.    0.5   1.   .0001 0.2  .00001    100

NORM

STOP
```

This file defines the instruction sequence for establishing the direct steady-state solution and/or the unsteady integration through time. Full printout is requested under *PRIN*, and defining NBUG(5) = 1 requests that the solution be output after every pass through subroutine FORMIT. For true unsteady solutions, setting NBUG(5) = $p$ specifies solution output after $p$ integration time steps. Under *MATL*, $k_L = 1. = k_R$ will yield Pe = 1 for (8.36). The nodal initial-condition $\{Q\}^{n=0}$, following *INIT*, which denotes "*INIT*ial condition," is the linear interpolation of the Dirichlet boundary conditions given in *DIRI*. The *command name UVEL* defines option "1," which selects a uniform constant for $u$; the magnitude definition (1.0) follows. The solution instruction is *SOLV*, hence the direct steady-state of (8.36) will be determined.

However, following *SOLV* (and indented a few columns to preclude its being "read" as a command in READIT) is the time integration definition sequence for *INTE*. The first four entries are

TZERO = 0.0: The solution initial time $t_0$.
TSTOP = 50.0: The time at which integration will be stopped.
TSTEP = 0.5: The maximum integration time-step $\Delta t$.
THETA = 1.0: The implicitness factor $\theta$.

The next four entries key integration procedure options for adjusting $\Delta t$ and determining whether a steady state has been achieved prior to reaching TSTOP. Specifically:

DQMIN = 0.0001: If all entries in $\{\Delta Q\}$ are less than this, define that steady state has been achieved.
DQMAX = 0.2: The maximum $\{\Delta Q\}$ allowed before decreasing $\Delta t$ to maintain accuracy.
TSTEPI = 0.0001: Initial $\Delta t$ to start the solution; it will grow to TSTEP under control of DQMAX.
NTSMAX = 100: The maximum number of integration time steps for terminating the solution process.

Execute this file, and the following output will appear on your monitor.

```
TITL  *****CODE EXERCISE 8.1:STEADY-STATE,M=4,PE=1*****
TYPE  K  N  NNODEL    REFL     NPR   NTRAN  NAXI
      1  1    2        1.0      1      0      0
PRIN  NBUG1  NBUG2   NBUG3   NBUG4   NBUG5  NBUG6
      1       1        1       1       1      0
GRID  NI   NPRI  NTRAN
      1  1  0
      XL   XR1   PR1   XR2   PR2   XR3   PR3
        1.0000   2.0000   1.0000
      NEM1    NEM2    NEM3
       4

      ELEMENT NODE CONNECTION ARRAY.(MEL)

          ELEMENT   NODE1   NODE2   NODE3   NODE4
              1        1       2
              2        2       3
              3        3       4
              4        4       5

      NODE COORDINATES(X,REAL VARIABLES)
      0.10000000E+01 0.12500000E+01 0.15000000E+01 0.17500000E+01
      0.20000000E+01

MATL  MACRO NODAL CONDUCTIVITIES
       1 0.10000E+01 0.10000E+01
      AK: UNIFORM

DIRI  NFIX:  2
      JFIX:    1  5
           0.00000    1.00000

INIT  NFIX:  5
      JFIX:    1   2  3  4  5
           0.00000  0.25000  0.50000  0.75000  1.00000

UVEL  SELECTION  UCON    PECLET
         1       1.00000  1.00000

FORM:  SYSTEM MATRIX  [K + H]; COLUMN FORMAT
           0.00       1.00      0.00
          -4.50       8.00     -3.50
          -4.50       8.00     -3.50
          -4.50       8.00     -3.50
           0.00       1.00      0.00

       DATA MATRIX  {b}  ROW FORMAT
           0.00       0.00      0.00      0.00      1.00

SOLV NODAL SOLUTION (REAL VARIABLES)
   0.00000000E+00 0.16487300E+00 0.37696859E+00 0.64948487E+00
   0.10000000E+01

NORM F.E. SOLN. ENERGY SEMI-NORM: E(QH)=0.87961592E+08
```

This output contains all debug and is quite familiar. Under *UVEL*, the code has indeed verified that $Pe = 1$ for the input data. Under *FORM*, note the system matrix $[K + H]$ is now nonsymmetric (due to $[U]$). The steady-state nodal solution $\{Q\}$ is a smooth interpolation of the endpoint Dirichlet data, and its energy seminorm has been computed and output following *NORM*.

The next step is to conduct a mesh study, to quantify the important issue of asymptotic convergence and accuracy estimation. Before proceeding, for completeness at this step, indent *FORM* and *SOLV* five spaces, bring *INTE* to left justification, redefine NBUG(5) = 100, and reexecute the input file. The output will appear identical except for the replacement of the *SOLV* field with

```
INTE    TZERO  TSTOP  TSTEP  THETA  DQMIN  DQMAX  TSTEPI   NTSMAX
         0.0    50.0   0.5    1.0  0.0001   0.2  0.00001   100

INTE    REACHED STEADY STATE
        TIME      #STEPS
        32.94       68

INTE     NODAL SOLUTION (REAL VARIABLES)
        0.00000000E+00 0.16487300E+00 0.37696859E+00 0.64948787E+00
        0.10000000E+01

NORM    F.E. SOLN. ENERGY SEMI-NORM: E(QH)=0.87961592E+08
```

Following input data reflection, and for the given definitions, *LEARN.FE* achieved the steady-state solution after 68 time-steps whereupon $t_f = t^{69} =$ 32.94 s. The resultant nodal solution $\{Q\}^{n+1}$ is indeed identical to the direct steady-state solution to PC precision.

**Code Exercise 8.2.** Determine the $M \geq 4$ uniform mesh required for the $k = 1$ basis finite element solution to be accurate to PC single precision, for the Peclet problem (8.33) for Pe = 0.1, 1.0, and 10.0.

The asymptotic error statement (4.91*a*) given in Chap. 4 remains valid, that is, at steady state:

$$\|\text{error}^h(t_f)\| \simeq C_k l_e^{2k}(c_1\|\text{data}\|_\Omega + c_2\|\text{data}\|_{\partial\Omega}) \qquad (4.91a)$$

where the notation $\|\cdot\|$ indicates "norm," which can be a specific nodal value or the energy seminorm. Further, you may utilize the exact solution (8.35) to absolutely verify your computational data predictions. The sequence of tests for $M \geq 4$ and Pe = 0.1, 1, 10 can be efficiently executed by suitable modification to the following base input file.

```
TITL
***** CODE EXERCISE 8.2; STEADY STATE,M=8,PE=1,ADVEC/DIFF *****

TYPE    [ K N NODEL    REFL   NPR NTRAN   NAXI ]
   1     1   2   1.   1

PRIN    [ NBUG(*) ]
   1     0  0 3  0  0

GRID    [ FOR N(XL XR(*) PR(*)), NEM(PR*) ]
          1.00  2.00    1.00
          8
MATL   [FOR NPR (KL(*) PR(*)), NEM(PR*)]
   1    1.00     1.00
INIT   [(NINIT, CODE),QZERO(*)]
         9  0   2
        0.

DIRI   [NDIRI, NODE(*),QB(*)]
     2 0 0
     1     9

    0.0    1.0

UVEL   [OPTION,MAG]
  1     1.0

FORM

SOLV

NORM

STOP
```

The data string following *INIT* utilizes the input repeat option to set $\{Q\}^{n=0}$ to zero. The subsequent *DIRI* specification overrides *INIT* data at any fixed node; hence, minimal user data specification is required for convergence studies.

The results of these experiments will verify that the finite element algorithm solutions behave predictably well for all $\mathrm{Pe} \leq 10$ on quite modest uniform meshes. Specifically, PC single-precision significance is predictable in $\|q^h\|_E$ for the $M \approx 32$ solution at $\mathrm{Pe} = 10$. Further, round-off error pollutes data obtained for $M > 16$ at $\mathrm{Pe} = 0.1$, hence the corresponding linear basis solutions are identical with the Lagrange interpolate of the exact solution (8.35) to PC precision.

The appropriateness of using modest uniform meshes changes quite dramatically, however, as the Peclet number is substantially increased. The following code exercise suggests the definitive computational experiments.

**Code Exercise 8.3.** Using the $k = 1$ basis algorithm on an $M = 20$ uniform mesh, generate steady-state solutions for the Peclet problem definition (8.34) for $\mathrm{Pe} = 10$, 100, and 1000.

This solution sequence is easily executed via modification and stacking of Code Exercise 8.2 input files. Set NBUG(4) = 4 to obtain integer print, and closely examine the nodal solution arrays, and thereafter request plots of each solution. You will note a distressing trend as the Peclet number becomes larger. Figure 8.2 graphs the solution nodal data, plotted as symbols for Pe = 10, 100 and Pe = 1000, in comparison to the analytical solution (8.35) shown as the line. The accurate solution for Pe $\leq$ 10 on an $M = 20$ uniform discretization has turned to absolute rubbish at Pe $\geq$ 100, i.e., a wild but bounded oscillation with period equal to $2l_e$. Note that the approximate solutions at Pe = 100 and 1000 are *not unstable*, as sometimes incorrectly stated in the literature. The "2 $\Delta x$" wave pattern is informing you that the selected discretization is incapable of supporting the solution

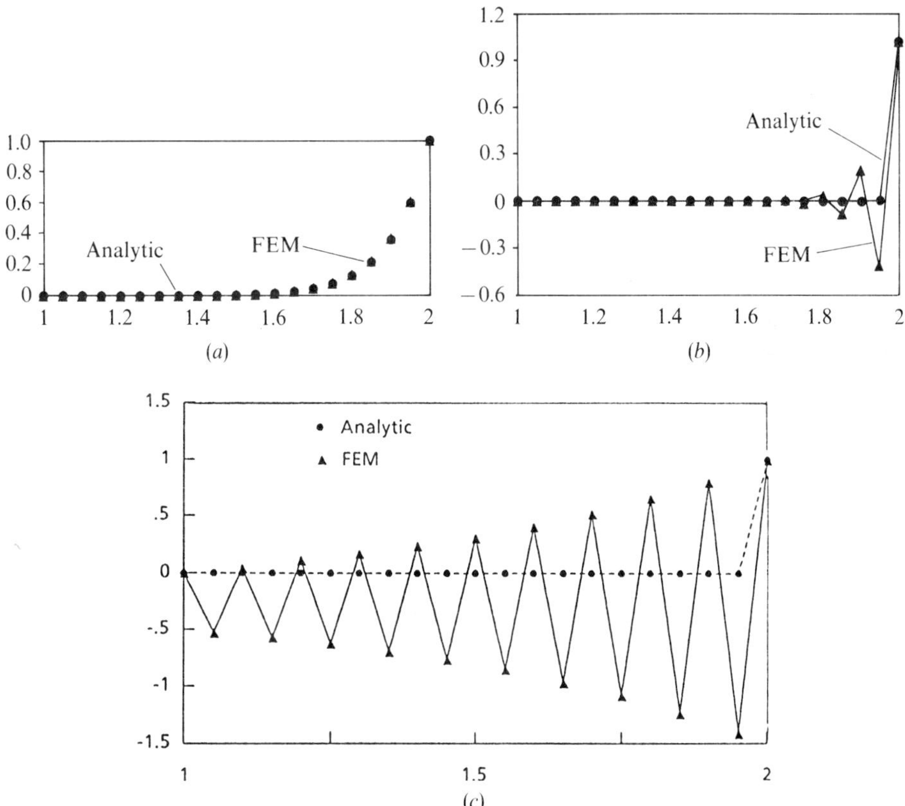


**FIGURE 8.2**
Nodal solutions for (8.33), $k = 1$, $M = 20$ uniform mesh. (*a*) Pe = 10, (*b*) Pe = 100, (*c*) Pe = 1000.

gradients, in accordance with "Don't Suppress the Wiggles; They're Telling You Something," the descriptive title of an article by Gresho and Lee (1979) on the subject.

As a consequence, the following extension to the code exercise is appropriate.

**Code Exercise 8.4.** Using the progression ratio option under *GRID*, determine whether a nonuniform discretization $\Omega^h$ of $0 \leq x \leq 1$ containing $M = 20$ elements can produce an "adequately" accurate $k = 1$ basis approximate solution for the model problem at Pe = 100. Then determine the uniform mesh refinements $M > 20$, e.g., $M = 40$, 80, required to accurately resolve the Pe = 100 problem solution. Compare the PC execution times for these various solutions, and then continue the computational experiment for Pe = 200, 400, ..., 1000. Then form your conclusion regarding the requirements for appropriate discretizations for the "convection-diffusion" problem statement (8.33).

Figure 8.3 illustrates the $M = 20$ nodal solutions you will obtain for *GRID* inputs PR = 0.8 and PR = 0.7. You should consider repeating these last three *LEARN.FE* code exercises using the $k = 2$ basis algorithm, which is easily accessible under command *TYPE*.

## 8.5 UNSTEADY CONDUCTION WITH BOUNDARY HEAT CONVECTION

The transient heat conduction problem with surface heat convection is an important subclass of (8.1) to (8.3) that results for the definitions

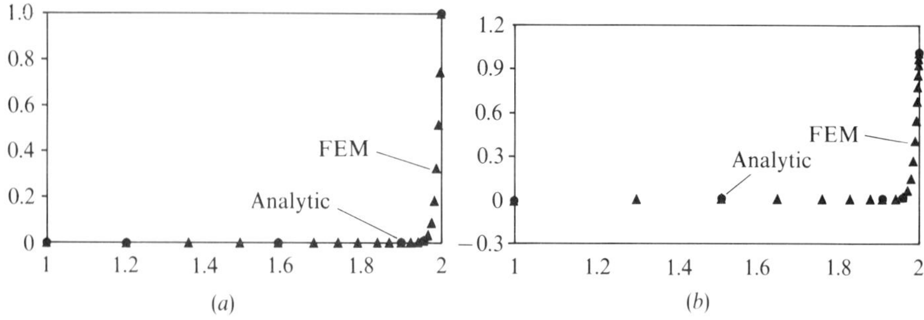


**FIGURE 8.3**
Nodal solutions for (8.33) for Pe = 100, $k = 1$, $M = 20$ nonuniform mesh. (*a*) PR = 0.8, (*b*) PR = 0.7.

$u \equiv 0 \equiv s$. For a model problem, consider conduction through an axisymmetric pipe heated on the interior by a hot fluid; recall Section 4.6. The associated form for (8.1) to (8.3) is

$$L(T) = \frac{\partial T}{\partial t} - \frac{1}{r}\frac{\partial}{\partial r}\left(\bar{k}r\frac{\partial T}{\partial r}\right) = 0 \qquad \text{on } \Omega \tag{8.37}$$

$$l(T) = \bar{k}\frac{dT}{dn} + \bar{h}(T - T_r) = 0 \qquad \text{on } \partial\Omega_1 \tag{8.38}$$

$$T(r_b) = T_b \qquad \text{on } \partial\Omega_2 \tag{8.39}$$

and again recall that $\bar{k} \equiv k/\rho c_p$ and $\bar{h} = h/\rho c_p$.

The linear basis weak statement (8.22) to (8.24) for uniform conductivity $\bar{k}$ is

$$\begin{aligned}
&([M] + \theta\,\Delta t[K + H])\{\Delta Q\} - \Delta t(\{b\} - [K + H]\{Q\}^n) \\
&= S_e\Bigg(\bigg(l_e\{R\}_e^T[A3000L] + \theta\,\Delta t\bigg(\frac{\bar{k}}{l_e}\{R\}_e^T[A3011L] \\
&\quad + \bar{h}r_1[A0L]\delta_{e1}\bigg)\bigg)\{\Delta Q\}_e \\
&\quad - \Delta t\bigg(\big(\bar{h}r_1[A0L]\{TR\}_e\delta_{e1} - \bigg(\frac{\bar{k}}{l_e}\{R\}_e^T[A3011L] \\
&\quad + \bar{h}r_1[A0L]\delta_{e1}\bigg)\{Q\}_e^n\big)\bigg)\Bigg)
\end{aligned} \tag{8.40}$$

The master matrix library for (8.40) is given in (8.27) to (8.28) with replacement of $\{K\}_e^T$ by $\{R\}_e^T$ in (8.28*c*). The algorithm statement (8.40) is valid for any discretization, and requires definition of the thermal data and the integration parameters for a specific analysis.

**Code Exercise 8.5.** Determine the early time transient evolution of the temperature distribution in the thick-walled circular cylinder problem defined in Code Exercise 4.5. Assume the isothermal initial condition $\{Q\}^{n=0} = T_b\{1\}$ and the geometric and thermal data specifications are $\bar{k} = 10$, $\bar{h} = 20$, $T_r = 1500$, $T_b = 306.86$, $r_1 = 1.0$, and $r_2 = 2.0$ in arbitrary consistent units. Select a (nonuniform) discretization $\Omega^h$ that you think is appropriate, define

$\theta = \frac{1}{2}$ for a time-accurate solution, and select a $\Delta t$ such that you reach $t_f = 0.01$ in a reasonable number of steps. Then conduct a range of experiments about these specifications, and compare your results to estimate solution accuracy at $t_f$. The base input file for an $M = 8$, $k = 1$ simulation for this exercise is:

```
TITL
***** CODE EXERCISE 8.5; TRANSIENT,M=8,CONDUCTION *****

TYPE    [ K N NODEL  REFL  NPR  NTRAN  NAXI ]
    1   1  2    1.  1    0     1

PRIN    [ NBUG(*) ]
    1   0  0 3  1

GRID     [ FOR N(XLXR(*) PR(*)),NEM(PR*) ]
           1.00  2.00   1.1
            8

MATL     [FOR NPR (KL(*), KR(*))]
   1       1.00    1.00
INIT     [(NINIT,CODE),QZERO(*)]
           9  0  2
           306.8625

DIRI  [NDIRI,NODE(*),QB(*)]
   1
   9
   306.8625

ROBN [NROB, NODE(*),(H(*),TR(*))]
   1
   1
           20.0          1500.0

INTE [TZERO  TSTOP  TSTEP  THETA  DQMIN  DQMAX  TSTEPI  NTSMAX]
        0.0   0.001 0.0001   0.5  0.0001   0.2   0.0001 100

STOP
```

The definitions to note above include NAXI = 1, for the axisymmetric geometry, and NBUG(5) = 1, which will produce in *INTE* output at every time step. Further, NBUG(4) = 3 will produce integer map output which is easy to view. The $M = 8$ nonuniform grid defines PR = 1.1; hence the mesh is refined toward the convection surface. Under *INTE*, TSTEP = 0.0001 defines $\Delta t = 0.0001$; hence 10 time steps will occur in reaching $t_f$. Further, TSTEPI = TSTEP; hence the code will execute at a fixed $\Delta t$ independent of the DQMAX and DQMIN input.

This code exercise is rather open-ended, but the specification is appropriate for what you might encounter in a "real situation." Your thinking regarding $\Omega^h$ should utilize the knowledge base established in conducting the previous several *LEARN.FE* code exercises. The maximum inner wall heat flux will occur at $t_o + \Delta t$, that is, the first time step. Unless you select a fine mesh adjacent to the cylinder inner wall, your approximate solution will manifest the "$2\,\Delta x$" error mode so evident in the previous steady model problem solutions. Actually, an analytical exercise (Baker 1983, Chap. 4) will show that, since the finite element algorithm (8.24) is *high-order-accurate*, that at least one $2\,\Delta x$ wave will always occur at $t = t_o + \Delta t$. However, a sufficiently refined $\Omega^h$ will render the amplitude negligible in an engineering sense. Thus, you can always obtain an "engineering-accurate" solution for your requirements. All it takes is sufficient computer resources to admit execution using an *adequate discretization* $\Omega^h$ and sufficiently small $\Delta t$.

# B. TWO-DIMENSIONAL PROBLEMS

## 8.6 ALGORITHM STATEMENT FOR CONVECTION-DIFFUSION

The $n$-dimensional generalization of the problem statement (8.1) to (8.4) is

$$L(T) = \frac{\partial T}{\partial t} + (\mathbf{u} \cdot \nabla)T - \nabla(\bar{k}\nabla T) - s = 0 \qquad \text{on } \Omega \qquad (8.41)$$

$$l(T) = \bar{k}\nabla T \cdot \mathbf{n} + \bar{h}(T - T_r) = 0 \qquad \text{on } \partial\Omega_1 \qquad (8.42)$$

$$T(\mathbf{x}_b) = T_b \qquad \text{on } \partial\Omega_2 \qquad (8.43)$$

$$T(\mathbf{x}, t = 0) = T_o \qquad \text{on } \Omega \bigcup \partial\Omega \qquad (8.44)$$

where $\mathbf{u}(\mathbf{x}, t) = u(\mathbf{x}, t)\mathbf{i} + v(\mathbf{x}, t)\mathbf{j}$ is the two-dimensional imposed velocity field and $\bar{k}$, $\bar{h}$, and $s$ retain their prior definitions.

An approximate solution for (8.41) to (8.44) has the form

$$T(\mathbf{x}, t) \simeq T^N(\mathbf{x}, t) = \sum_{j=1}^{N} \psi_j(\mathbf{x})Q_j(t) \qquad (8.45)$$

The finite element procedure introduces the discretization $\Omega^h$ of $\Omega$; hence the form of (8.45) becomes

$$T(\mathbf{x}, t) \simeq T^h(\mathbf{x}, t) = \bigcup_e T_e(\mathbf{x}, t) \tag{8.46}$$

$$T_e(\mathbf{x}, t) = \{N_k(\boldsymbol{\eta})\}^T\{Q(t)\}_e \tag{8.47}$$

The finite element approximation to the weak statement written for (8.41) to (8.44) is the following modest extension on (8.6) and (8.10).

$$\begin{aligned} WS^h = S_e\Bigg(&\int_{\Omega_e} \{N_k\}\{N_k\}^T\, d\tau\{Q\}'_e + \int_{\Omega_e} \{N_k\}\mathbf{u}_e \cdot \nabla\{N_k\}^T\, d\tau\{Q\}_e \\ &+ \int_{\Omega_e} \nabla\{N_k\} \cdot \bar{k}_e \nabla\{N_k\}^T\, d\tau\{Q\}_e - \int_{\Omega_e} \{N_k\}\{N_k\}^T\, d\tau\{S\}_e \\ &+ \int_{\partial\Omega_e \cap \partial\Omega_1} \{N_k\}\bar{h}_e\{N_k\}^T\, d\sigma\{Q - TR\}_e\Bigg) = \{0\} \end{aligned} \tag{8.48}$$

Comparing (8.48) to (8.10), and recalling the ease with which $n$-dimensional finite element bases are constructed, very little is involved that is "new." Once the modest hurdles in evaluating the new convection master matrix integrals are cleared, (8.48) is no more than the matrix ODE system

$$\begin{aligned} WS^h &= S_e([M]_e\{Q\}'_e + ([U]_e + [V]_e + [K]_e + [H]_e)\{Q\}_e - \{b\}_e) \\ &= [M]\frac{d\{Q\}}{dt} + [U + V + K + H]\{Q\} - \{b\} = \{0\} \end{aligned} \tag{8.49}$$

This statement is also a modest extension on the one-dimensional statement (8.14), and the previous discussion on ODE methodology is directly applicable. Thus, the algorithm global matrix statement (8.22) implemented with (8.49) is

$$\begin{aligned} S_e(([M]_e + \theta\Delta t[U + V + K + H]_e)\{\Delta Q\}_e \\ -\Delta t(\{b\}_e - [U + V + K + H]_e\{Q\}^n_e)) = \{0\} \end{aligned} \tag{8.50}$$

Comparing (8.50) and (8.22), the extension to the three-dimensional statement conceptually involves no more than appending the element matrix $[W]_e$ for the third velocity component. However, as can occur in any multidimensional steady-state problem, the global matrix defined in (8.50), and its three-dimensional extension, will become unacceptably large for a PC. Hence, the issue of numerical iterative methods to approximately solve (8.50) again becomes the major focus. A uniquely efficient procedure is developed in a subsequent section (Sec. 8.9) that involves only one-dimensional matrices coupled with a grid-sweeping *ADI*-type procedure highly suited to a PC implementation.

## 8.7 MASTER MATRIX LIBRARY

For completeness, and prior to consideration of approximate matrix solution methods, the linear triangle and bilinear quadrilateral basis master matrices are developed and collected in this section. The Fortran nomenclature base form for (8.50) for any two-dimensional basis implementation, and assuming all fluid convection and thermal data do not require interpolation, is

$$\begin{aligned} S_e((DET_e[B200d] &+ \theta\,\Delta t(\bar{u}_e[B201d] + \bar{v}_e[B202d]_e \\ &+ \bar{k}_e DET_e^{-1}[B2JJd]_e + \bar{h}_e l_e[A200d]))\{\Delta Q\}_e \\ &- \Delta t(DET_e[B200d]\{S\}_e + \bar{h}_e l_e[A200d]\{TR\}_e \\ &- (\bar{u}_e[B201d] + \bar{v}_e[B202d] + \bar{k}_e DET_e^{-1}[B2JJd]_e \\ &+ \bar{h}_e l_e[A200d])\{Q\}_e^n))) \equiv \{0\} \end{aligned} \tag{8.51}$$

The remaining matrix parameter $d$ then takes the appropriate label dependent on the specific basis choice. Should the fluid convection velocity or thermal conductivity be distributed on $\Omega$, and hence require interpolation on $\Omega_e$ (or boundary convection coefficient on $\partial\Omega_1$), the appropriate replacements in (8.51) are

$$\begin{aligned} \bar{u}_e[B201d] &\rightarrow \{U\}_e^T[B3001d] \\ \bar{v}_e[B202d] &\rightarrow \{V\}_e^T[B3002d] \\ \bar{k}_e[B2JJd] &\rightarrow \{K\}_e^T[B30JJd] \\ \bar{h}_e[A200d] &\rightarrow \{H\}_e^T[A3000d] \end{aligned} \tag{8.52}$$

All except the fluid convection master matrices already exist; hence the added formulation requirements for the generalized problem class are modest.

**Linear Basis Triangular Element**

In this case, the base fluid convection matrix library expression for the appropriate terms in (8.51) is

$$\bar{u}_e[B201L]_e + \bar{v}_e[B202L]_e = \bar{u}_j^e[B20JL]_e \tag{8.53}$$

where $1 \leq (j, J) \leq 2 = n$ is a summation index. The master matrix definition is then

$$\begin{aligned}[B20JL]_e &\equiv \int_{\Omega_e} \{N_1(\zeta_i)\}\nabla\{N_1(\zeta_i)\}^T \, d\tau \\ &= \int_{\Omega_e} \{N_1(\zeta_i)\} \frac{\partial\{N_1(\zeta_i)\}^T}{\partial\zeta_i} \left[\frac{\partial\zeta_i}{\partial x_j}\right]_e d\tau \\ &= \left[\frac{\partial\zeta_i}{\partial x_j}\right]_e \int_{\Omega_e} \{N_1(\zeta_i)\} \, d\tau \, \frac{\partial\{N_1\}^T}{\partial\zeta_i}\end{aligned} \tag{8.54}$$

realizing that $\{N_1\}$ is a linear function of $\zeta_i$ and using the chain rule. The integrand remaining in (8.54) is easy to evaluate, yielding $A_e\{B10B\} = A_e\{1/3, 1/3, 1/3\}^T$.

The derivative of $\{N_1\}^T$ by each $\zeta_i$ is one or zero, as given in (5.27), and the metric data array $(\partial\zeta_i/\partial x_j)_e$ is the $2 \times 3$ matrix of nodal coordinate differences given in (5.29). Recalling the pseudo-Kronecker delta matrix $[\delta_{ik}]$ developed in (5.32), we are led to the definition

$$\begin{aligned}\frac{\partial\{N_1\}^T}{\partial\zeta_i} &= \begin{cases}\{1 \quad 0 \quad 0\} & \text{for } i = 1 \\ \{0 \quad 1 \quad 0\} & \text{for } i = 2 \\ \{0 \quad 0 \quad 1\} & \text{for } i = 3\end{cases} \\ &\equiv \{\delta_i\}^T\end{aligned} \tag{8.55}$$

with the definition

$$\delta_i \equiv \begin{cases}1 & \text{for position } i \text{ in the row matrix} \\ 0 & \text{for all other matrix locations}\end{cases} \tag{8.56}$$

Combining (5.29) and (5.121) leads to the definition

$$\begin{aligned}\left[\frac{\partial\zeta_i}{\partial x_j}\right]_e &= \begin{bmatrix} Y_2 - Y_3 & X_3 - X_2 \\ Y_3 - Y_1 & X_1 - X_3 \\ Y_1 - Y_2 & X_2 - X_1 \end{bmatrix}_e (2A_e)^{-1} \\ &\equiv \frac{ZETAIJ_e}{2A_e} \qquad \text{for } 1 \leq I \leq 3 \quad \text{and} \quad 1 \leq J \leq 2\end{aligned} \tag{8.57}$$

Substituting into (8.54) then produces the convection master matrix statement

$$
\begin{aligned}
[B20JL]_e &\equiv A_e ZETAIJ_e (2A_e)^{-1}\{B10L\}\{\delta_i\}^T \\
&= \tfrac{1}{2} ZETAIJ_e\{B10L\}\{\delta_i\}^T
\end{aligned}
\tag{8.58}
$$

The triangle basis convection master matrix for an interpolation of the fluid velocity field distribution [i.e., (8.52)] involves the integral

$$
\begin{aligned}
[B300JL]_e &\equiv \int_{\Omega_e} \{N_1\}\{N_1\}\nabla\{N_1\}^T \, d\tau \\
&= (2A_e)^{-1} ZETAIJ_e \int_{\Omega_e} \{N_1\}\{N_1\} \, d\tau \{\delta_i\}^T
\end{aligned}
\tag{8.59}
$$

The product of the two column matrices in (8.59) is not a "regular" matrix product definition, but instead is a *column hypermatrix* of degree 1. The integrals are easy to form:

$$
\begin{aligned}
\int_{\Omega_e} \{N_1\}\{N_1\} \, d\tau &= A_e \int_{\Omega_e} \begin{Bmatrix} \zeta_1 \\ \zeta_2 \\ \zeta_3 \end{Bmatrix} \begin{Bmatrix} \zeta_1 \\ \zeta_2 \\ \zeta_3 \end{Bmatrix} d\tau \\
&= A_e \int_{\Omega_e} \begin{Bmatrix} \{N_1\}\zeta_1 \\ \{N_1\}\zeta_2 \\ \{N_1\}\zeta_3 \end{Bmatrix} d\tau \\
&= \frac{A_e}{6} \left\{ \begin{Bmatrix} 2 \\ 1 \\ 1 \end{Bmatrix} \begin{Bmatrix} 1 \\ 2 \\ 1 \end{Bmatrix} \begin{Bmatrix} 1 \\ 1 \\ 2 \end{Bmatrix} \right\} \\
&\equiv A_e\{B200L\}
\end{aligned}
\tag{8.60}
$$

The three (column matrix) entries in $\{B200L\}$ are given above the last line in (8.60). Each is premultiplied by $\{UJ\}_e^T$, the transpose of the nodal values of the convection velocity vector $u_j^e$ on $\Omega_e$, which reduces the expression in (8.60) to a (regular) column matrix. The Fortran master matrix statement for this generalization is

$$
[B300JL]_e = \tfrac{1}{2} ZETAIJ_e\{B200L\}\{\delta_i\}^T \tag{8.61}
$$

which is a modest notational (and DO loop) modification to (8.58).

**Bilinear Basis Quadrilateral Element**

For the quadrilateral, it is also convenient to combine the two fluid convection terms in (8.51), yielding

$$\bar{u}_e[B201B]_e + \bar{v}_e[B202B]_e = \bar{u}_j^e[B20JB]_e \tag{8.62}$$

and $1 \le (j, J) \le 2 = n$. The master matrix definition is thus

$$\begin{aligned}[B20JB]_e &\equiv \int_{\Omega_e} \{N_1^+(\eta_i)\}\nabla\{N_1^+(\eta_i)\}^T\, d\tau \\ &= \int_{\Omega_e} \{N_1^+\}\frac{\partial\{N_1^+\}^T}{\partial\eta_i}\left[\frac{\partial\eta_i}{\partial x_j}\right]_e \det{}_e\, d\eta_1\, d\eta_2\end{aligned} \tag{8.63}$$

The derivative column matrix and metric data matrix are defined in (6.28), (6.29), and (6.32), and each is a linear function of the $\eta_i$ coordinate system. Thus, for completely general element shapes, the entries in (8.63) must be evaluated using an appropriate numerical quadrature rule, as detailed in Section 6.4, Bilinear Basis. Alternately, for $\Omega_e$ differing only slightly from a parallelepiped, the simplified formulation discussed in Section 6.4, Bilinear Basis Simplified, is appropriate. Here, the metric data matrix and the transformation matrix determinant are evaluated at the element centroid—hence extracted from the integrand in (8.63). Recalling the definitions given in (6.44) to (6.45), the simplified form of (8.63) is

$$\begin{aligned}[B20JB]_e &\simeq \overline{\det{}_e\left[\frac{\partial\eta_i}{\partial x_j}\right]_e} \int_{\Omega_e} \{N_1^+\}\frac{\partial\{N_1^+\}^T}{\partial\eta_i}\, d\eta_1\, d\eta_2 \\ &= ETAIJ_e \int_{\Omega_e} \{N_1^+\}\frac{\partial\{N_1^+\}^T}{\partial\eta_i}\, d\eta_1\, d\eta_2\end{aligned} \tag{8.64}$$

and $\overline{\det_e} \simeq DET_e$ cancels out. The integrals remaining in (8.64) are readily evaluated; hence, (8.64) yields

$$[B20JB]_e = ETAIJ_e[B20IB], \qquad \text{for } 1 \le I \le 2 \tag{8.65a}$$

and the two *element-independent* matrices $[B20IB]$ in (8.65*a*) are

$$[B201B] = \frac{1}{12}\begin{bmatrix} -2 & 2 & 1 & -1 \\ -2 & 2 & 1 & -1 \\ -1 & 1 & 2 & -2 \\ -1 & 1 & 2 & -2 \end{bmatrix} \tag{8.65b}$$

$$[B202B] = \frac{1}{12}\begin{bmatrix} -2 & -1 & 1 & 2 \\ -1 & -2 & 2 & 1 \\ -1 & -2 & 2 & 1 \\ -2 & -1 & 1 & 2 \end{bmatrix} \tag{8.65c}$$

which completes the formulation of (8.62).

The extension to interpolation of the fluid velocity distribution is direct when the simplified metric data procedure is used. Specifically, combining the two velocity expressions in (8.52) yields the master matrix definition

$$\begin{aligned} [B300JB]_e &\equiv \int_{\Omega_e} \{N_1^+\}\{N_1^+\}\frac{\partial\{N_1^+\}}{\partial\eta_i}\left[\frac{\partial\eta_i}{\partial x_j}\right]_e \det_e d\eta_1\, d\eta_2 \\ &\simeq ETAIJ_e \int_{\Omega_e} \{N_1^+\}\{N_1^+\}\frac{\partial\{N_1^+\}}{\partial\eta_i}\, d\eta_1\, d\eta_2 \\ &= ETAIJ_e[B300IB] \end{aligned} \tag{8.66}$$

The *element-independent* master matrices defined in (8.66) are the following degree 1 hypermatrices.

$$[B3001B] = \frac{1}{36}\begin{bmatrix} -\begin{Bmatrix}6\\3\\1\\2\end{Bmatrix} & \begin{Bmatrix}6\\3\\1\\2\end{Bmatrix} & \begin{Bmatrix}2\\1\\1\\2\end{Bmatrix} & -\begin{Bmatrix}2\\1\\1\\2\end{Bmatrix} \\ -\begin{Bmatrix}3\\6\\2\\1\end{Bmatrix} & \begin{Bmatrix}3\\6\\2\\1\end{Bmatrix} & \begin{Bmatrix}1\\2\\2\\1\end{Bmatrix} & -\begin{Bmatrix}1\\2\\2\\1\end{Bmatrix} \\ -\begin{Bmatrix}1\\2\\2\\1\end{Bmatrix} & \begin{Bmatrix}1\\2\\2\\1\end{Bmatrix} & \begin{Bmatrix}1\\2\\6\\3\end{Bmatrix} & -\begin{Bmatrix}1\\2\\6\\3\end{Bmatrix} \\ -\begin{Bmatrix}2\\1\\1\\2\end{Bmatrix} & \begin{Bmatrix}2\\1\\1\\2\end{Bmatrix} & \begin{Bmatrix}2\\1\\3\\6\end{Bmatrix} & -\begin{Bmatrix}2\\1\\3\\6\end{Bmatrix} \end{bmatrix} \tag{8.67a}$$

$$[B3002B] = \frac{1}{36}\begin{bmatrix} -\begin{Bmatrix}6\\2\\1\\3\end{Bmatrix} & -\begin{Bmatrix}2\\2\\1\\1\end{Bmatrix} & \begin{Bmatrix}2\\2\\1\\1\end{Bmatrix} & \begin{Bmatrix}6\\2\\1\\3\end{Bmatrix} \\ -\begin{Bmatrix}2\\2\\1\\1\end{Bmatrix} & -\begin{Bmatrix}2\\6\\3\\1\end{Bmatrix} & \begin{Bmatrix}2\\6\\3\\1\end{Bmatrix} & \begin{Bmatrix}2\\2\\1\\1\end{Bmatrix} \\ -\begin{Bmatrix}1\\1\\2\\2\end{Bmatrix} & -\begin{Bmatrix}1\\3\\6\\2\end{Bmatrix} & \begin{Bmatrix}1\\3\\6\\2\end{Bmatrix} & \begin{Bmatrix}1\\1\\2\\2\end{Bmatrix} \\ -\begin{Bmatrix}3\\1\\2\\6\end{Bmatrix} & -\begin{Bmatrix}1\\1\\2\\2\end{Bmatrix} & \begin{Bmatrix}1\\1\\2\\2\end{Bmatrix} & \begin{Bmatrix}3\\1\\2\\6\end{Bmatrix} \end{bmatrix} \tag{8.67b}$$

## 8.8 TENSOR PRODUCT FACTORIZATION ALGORITHM

Equation (8.50) is the assembly expression of the two-dimensional finite element algorithm for (8.41) to (8.44). Upon realizing that the fluid convection, diffusion, and boundary convection contributions can each be resolved into components parallel to the global coordinate system $x_i$, we can accordingly write (8.50) as

$$\begin{aligned} S_e(([M]_e &+ \theta\,\Delta t[UX + UY + KX + KY + HX + HY]_e)\{\Delta Q\}_e \\ &- \Delta t(\{b_x + b_y\}_e - [UX + UY + KX + KY + HX + HY]_e\{Q\}_e^n)) = \{0\} \end{aligned} \tag{8.68}$$

On assembly over all $\Omega_e$ of $\Omega^h$, (8.68) becomes

$$([M] + \theta\,\Delta t[UX + UY + KX + KY + HX + HY])\{\Delta Q\} = -\Delta t\{R\}^n \tag{8.69}$$

where all the *data* terms, that is, those multiplied by $\Delta t$ in the second expression in (8.68), have been combined into the *residual* $\{R\}^n$.

The global rank matrix premultiplying $\{\Delta Q\}$ in (8.69) can be very large for a sufficiently dense mesh $\Omega^h$: Hence, a direct solution procedure will be extremely slow on the PC. Recall from Chap. 5 that the discussed *stationary* or *point* iteration procedures involved resolution of the matrix $[A]$ into a *sum* of matrices, for example, the lower $[L]$ and upper $[U]$ triangular forms plus the diagonal $[D]$. For the unsteady problem class now being considered, the appearance of the global *mass matrix* $[M]$ in (8.69) facilitates an approximate resolution of the system matrix $([M] + \theta\,\Delta t[\,\cdot\,])$ into a *product* of reduced rank matrices that is particularly amenable to an efficient "grid-sweeping" iteration procedure. Specifically, letting $[A]$ denote the system matrix in (8.69), the suggested resolution is

$$\begin{aligned}[A] &\equiv [M] + \theta\,\Delta t[UX + UY + KX + KY + HX + HY]\\ &\simeq [AX][AY]\end{aligned} \tag{8.70}$$

The product $[AX][AY]$ in (8.70) is indeed an approximation; in other words, it is inexact with a truncation error of order $O(\theta\,\Delta t)^2$. This error order is the same as that for the forward or backward Euler integration procedures [recall (8.15)], and is lower order than that of the trapezoidal rule (8.16). Hence, for relatively modest time steps $\Delta t$, one expects adequate performance as will become verified.

Combining (8.70) and (8.69) yields

$$[A]\{\Delta Q\} \simeq [AX][AY]\{\Delta Q\} = -\Delta t\{R\}^n \tag{8.71}$$

Making the definition $[AY]\{\Delta Q\} \equiv \{P\}$, (8.71) can be expressed as the two-step solution sequence

$$\begin{aligned}[AX]\{P\} &= -\Delta t\{R\}^n = -\Delta t\{R(Q^n)\}\\ [AY]\{\Delta Q\} &= \{P\}\\ \{Q\}^{n+1} &= \{Q\}^n + \{\Delta Q\}\end{aligned} \tag{8.72}$$

where $\{P\}$ is simply an intermediate (column matrix) array of data.

The resolution (8.70) and approximate solution sequence (8.72) is of little value if $[AX]$ and/or $[AY]$ is the same "size" as $[A]$. This is not the case, however, since the element mass matrix $[M]_e$ can be exactly resolved into factors $[MX]_e$ and $[MY]_e$, whereupon an excellent approximate construction for the factored matrices in (8.72) is

$$\begin{aligned}[AX] &\equiv S_e([AX]_e) = S_e([MX]_e + \theta\,\Delta t[UX + KX + HX]_e)\\ [AY] &\equiv S_e([AY]_e) = S_e([MY]_e + \theta\,\Delta t[UY + KY + HY]_e)\end{aligned} \tag{8.73}$$

In (8.73), every element matrix on the right side is formed using the *one-dimensional* master matrix library. Hence, both $[AX]$ and $[AY]$ are tridiagonal, for the linear basis formulation, and the computer resources required to equation-solve (8.72) are thus truly minimal.

The theoretical support behind the formulational structure (8.72) and (8.73) involves the concept of tensor (outer) matrix products, which lies well beyond the introductory level of this development. The tensor factorization (8.71) and (8.73) belongs to the family of *alternating-direction-implicit* (ADI) linear algebra procedures, which replace large sparse matrix operations with tridiagonal grid-sweeping methods.

For a quadrilateral region discretization $\Omega^h$, constituted of $I$ rows by $J$ columns of nodes, the (sparse) matrix $[A]$ in (8.70) is of the form

$$[A] = \left[ \; [0] \quad +J \quad [0] \quad -J \quad [0] \quad [0] \; \right] \qquad (8.73a)$$

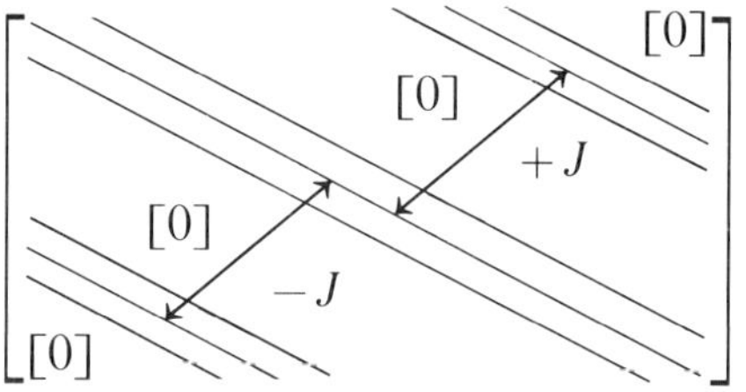

For columnwise node numbering, the length of the main diagonal in (8.73*a*) is $I \times J$, nodes $j-1$, $j$, and $j+1$ are coupled via the three main diagonals, while nodes $j-1 \pm J$, $j \pm J$, $j+1 \pm J$ are coupled across a submatrix of zeros of span $\pm J$, as indicated. The matrix approximate factorization (8.70), in concert with the grid-sweeping linear algebra procedure (8.72), decouples the $\pm J$ span matrix terms from those lying directly adjacent to the main diagonal.

Using a geometrical interpretation, (8.70) thus replaces $[A]$ above with the essential form

$$[AX][AY] \simeq \left[ \; [0] \quad [0] \; \right] \otimes \left[ \; [0] \quad [0] \; \right] \qquad (8.73b)$$

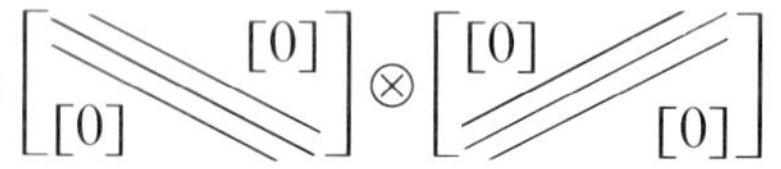

where the symbol "$\otimes$" denotes the outer (tensor) matrix product. The first step in (8.72) then involves solution of the tridiagonal form,

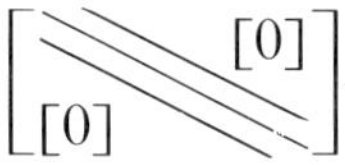

$$\left[ \; [0] \quad [0] \; \right] \{P\} = -\Delta t \{R\} \qquad (8.73c)$$

The data in $\{P\}$ is then shuffled, to exchange the intrinsic column ordering to row ordering, which operationally accounts for the tensor product $\otimes$ in (8.73*b*). The resultant solution of

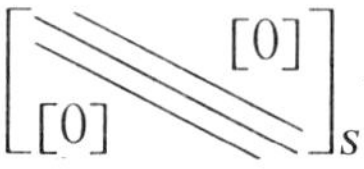

$$\left[ \; [0] \quad [0] \; \right]_S \{\Delta Q_S\} = \{P_S\} \qquad (8.73d)$$

then produces the solution matrix $\{\Delta Q\}$ ordered row-wise, as results from $\{P\}$ being so ordered, both of which are indicated by the subscripts $S$ in (8.73*d*).

As a final note, since node rows and columns have become essentially decoupled by the tensor factorization, the solution steps (8.73*c*) and (8.73*d*) forming (8.72) can be performed with matrices of main diagonal length $J$ and $I$, respectively, rather than $I \times J$. Hence, for an $11 \times 7$ nodal mesh, for example, then 11 matrix statements of the form (8.73*c*) would be solved first, either serially or in parallel, whereafter seven problem statements (8.73*d*) would be similarly solved following the column-row data reordering. Thus, the tensor matrix approximate factorization formulation, as a numerical linear algera procedure, is highly amenable to optimization for vector and/or parallel processing hardware architectures as they become available in the PC environment.

## 8.9 THE TIME-SPLITTING AND ADI APPROXIMATIONS

It is important to note, in the development of the tensor approximate factorization (TAF) procedure, that *no approximations* were made in forming the residual $\{R\}$ of the parent global matrix statement (8.69). Using the developed two-dimensional master matrix library, one can exactly form $\{R\}$ for any discretization $\Omega^h$. Therefore, the only approximation error lies in using (8.70) to replace the system matrix. The associated $O(\theta\, \Delta t)^2$ error can be totally eliminated if one recasts the parent statement (8.69) as a Newton iteration, at the expense of repeating the tridiagonal matrix solution sequence (8.72) a number of times at each time station $t^{n+1}$.

In our view, these complications lie beyond the scope of this introductory exposition. Alternatively, there are two further approximations for the TAF algorithm (8.70) to (8.72) that produce solution procedures highly amenable to the present-day PC environment. Both the *time-splitting* (TS) finite difference method of Yanenko (1971), and the familiar *alternating-direction-implicit* (ADI) finite difference procedure (Ames, 1977) can be directly interpreted within the TAF algorithm as simplifications to formation of the exact residual $\Delta t\{R\}^n$.

Comparing (8.68) and (8.69), the residual definition is

$$\{R\}^n = -\{b_x\} - \{b_y\} + [UX + UY + KX + KY + HX + HY]\{Q\}^n \tag{8.74}$$

In (8.74), all contributions are formed using full-dimensional element matrices, and $\{R\}^n$ contains contributions stemming from conservation law terms resolved parallel to the global coordinate system $x_i$. If one assumes (defines) that these contributions can be formed using the one-dimensional master matrix library, the resultant element-level residual contributions formed parallel to the $x_i$ coordinate directions are

$$\begin{aligned}\{RX\}_e &= ([UX]_e + [KX]_e + [HX]_e)\{Q\}_e - \{b_x\}_e \\ &= (\bar{u}_e[A201L] + \bar{k}_e l_e^{-1}[A211L] + \bar{h}_e[A0L])\{Q\}_e \\ &\quad - l_e[A200L]\{S\}_e - \bar{h}_e[A0L]\{TR\}_e \end{aligned} \tag{8.75a}$$

$$\begin{aligned}\{RY\}_e &= ([UY]_e + [KY]_e + [HY]_e)\{Q\}_e - \{b_y\}_e \\ &= (\bar{v}_e[A201L] + \bar{k}_e l_e^{-1}[A211L] + \bar{h}_e[A0L])\{Q\}_e \\ &\quad - l_e[A200L]\{S\}_e - \bar{h}_e[A0L]\{TR\}_e \end{aligned} \tag{8.75b}$$

The linear basis one-dimensional element library from (8.26) has been used for illustration in (8.75), and the hypermatrix forms (8.28) could replace the base forms if needed. The element measures $l_e$ appearing in (8.75) correspond to the appropriate coordinate direction, and the conductivity $k_e$ could certainly be given a directional character if desired. Finally, $\{R\} \simeq S_e(\{RX\}_e + \{RY\}_e)$ yields the resultant approximation to the multi-dimensional residual, if it were formed. However, this is not done in either the TS or ADI method, since the resolved approximations are always coupled with the resolved matrix factorization (8.73) on node rows and columns as appropriate.

As the final preparatory step, the base form of the tensor matrix factorization in (8.73) involves the one-dimensional master matrix library as follows.

$$\begin{aligned}[AX]_e &= [MX]_e + \theta\,\Delta t[UX + KX + HX]_e \\ &= l_e[A200L] + \theta\,\Delta t(\bar{u}_e[A201L] + \bar{k}_e l_e^{-1}[A211L] + \bar{h}_e[A0L]) \end{aligned} \tag{8.76a}$$

$$\begin{aligned}[AY]_e &= [MY]_e + \theta\,\Delta t[UY + KY + HY]_e \\ &= l_e[A200L] + \theta\,\Delta t(\bar{v}_e[A201L] + \bar{k}_e l_e^{-1}[A211L] + \bar{h}_e[A0L]) \end{aligned} \tag{8.76b}$$

Then, adding subscript $(I, J)$ on the assembly operator to denote the direction of one-dimensional assembly, that is, parallel to the grid-sweep direction, the TS simplification to (8.72) is the sequence

$$
\begin{aligned}
S_{e,J}([AX]_e)\{P\} &= -\tfrac{1}{2}\Delta t S_{e,J}(\{RX(Q)^n\}_e) \\
\{Q\}^{n+1/2} &\equiv \{Q\}^n + \{P\} \\
S_{e,I}([AY]_e)\{\Delta Q\} &= -\tfrac{1}{2}\Delta t S_{e,I}(\{RY(Q)^{n+1/2}\}_e) \\
\{Q\}^{n+1} &\equiv \{Q\}^{n+1/2} + \{\Delta Q\}
\end{aligned}
\tag{8.77}
$$

Conversely, the ADI simplification to (8.72) is the sequence

$$
\begin{aligned}
S_{e,J}([AX]_e)\{P\} &= -\tfrac{1}{2}\Delta t S_{e,I}(\{RY(Q^n)\}_e) \\
\{Q\}^{n+1/2} &\equiv \{Q\}^n + \{P\} \\
S_{e,I}([AY]_e)\{\Delta Q\} &= -\tfrac{1}{2}\Delta t S_{e,J}(\{RX(Q^{n+1/2})\}_e) \\
\{Q\}^{n+1} &\equiv \{Q\}^{n+1/2} + \{\Delta Q\}
\end{aligned}
\tag{8.78}
$$

The distinctions between the two approximations to the TAF algorithm are now quite apparent. First, both utilize the intermediate solution data array $\{P\}$ to update the dependent variable at the half time interval. Then, this value $\{Q\}^{n+1/2}$ is used for a directional residual evaluation during the second solution sweep, in distinction to direct use of $\{P\}$ in (8.72). The TS procedure (8.77) assembles the factor matrix and the one-dimensional residual in the same direction, thus, subscripts $J$ or $I$ appear uniformly in each equation. In distinction, the ADI procedure (8.78) crosses $J$ and $I$ in each directional equation, and hence requires data for the directional residual that does not lie on the current sweep row or column. Finally, the length of the diagonal in $[AX]$ or $[AY]$ is always $I$ or $J$, not the product; hence either procedure is also amenable to vector or parallel processing considerations.

Considering all aspects of the finite element weak statement approximation for the two-dimensional conservation law system (8.41) to (8.44), the logical *LEARN.FE* implementation utilizes the time-split formulation (8.77) as a direct extension on the one-dimensional methodology. The next section highlights code use for representative problem statements.

## 8.10 PRACTICAL PROBLEMS IN TWO DIMENSIONS

As in the one-dimensional case, we select a representative steady state and unsteady problem statement to evaluate the *LEARN.FE* implementation for the time-split linear basis algorithm.

## Steady-State Source Dispersion

Consider an industrial process wherein a stack emits an effluent into the atmosphere at a steady rate. Assume that a steady wind is blowing across the top of the stack parallel to the $x$ coordinate axis. Further assume that the effluent diffuses only laterally, i.e., in the $y$-direction; hence the diffusion scalar coefficient parallel to the wind direction is zero. Figure 8.4 illustrates this problem statement. Set up the definition according to the TS algorithm statement (8.75) to (8.77) for the following data:

$$\bar{u}_e \equiv \bar{u} = 0.5 \text{ m/s}$$

$$\bar{v}_e \equiv \bar{v} = 0.$$

$$\bar{k}_x = 0.$$

$$\bar{k}_y = 0.02 \text{ m}^2/\text{h}$$

$$s = 100 \text{ g/(h} \cdot \text{m}^2)$$

**Code Exercise 8.6.** Access the base input file for this *LEARN.FE* code exercise, and the file following will appear on your monitor. It defines a suitable grid and integration parameter set for proceeding to the steady state via *INTE*. Compute the steady-state atmospheric distribution of the stack effluent, and compare the results to the exact solution [c.f., Baker (1983), p. 213]. Figure 8.5 summarizes this comparison for reference.

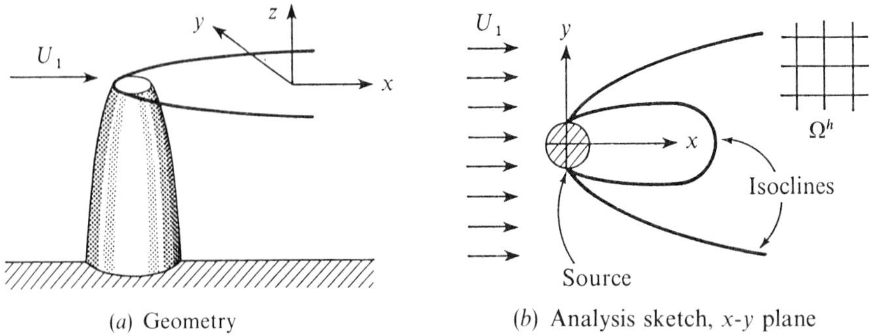


(*a*) Geometry (*b*) Analysis sketch, $x$-$y$ plane

**FIGURE 8.4**
Illustration of the steady-state dispersion problem.

```
TITL   ***** CODE EXERCISE 8.6, 2D GAUSSIAN PLUME   **********

TYPE    [ K   N   NNODEL   REFL   NPR   NTRAN   NAXI NZSTR ]
         1    2    4    1.    1    1    0    0    0

PRIN    [ NBUG(*) ]
         0       0       0       4      24       0

GRID    [ FOR N ( XL XR(*) PR(*) ),NEM(PR *) ]
        0.0000     20.0000      1.0000
        20
        0.0000     14.0000      1.0000
        14

MATL [FOR NPR (KL(*), KR(*))]
        1  0.10000  0.10000

UVEL [OPTION, MAG]
        1    1.0000

SORC [NSORC, JS(*), S(*)]
     9  0
   128 129 130 149 150 151 170 171 172
   25.   50.   25.
   50. 100.   50.
   25.   50.   25.

DIRI [(NDIR, NODE(*) , BCODE) , FACE,  QB(*)]
     0   0  3
     1
     0.
     3
     0.
     4
     0.

INTE   [TZERO TSTOP TSTEP THETA   DQMIN DQMAX TSTEPI NTSMAX]
           0.0 100.0   0.7   1.0  0.0001  10.0  0.001    300

STOP
```

## Unsteady Source Convection with No Diffusion

Consider the situation where a known initial distribution for some scalar field is moved by fluid convection around the two-dimensional plane in the absence of diffusion. The correct solution is known; that is, the distribution at any time $t^n$ remains unchanged from the initial data except for appearing at a different location in $\Omega$. A very demanding and highly informative example test specification is the "rotating cone" problem. The initial condition is a smooth bell-shaped curve rotated around its centroidal axis, such that it generates an axisymmetric distribution of concentration contours. This distribution then traverses a curved trajectory in the $x$–$y$

Finite element, $k = 1$

| | | | | | | | | | | | | | | | | | | | | |
|---|---|---|---|---|---|---|---|---|---|---|---|---|---|---|---|---|---|---|---|---|
| | | | | | | | | | | | | | | | | | | | 0 | 0 |
| | | | | | | | | | | | | | | | | | | 1 | 1 | 1 |
| | | | | | | | | | | | | 1 | 1 | 1 | 2 | 2 | 2 | 3 | 3 | 3 |
| | Source Region | | | | | | | 1 | 2 | 2 | 3 | 4 | 5 | 6 | 7 | 8 | 9 | 10 | 11 | 10 |
| | | | | 1 | 3 | 5 | 8 | 10 | 12 | 14 | 16 | 18 | 20 | 21 | 23 | 24 | 25 | 26 | 27 | 26 |
| | | 5 | 15 | 23 | 30 | 34 | 38 | 41 | 43 | 45 | 46 | 47 | 48 | 48 | 48 | 49 | 49 | 49 | 49 | 48 |
| 0 | 8 | 51 | 92 | 99 | 97 | 95 | 93 | 90 | 88 | 86 | 83 | 81 | 79 | 78 | 76 | 74 | 73 | 71 | 70 | 71 |
| 0 | 15 | 89 | 154 | 154 | 141 | 131 | 123 | 116 | 111 | 106 | 101 | 97 | 94 | 91 | 88 | 85 | 83 | 81 | 79 | 81 |
| U → 0 | 25 | 98 | 162 | 166 | 145 | 137 | 129 | 121 | 115 | 109 | 105 | 100 | 97 | 93 | 90 | 88 | 85 | 83 | 81 | Analytic |
| −2 | 23 | 90 | 150 | 160 | 148 | 138 | 130 | 123 | 117 | 111 | 107 | 102 | 99 | 95 | 92 | 89 | 87 | 84 | 82 | 81 |
| −1 | 13 | 51 | 89 | 102 | 100 | 98 | 95 | 93 | 90 | 88 | 86 | 84 | 82 | 80 | 78 | 76 | 75 | 73 | 72 | 71 |
| | 1 | 5 | 13 | 21 | 27 | 31 | 35 | 38 | 41 | 43 | 44 | 45 | 46 | 47 | 47 | 48 | 48 | 48 | 48 | 48 |
| | | | −1 | −1 | 0 | 2 | 5 | 7 | 10 | 12 | 14 | 16 | 18 | 19 | 21 | 22 | 23 | 24 | 25 | 26 |
| | | | | | | | | | 1 | 2 | 3 | 4 | 4 | 5 | 6 | 7 | 8 | 9 | 10 | 10 |
| | | | | | | | | | | | | 1 | 1 | 1 | 1 | 2 | 2 | 3 | 3 | 3 |
| | | | | | | | | | | | | | | | | | | 1 | 1 | 1 |
| | | | | | | | | | | | | | | | | | | | 0 | 0 |

Finite element, $k = 2$

**FIGURE 8.5**
Comparison of $M = 20 \times 14$ linear and quadratic basis nodal solutions for dispersion of an effluent, steady state.

plane by imposing a solid body rotation velocity field of the form $\mathbf{u}(x, y) = r\dot{\theta}\mathbf{e}_\theta$, where $r$ is the radial coordinate, $\dot{\theta}$ is the angular velocity, and $\mathbf{e}_\theta$ is the unit vector parallel to the angular direction.

Figure 8.6*a* gives a perspective view of such an initial distribution, as a solution surface plot on an $M = 32 \times 32$ uniform discretization of $\Omega$. It also coincides with the exact solution after any number of complete revolutions about the $x$–$y$ plane; that is, the distribution returns periodically to its original location. Figure 8.6*b* presents the surface plot of the linear basis TS algorithm solution, as obtained using a nondimensional time-step $\Delta t = 0.5$ to integrate over the exact elapsed time needed to return the analytical solution to the initial location. The finite element solution is quite good, even though a trailing dispersive error wake is evident, and the peak has not quite returned to its initial location. Other solution algorithms, such as a conventional second-order accurate finite difference procedure, will produce approximate solutions that are much poorer in comparison. The following code exercise will guide you in conducting a range of computational experiments for this demanding problem statement.

**Code Exercise 8.7.** Access the input file for this code exercise, which is set up to conduct the rotating cone experiment for the linear basis TS algorithm on a uniform quadrilateral element discretization of the unit square. The nondimensional time-step is $\Delta t = 0.25$, and the final time of $t_f = 16$ is exact for one-half revolution of the distribution. Execute this file, and record the nodal solution at $t_f$. Then double and halve $\Delta t$, reexecute the problem

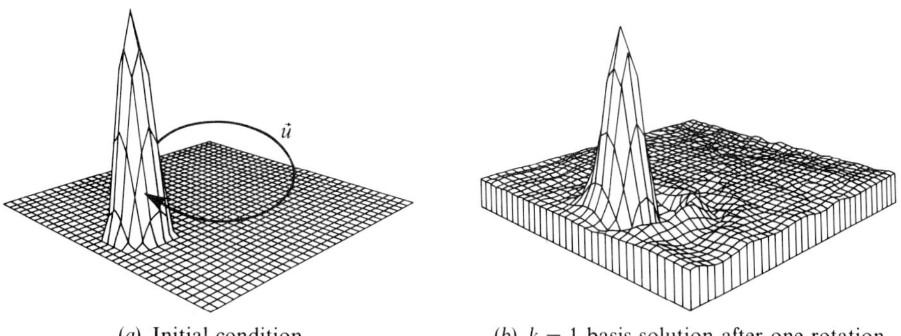


(*a*) Initial condition (*b*) $k = 1$ basis solution after one rotation

**FIGURE 8.6**
The rotating cone pure convection problem. (*a*) Initial condition; (*b*) $k = 1$ basis solution after one rotation, from Baker (1983, p. 212).

and compare the three solutions to the exact solution (the nodal initial condition). Hence, change the specification to $k = 2$ and repeat these three executions. Then draw conclusions about the need to limit nondimensional $\Delta t$ for fluid convection problems without diffusion, and the influence on accuracy that the more complete trial space exerts for this type of problem.

```
TITL
***  CODE EXERCISE 8.7, 2D ROTATING CONE, HALF TURN, LINEAR BASIS ***

TYPE     [ K   N   NNODEL   REFL   NPR   NTRAN   NAXI  NZSTR ]
           1   2    4    1.    1    1    0    0    0

PRIN     [ NBUG(*) ]
           0    0    0    4    16    0

GRID     [ FOR N ( XL XR(*) PR(*) ), NEM(PR *) ]
         0.0000      1.0000       1.0000
         32
         0.0000      1.0000       1.0000
         32

UVEL  [(OPTION, MAG), CENTROID]
         3        0.2
         0.5000       0.5000

INIT  [NINIT, NODE(*),QZERO(*)]
        45   0    0 0.00000
      434 435 436 437 438
      466 467 468 469 470 471 472
      499 500 501 502 503 504 505
      532 533 534 535 536 537 538
      565 566 567 568 569 570 571
      598 599 600 601 602 603 604
      632 633 634 635 636
      02.      10.      15.      10.      02.
      02.      20.      41.      50.      41.      20.      02.
      10.      41.      72.      85.      72.      41.      10.
      15.      50.      85.     100.      85.      50.      15.
      10.      41.      72.      85.      72.      41.      10.
      02.      20.      41.      50.      41.      20.      02.
      02.      10.      15.      10.      02.

DIRI  [(NDIR, NODE(*) , BCODE) , FACE,  QB(*)]
      0    0  4
      1
      0.
      2
      0.
      3
      0.
      4
      0.

INTE    [TZERO TSTOP TSTEP THETA      DQMIN DQMAX TSTEPI NTSMAX]
            0.0   16.0   0.25    0.5 0.000001 200.0    0.25       64

STOP
```

## 8.11 SUMMARY

This draws to a close the topical coverage in this text. It is your authors' sincere hope you have found this material, and the presentation format, of significant help in establishing a feeling of comfort with "the finite element method." The material presented here has remained introductory throughout; hence certain issues of more theoretical nature have been passed over. The reader interested in such matters, as well as in the extension to more comprehensive problem statements, should now possess the knowledge base needed to approach the literature with confidence. We certainly hope so!

## 8.12 ADDITIONAL REFERENCES

Ames, W. F.: "Numerical Methods for Partial Differential Equations," Academic, New York (1977)

Gresho, P. M. and R. L. Lee, (1979): "Don't Suppress the Wiggles, They're Telling You Something," in T. J. R. Hughes (ed.), *Finite Element Methods for Convection Dominated Flow*, Vol. AMD 41, American Society of Mechanical Engineers, New York.

Yanenko, N. N. (1971): *The Method of Fractional Steps*, Springer-Verlag, New York.

# APPENDIX

# *LEARN.FE* KEY SUBROUTINE LISTINGS

MAIN
READIT
ELMAT
FORMIT
BLOCK DATA

```
C     ...................................................................
C     .                                                                 .
C     .                         LEARN . FE                              .
C     .                                                                 .
C     .        ***    SINGLE VARIABLE FINITE ELEMENT CODE     ***       .
C     .        ***         FOR LINEAR ELLIPTIC AND            ***       .
C     .        ***    INITIAL - BOUNDARY VALUE PROBLEMS       ***       .
C     .                                                                 .
C     .    FEATURES (V6.5, 10/90) :                                     .
C     .                                                                 .
C     .       **  COMMAND NAME DRIVEN                                   .
C     .       **  ONE AND TWO DIMENSIONAL GEOMETRIES                    .
C     .       **  LINEAR, QUADRATIC AND CUBIC BASES IN 1-D              .
C     .       **  ISOPARAMETRIC GRID TRANSFORMATIONS                    .
C     .       **  TRIANGLE AND QUADRILATERAL BASES IN 2-D               .
C     .       **  AUTOMATED NON-UNIFORM BLOCK DISCRETIZATIONS           .
C     .       **  DIRICHLET, ROBIN AND NEUMANN BOUNDARY CONDITIONS      .
C     .       **  VARIABLE MATERIAL PROPERTIES                          .
C     .       **  MULTIPLE SOURCE DISTRIBUTIONS                         .
C     .       **  IMPOSED FLUID VELOCITY CONVECTION FIELDS              .
C     .       **  BANDED NON-SYMMETRIC EQUATION SOLVERS                 .
C     .       **  UNSTEADY SOLUTIONS VIA MATRIX TENSOR PRODUCTS         .
C     ...................................................................
C
C     LEARN.FE IS A COPYRIGHTED PRODUCT DISTRIBUTED WITH THE TEXTBOOK
C
C                       " FINITE ELEMENTS 1-2-3 "
C
C     WRITTEN BY A.J. BAKER AND D.W. PEPPER AND PUBLISHED BY McGRAW-HILL
C     IN 1990.  FOR WIDEST COMPATIBILITY, THE DISTRIBUTION VEHICLE IS
C     THE ENCLOSED LOW-DENSITY (360 KBYTE) 5.25 INCH DISKETTE CONTAINING
C     THE EXECUTABLE CODE, THE GRAPHICS SYSTEM, THE FULL COMPLEMENT OF
C     CODE EXERCISE INPUT FILES AND SELECTED CODE EXECUTION OUTPUT FILES.
C
C     LEARN.FE IS WRITTEN IN ANSI STANDARD FORTRAN 77, AND WILL OPERATE
C     DIRECTLY ON ANY PC OR PC-COMPATIBLE COMPUTER WITH 640 KBYTE MEMORY,
C     A MATH CO-PROCESSOR, AND DISK OPERATING SYSTEM DOS 3.0 OR HIGHER.
C
C     THE LOW DENSITY DISK FORMAT DOES NOT ALLOW FOR INCLUSION OF THE
C     LEARN.FE SOURCE CODE.  THIS APPENDIX CONTAINS SOURCE LISTINGS FOR ALL
C     SUBROUTINES THAT YOU WOULD UPDATE OR MODIFY FOR CODE EXERCISES OR FOR
C     CODE CAPABILITY EXTENSION.  TO ACCOMPLISH CODE EXTENSION OF COURSE
C     REQUIRES THAT YOU HAVE A FORTRAN COMPILER AND EDIT CAPABILITY.
C
C     FOR USERS INTERESTED IN SUBSTANTIAL EXTENSION, THE LEARN.FE SOURCE
C     CODE IS AVAILABLE WITH FULLY PAID LICENSE FROM COMCO*.  COMPLETE THE
C     REQUEST AND DISCLAIMER FORM CONTAINED WITH THE DISKETTE AND MAIL IT
C     TO COMCO, OR MAKE A DIRECT CONTACT. FOR USERS WITH A MAC II COMPUTER,
C     THE TAILORED EXECUTABLE LEARN.FE VERSION IS AVAILABLE FROM COMCO.
C
C     YOUR AUTHORS HOPE THE COMBINATION OF "FINITE ELEMENTS 1-2-3" AND
C     THE LEARN.FE CODE PROVIDE FOR YOUR ACQUISITION OF FINITE ELEMENT
C     ANALYSIS LITERACY.  WE WELCOME YOUR COMMENTS AND SUGGESTIONS.
C
C                 * COMPUTATIONAL MECHANICS CORP. (COMCO)
C                       601 CONCORD ST, SUITE LL-C
C                       KNOXVILLE, TENNESSEE  37919
C                          VOICE: (615) 546-3664
C                            FAX: (615) 546-7463
C
C
C
C
C
C
```

```
C***********************   LEARN.FE COMMAND NAME NOMENCLATURE   ********************
C
C     THE FOLLOWING ORDER IS THAT OF APPEARANCE IN THE CODE EXERCISES
C
C     TITL
C           READS ARRAY IN A-FORMAT AND THEN OUTPUTS IT
C
C     TYPE [ K    N   NNODEL   REFL    NPR   NTRAN   NAXI   NZSTR  ]
C           K.........FINITE ELEMENT BASIS DEGREE
C           N.........DIMENSION OF PROBLEM (N=1 OR N=2 ONLY)
C           NNODEL....# OF NODES / ELEMENT (DEPENDS ON K AND N)
C           REFL......AVAILABLE FOR NON-D LENGTH SCALE
C           NPR.......# OF MACRO DISCRETIZATION PROGRESSION RATIOS
C           NTRAN....INDEX DEFINING PROBLEM GEOMETRY
C                    = 0 = RECTANGULAR CARTESIAN
C                    = 1 = GENERAL STRAIGHT-SIDED QUADRILATERAL
C                    = 2 = PARABOLIC CURVED-SIDED QUADRILATERAL
C                    = 3 = REGION BETWEEN TWO CONCENTRIC CIRCULAR ARCS,
C                          GRID FORMED OF RADIAL AND CIRCUMFERENTIAL ARCS.
C                          (SEE NOTES UNDER "GRID" ON DATA STRUCTURES)
C           NAXI......(0,1) SWITCH FOR RECTILINEAR / AXISYMMETRIC PROBLEM
C           NZSTR.....(0,1) SWITCH FOR FINNED TUBE HEAT EXCHANGER
C
C     PRIN [ NBUG(*) ]
C           NBUG(*)........OUTPUT PRINT CONTROLS
C                (1)... =1  PRINT 2D DIRICHLET BOUNDARY ARRAYS
C                (2)... =1  PRINT SYSTEM MATRICES (SMAT AND F)
C                (3)... =1  PRINT ELEMENT-NODE CONNECTION TABLE (MEL)
C                (4)... =0  FLOATING POINT FORMAT
C                (4)... =1  EXPONENTIAL FORMAT
C                (4)... =2  INTEGER FORMAT
C                (4)... =3  FLOATING POINT & INTEGER FORMATS
C                (5)    =X  TRANSIENT OUTPUT PRINTED EVERY "X" TIMESTEPS, PLUS AT
C                           TERMINATION (STEADY STATE, MAX TIMESTEPS, MAX TIME)
C                (6)    =1  CODE DEBUG PARAMETER, PRINTS EVERY STEP IN FORMIT
C                           CAUTION - PRODUCES LARGE OUTPUT FILES
C
C     GRID [ FOR N (XL XR(*)  PR(*)) , NEM(PR*) ]
C           N.........DIMENSION OF THE PROBLEM STATEMENT
C           XL........X COORDINATE OF THE DOMAIN LEFT END
C           XR(*).....X NODAL COORDINATE OF THE NEXT RIGHT MACRO DOMAIN
C                       (*=1,2 OR 3 FOR N = 1 AND 1 OR 2 FOR N = 2)
C           PR(*).....MESH GENERATION PROGRESSION FACTOR FOR NEXT MACRO DOMAIN
C                       (*= SAME VALUE AS FOR XR(*))
C           NEM(PR*)..NUMBER OF FINITE ELEMENTS TO BE GENERATED IN EACH MACRO
C                       DOMAIN FOR PR(*) DEFINITION
C
C           NOTE: THIS DATA STRING IS REPEATED FOR N = 2 PROBLEMS WITH XL, XR(*), ETC,
C                 INTERPRETATIONS CHANGED TO YL, YU(*), ETC. COORDINATES SPECIFYING
C                 GEOMETRY ARE INPUT IN THE BASE GLOBAL CARTESIAN REFERENCE FRAME.
C                 IN CASE OF A NON-CARTESIAN RECTANGULAR OR CURVE-SIDED DOMAIN
C                 GEOMETRY (NTRAN > 0), ADDITIONAL MACRO NODAL DATA REQUIRED INCLUDE
C                 COORDINATES OF THE MACRO VERTEX NODES (NTRAN = 1) AND MID-SIDE NODES
C                 (NTRAN = 2,3).  THESE ARE INPUT TO XC1-XC8 AND YC1-YC8 ARRAYS. NODAL
C                 ORDERING IS COUNTERCLOCKWISE AND DEFINITIONS ARE:
C
C                   XC1 - XC4    = X-COORDINATES OF VERTEX NODES
C                   YC1 - YC4    = Y-COORDINATES OF VERTEX NODES
C                   XC5 - XC8    = X-COORDINATES OF MID-SIDE NODES
C                   YC5 - YC8    = Y-COORDINATES OF MID-SIDE NODES
C
C     EXIT
C           RETURNS CODE EXECUTION TO TOP OF READIT FOR NEXT INPUT DATA FILE
C
C     STOP
C           HALTS CODE EXECUTION
```

```
C
C     MATL [ FOR NPR, KL(*) KR(*) ]       (FOR N = 1)
C           NPR.......THE NUMBER OF PROGRESSION FACTORS IN MESH
C           KL(*).....DIFFUSION COEFFICIENT FOR LEFT MACRO NODE
C           KR(*).... DIFFUSION COEFFICIENT FOR RIGHT MACRO NODE
C
C     MATL [ FOR MACRO(I), KI(*) ]       (FOR N = 2)
C           MACRO(I)..THE NUMBER OF MACRO ELEMENTS ( 1 OR 2 )
C           KI(*).....DIFFUSION COEFFICIENT IN MACRO ELEMENT
C
C     ROBN [ NROB, NODE(*), H(*)  TR(*) ]
C           NROB......THE NUMBER OF ROBIN (VARIBLE FLUX) B.C.
C           NODE(*)...NODE NUMBER FOR B.C. APPLICATION
C           H(*)......HEAT TRANSFER COEFFICIENT
C           TR(*).....HEAT EXCHANGE REFERENCE TEMPERATURE
C
C           NOTE:  SEVERAL DATA REPEAT OPTIONS ARE AVAILABLE TO REDUCE
C                  THE AMOUNT OF INPUT SPECIFICATION, SEE READIT LOOP.
C
C     DIRI [ NDIR, NODE(*), QB(*) ]
C           NDIR......THE NUMBER OF DIRICHLET (FIXED) B.C.
C           NODE(*)...NODE NUMBER FOR B.C. APPLICATION
C           QB(*).....THE VALUE TO BE ASSIGNED TO NODAL VARIABLE
C
C           NOTE:  SEVERAL DATA REPEAT OPTIONS ARE AVAILABLE TO REDUCE
C                  THE AMOUNT OF INPUT SPECIFICATION, SEE READIT LOOP.
C
C     NEUM [ NNEU, NODE(*), FF(*) ]
C           NNEU......THE NUMBER OF NEUMANN (FIXED FLUX) B.C.
C           NODE(*)...NODE NUMBER FOR B.C. APPLICATION
C           FF(*).....THE VALUE OF FIXED FLUX TO BE ASSIGNED
C
C     FORM
C           FORM THE PROBLEM SYSTEM MATRIX AND DATA COLUMN MATRIX FOR
C             A STEADY-STATE SOLUTION
C
C     SOLV
C           SOLVE THE SYSTEM MATRIX STATEMENT USING AN L-U DIRECT METHOD
C
C     NORM
C           COMPUTE THE ENERGY SEMI-NORM OF THE APPROXIMATE SOLUTION
C
C     SORC [ NSORC OPTION(*), CENTROID, PEAK  HALFWIDTH ]
C           NSORC.......THE NUMBER OF SOURCES, DEPENDENT UPON OPTION
C           OPTION.....=0 READ NODES AND CORRESPONDING SORC(*)
C                      =1 SINE-WAVEDISTRIBUTION
C                      =2 GAUSSIAN  DISTRIBUTION
C           CENTROID......X AND Y COORDINATE OF SOURCE PEAK
C           PEAK..........SOURCE AMPLITUDE
C           HALFWIDTH.....SOURCE HALFWIDTH
C
C           NOTE:  SEVERAL OPTIONS EXIST TO SIMPLIFY SOURCE DATA INPUT,
C                  INCLUDING INTERROGATION OF ANALYTICAL FUNCTIONS AFTER
C                  GRID GENERATION, SEE THE READIT INPUT STREAM.
C
C     GAUS [ CONVERGENCE ]
C           SOLVE THE SYSTEM MATRIX STATEMENT ITERATIVELY USING GAUSS-SEIDEL
C             TO CONVERGENCE SPECIFICATION
C
C     INIT [ (NINIT CODE(*) OPTION), NODES(*), QZERO(*) ]
C           NINIT.....>0 THE NUMBER OF NODAL DATA TO BE INSERTED
C           CODE(*)...=0 FOR USER SPECIFICATION
C                     =1 FOR N=2 ROTATING CONE
C           OPTION....=0 USER SPECIFICATION OF NODAL QZERO
C                     =1 INPUT NONZERO VALUES ONLY FOR QZERO
C                     =2 UNIFORM INITIAL CONDITION
```

```
C          NODES(*).... NODAL NUMBERS FOR DATA SPECIFICATION
C          QZERO(*).... NODAL INITIAL CONDITION
C
C     UVEL [ OPTION, MAG ]
C          OPTION....=1 FLOW PARALLEL TO X DIRECTION ONLY
C                    =2 FLOW PARALLEL TO Y DIRECTION ONLY
C                    =3 SOLID BODY ROTATION FLOW FIELD
C          MAG......... MAGNITUDE OF THE VELOCITY INPUT FIELD
C
C     INTE [ TZERO  TSTOP  TSTEP  THETA  DQMIN  DQMAX  TSTEPI  NTSMAX ]
C          TZERO.... INITIATION TIME FOR AN UNSTEADY SOLUTION
C          TSTOP.....THE TIME TO STOP THE INTEGRATION PROCESS
C          TSTEP.....THE MAXIMUM ALLOWED INTEGRATION TIME STEP
C          THETA.....TIME INTEGRATION IMPLICITNESS FACTOR
C                    =0.5 (1.0) FOR SECOND (FIRST) ORDER ACCURACY
C          DQMIN.....DEFINES STEADY STATE WHEN ALL INCREMENTS LESS THAN THIS
C          DQMAX.....REDUCE TIME STEP SIZE IF ANY INCREMENT EXCEEDS THIS
C          TSTEPI....INITIAL TIME INTEGRATION STEP SIZE.  IF TSTEPI = TSTEP,
C                    THEN TIME INTEGRATION PERFORMED AT FIXED TIME STEP
C                    INDEPENDENT OF ALL CONTROLS.
C          NTSMAX....THE MAXIMUM NUMBER OF TIME STEPS TO TERMINATE EXECUTION
C
C*****************************************************************************
*
C
C************* ADDITIONAL USEFUL INFORMATION REGARDING CODE VARIABLES *************
*
C
C
C     NFIX......# OF DIRICHLET BOUNDARY NODES
C     JFIX(*)...DIRICHLET BOUNDARY NODE NUMBERS KEYED TO DATA
C     NNEU(*)...# OF NEUMANN BOUNDARY NODES
C     JNEU(*)...NEUMANN BOUNDARY NODE NUMBERS KEYED TO DATA
C     NROB......# OF ROBIN BOUNDARY NODES
C     JROB(*)...ROBIN BOUNDARY NODE NUMBERS KEYED TO DATA
C     INPR......PROGRESSION RATIO NUMBER
C     NBAND.....BANDWIDTH OF SYSTEM MATRIX (SMAT)
C     NDOFS.....DEGREES OF FREEDOM IN THE SYSTEM
C     NNODES....TOTAL NUMBER OF NODES
C     NDOFN.....# OF DEGREES OF FREEDOM / NODE
C     NDOFE.....DEGREES OF FREEDOM / ELEMENT
C     NELES.....NUMBER OF ELEMENTS IN THE SYSTEM DISCRETIZATION.
C     NELE(*)...# OF ELEMENTS IN DISCRETIZATION / DIMENSION
C     NNOD(*)...# OF NODES / DIMENSION
C     AK(*).....NODAL THERMAL CONDUCTIVITY ARRAY
C     AU(*).....NODAL X DIRECTION FLUID VELOCITY
C     AV(*).....NODAL Y DIRECTION FLUID VELOCITY
C     EMAT(*,*).ELEMENT MATRIX LOCAL TO FORMIT
C     F(*)......GLOBAL DATA MATRIX
C     SMAT(*,*).ASSEMBLED SYSTEM MATRIX
C     MEL(*,*)..ELEMENT NODE CONNECTION ARRAY
C     ENORMH....ENERGY SEMI-NORM OF F.E. SOLUTION
C     TOL.......GAUSS-SEIDEL ITERATIVE CONVERGENCE TOLERANCE
C
```

```
C**********************************************************************
C     MAIN                                                            *
C                                                                     *
C     FUNCTION:  INITIALIZE NAMED COMMON ARRAYS AND DIRECT SOLUTION   *
C                 SEQUENCE FOR AS MANY INPUT FILES AS DESIRED         *
C**********************************************************************

      PROGRAM LEARN
      IMPLICIT REAL(A-H,O-Z)
      REAL*4 DURATION,START
      CHARACTER*80 TITLE
      CHARACTER FIN*15, FOUT*15
      COMMON/BIGMT/SMAT(5450),F(1090),Q(1090),JINI(1090),
     *XG(1090),YG(1090),AK(1090),AU(1090),AV(1090),SORC(1090),
     *XL(1090),JFIX(1090),MEL(4800)
      COMMON/CONST/K,N,NAXI,NNODEL,NBAND,NDOFS,NDOFN,MSW,NTRAN,INTEG,
     *NELES,NNODES,NSORC,MO,NPR,NFIX,INNEU,JK,IFNNEU,JNEU(68),
     *NNOD(3),NELE(3),NBUG(6),NEM(2,3),NPRA(2),NNEU(0:4),
     *DELTAT,TRNFLG,THETA,PREMAC,ENORMH,ZSTAR,
     *ANUMAN(2,68),REFL,XI(2),XF(2,3),PR(2,3)
      COMMON/MATX/A3000L(2,2,2),A3011L(2,2,2),A3000Q(3,3,3),
     *A3011Q(3,3,3),A3011C(2,4,4),A3011H(2,4,4),
     *A3001L(2,2,2),A3001Q(3,3,3),A200C(4,4),
     *B211B(4,4),B222B(4,4),B212B(4,4),B221B(4,4),
     *B200L(3,3),B200B(4,4)
      COMMON/COORDN/XC1(9),XC2(9),XC3(9),XC4(9),XC5(9),XC6(9),
     *XC7(9),XC8(9),YC1(9),YC2(9),YC3(9),YC4(9),
     *YC5(9),YC6(9),YC7(9),YC8(9)

      DATA IGOHON/1/

C     THE FOLLOWING COMMANDS, AND THOSE BEFORE "STOP", MAY BE
C     CHANGED OR COMMENTED DEPENDENT UPON SPECIFIC COMPUTER SYSTEM.


      OPEN (UNIT=5,FILE=FIN,STATUS='UNKNOWN')
      OPEN (UNIT=6,FILE=FOUT,STATUS='UNKNOWN')
C     START THE CLOCK TIMING AT ZERO

C     START = SECNDS(0.0)

C     LEARN F.E. IS WRITTEN FOR EITHER SINGLE OR DOUBLE PRECISION. IT
C     CAN BE CHANGED TO DOUBLE PRECISION
C     BY CHANGING ALL 'REAL' TO 'DOUBLE PRECISION' OR 'REAL*8' ON SOME MACHINES
C         AND CHANGING ALL 'E0' CONSTANTS TO 'D0' CONSTANTS
C     ALSO CHANGE DATA STATEMENT FOR COMMON VARIABLE 'PREMAC' TO
C     APPROXIMATE MACHINE PRECISION (DEFAULT 1.0E-07)


C     THE DIMENSIONED SETTINGS OF THE PROGRAM ARE:
C          1.  1090  SYSTEM DEGREES OF FREEDOM (NDOFS)
C              (20*20) LINEAR 2-D GRID=(21X21) NODES
C          2.  3270 SMAT (SYSTEM MATRIX) . THE LARGEST 2-D DIRECT SOLVE
C              AVAILABLE IS M=32X32 TRIANGLE OR 16X16 QUADRATIC
C              QUADRILATERAL ELEMENTS. REQUIRED SMAT SIZE
C              IS (TOTAL # OF NODES)*(3+2*(# OF XNODES)) OR
C              (16+1)*(16+1)*(3+2*(16+1)) FOR 16X16.THEN ROUND SOLUTION
C              UP TO THE NEAREST MULTIPLE OF 15. SMAT DIMENSION CAN BE
C              EXPANDED IF COMPUTER MEMORY IS SUFFICIENT.
C              REQUIRED SMAT SIZE IN 1-D OR 2-D TIMESPLIT IS 5*
C              MAX(# OF XNODES,# OF YNODES).
C          3.   4800 MEL ENTRIES (ELEMENT CONNECTION ARRAY, 20*14*4)
C          4.   68  NEUMANN   B.C. NODES (JNEU)
C          5.   68  DIRICHLET B.C. NODES (JFIX)
C          6.   68  ROBIN B.C.NODES (JROB)
```

```
C         1.    ESTABLISH FINITE ELEMENT MASTER MATRIX LIBRARY.

      CALL ELMAT

C     THE INTEGER IGOHON IS A CONTINUATION SWITCH:
C         IGOHON =1 PROGRAM CYCLES THROUGH DATA FILES
C         IGOHON =0 AT END OF LAST DATA FILE
C
C         2.     FOR EACH DATA CASE:

 101   IF (IGOHON .EQ. 1) THEN

C         3. INITIALIZE NAMED COMMON

             TITLE=' '

C         4. READ DATA, FILL COMMON AND PERFORM SOLUTION

             CALL READIT(IGOHON,ISOLV,IPLOT)


             IF (IPLOT.EQ.1) CALL PLOTIT
       GO TO 101
       END IF
C      DURATION=SECNDS(START)
C      WRITE(0,*)'DURATION OF PROGRAM =',DURATION,'SECONDS'
       CLOSE (UNIT=6)
       CLOSE (UNIT=5)
C      PAUSE
       STOP
      END
```

```
      SUBROUTINE READIT(IGOHON,ISOLV,IPLOT)
C***********************************************************************
C     DEFINE PROBLEM STATEMENT USING PROGRAM EXECUTION                 *
C     COMMAND NAME SEQUENCE                                            *
C***********************************************************************
      IMPLICIT REAL(A-H,O-Z)

      COMMON/BIGMT/SMAT(5450),F(1090),Q(1090),JINI(1090),
     *XG(1090),YG(1090),AK(1090),AU(1090),AV(1090),SORC(1090),
     *XL(1090),JFIX(1090),MEL(4800)
      COMMON/CONST/K,N,NAXI,NNODEL,NBAND,NDOFS,NDOFN,MSW,NTRAN,INTEG,
     *NELES,NNODES,NSORC,MO,NPR,NFIX,INNEU,JK,IFNNEU,JNEU(68),
     *NNOD(3),NELE(3),NBUG(6),NEM(2,3),NPRA(2),NNEU(0:4),
     *DELTAT,TRNFLG,THETA,PREMAC,ENORMH,ZSTAR,
     *ANUMAN(2,68),REFL,XI(2),XF(2,3),PR(2,3)
      COMMON/MATX/A3000L(2,2,2),A3011L(2,2,2),A3000Q(3,3,3),
     *A3011Q(3,3,3),A3011C(2,4,4),A3011H(2,4,4),
     *A3001L(2,2,2),A3001Q(3,3,3),A200C(4,4),
     *B211B(4,4),B222B(4,4),B212B(4,4),B221B(4,4),
     *B200L(3,3),B200B(4,4)
      COMMON/COORDN/XC1(9),XC2(9),XC3(9),XC4(9),XC5(9),XC6(9),
     *XC7(9),XC8(9),YC1(9),YC2(9),YC3(9),YC4(9),
     *YC5(9),YC6(9),YC7(9),YC8(9)

      CHARACTER*4 NAMES(19),NAME
      CHARACTER*80 TITLE
C****************************************************
C    COMMAND NAMES
      DATA NAMES / 'TITL', 'TYPE', 'PRIN',
     *'GRID', 'MATL', 'DIRI',
     *'ROBN', 'NEUM', 'FORM',
     *'SORC', 'SOLV', 'INTE',
     *'NORM', 'GAUS', 'INIT',
     *'UVEL', 'GRAF', 'EXIT',
     *'STOP'/
C****************************************************
      WRITE(6,9520)
      WRITE(6,*) ('*',I=1,76)
      WRITE(6,9420)
      WRITE(6,*) ('*',I=1,76)
      IFAIL - 0
      ISOLV=0
      IPLOT=0
      IENERG=0
      INNEU=0
      IFNNEU=0
      NSORC=0
      NFIX=0
      ZSTAR=0.
      ENORMH=0.
      DO 12 I=1,4800
      MEL(I)=0
 12   CONTINUE
      DO 11 I=1,1090
      AK(I)=0.
      AU(I)=0.
      AV(I)=0.
      SORC(I)=0.
      XG(I)=0.
      YG(I)=0.
      F(I)=0.
      Q(I)=0.
 11   CONTINUE
      DO 2 I=1,68
      JNEU(I)=0.
      DO 2 J=1,2
```

```
      ANUMAN(J,I)=0.
 2    CONTINUE
      NNEU(0)=0
      NNEU(1)=0
      NNEU(2)=0
      NNEU(3)=0
      NNEU(4)=0
      NAME=NAMES(1)
 102  IF ((NAME .NE. NAMES(18)) .AND. (NAME .NE. NAMES(19))) THEN
      READ(5,9500)  NAME
      IF (NAME .EQ. '    ') GOTO 102
      IF (NAME .EQ. NAMES(1)) THEN
C**********************************************************
C              1. READ TITLE.(TITL).
C**********************************************************

      READ(5,'(A80)')TITLE
      WRITE(6,9700) NAME,TITLE

      ELSE IF (NAME .EQ. NAMES(2)) THEN
C**********************************************************
C              2. READ PROBLEM AND SOLUTION DEFINITIONS .(TYPE).
C**********************************************************
      READ(5,*)K,N,NNODEL,REFL,(NPRA(I),I=1,N),NTRAN,NAXI,MSW
      IF (N .EQ. 1) THEN
      WRITE(6,*)
     .' TYPE:    K  N  NNODEL  REFL  NPR1 NTRAN  NAXI  NZTR'
      WRITE(6,9533)K,N,NNODEL,REFL,NPRA(1),NTRAN,NAXI,MSW
      ELSE
      WRITE(6,*)
     .' TYPE:    K  N  NNODEL  REFL  NPR1  NPR2  NTRAN  NAXI  NZTR'
      WRITE(6,9535)K,N,NNODEL,REFL,NPRA(1),NPRA(2),NTRAN,NAXI,MSW
      ENDIF
      IF (K.EQ.4) THEN
      NDOFN=2
      ELSE
      NDOFN=1
      ENDIF


      ELSE IF (NAME .EQ. NAMES(3)) THEN
C*****************************************************
C               3.  READ INPUT/OUTPUT PRINT CONTROL PARAMETERS.(PRIN).
C                   DEFAULT: NBUG(*)=0
C*****************************************************

      READ(5,*)(NBUG(I),I=1,6)
      WRITE(6,*) ' PRIN:  NBUG1  NBUG2  NBUG3  NBUG4  NBUG5  NBUG6'
      WRITE(6,9534)(NBUG(I),I=1,6)
      NBUG7=0
      IF (NBUG(2) .EQ. 1) NBUG7=1
      IF (NBUG(4) .EQ. 2) THEN
C                  NBUG(4)=0
      INTEG=1
      ELSE IF (NBUG(4) .EQ. 3) THEN
C                  NBUG(4)=0
      INTEG=2
      ELSE
      INTEG=0
      END IF

      ELSE IF (NAME .EQ. NAMES(4)) THEN
C*******************************************************
C               4. READ SOLUTION DOMAIN GEOMETRY AND DISCRETIZATION
C                  MACROELEMENTS.(GRID).
C*******************************************************
```

```
      DO 410 I=1,N
      WRITE(6,*) '  GRID:        NI  NPRI NTRAN'
      WRITE(6,9532) I,NPRA(I),NTRAN
      IF(I.EQ.1)WRITE(6,*)
     *'                XL       XR1       PR1       XR2',
     *'       PR2      XR3      PR3'
      IF(I.EQ.2)WRITE(6,*)
     *'                YL       YU1       PR1       YU2',
     *'       PR2      YU3      PR3'
      READ(5,*)XI(I),(XF(I,J),PR(I,J),J=1,NPRA(I))
      WRITE(6,9552)XI(I),(XF(I,J),PR(I,J),J=1,NPRA(I))
      READ(5,*)(NEM(I,J),J=1,NPRA(I))
      WRITE(6,*) '               NEM1  NEM2  NEM3'
  410 WRITE(6,9532)(NEM(I,J),J=1,NPRA(I))

C     DEFINE REFERENCE LENGTH SCALE
      IF(REFL.EQ.0.) REFL=1.
C
C    ISOPARAMETRIC GRID DEFINITION SEQUENCES (NTRAN > 0)
C


C
C---- INPUT CORNER NODE COORDINATES. ( NTRAN = 1 )
C
      IF ( NPRA(1) .EQ. 1 ) NPRA1=1
      IF ( NPRA(1) .GE. 2 ) NPRA1=2
      DO 420 I=1,NPRA1
      J=1
      LI=(I-1)*NPRA(1)+J
      IF (NTRAN .GE. 1 .AND. NTRAN .LE. 2) THEN
      READ(5,*) XC1(LI),XC2(LI),XC3(LI),XC4(LI)
      READ(5,*) YC1(LI),YC2(LI),YC3(LI),YC4(LI)
      WRITE(6,9402)
      WRITE(6,9551) XC1(LI),XC2(LI),XC3(LI),XC4(LI)
      WRITE(6,9406)
      WRITE(6,9551) YC1(LI),YC2(LI),YC3(LI),YC4(LI)
      XC1(LI)=XC1(LI)/REFL
      XC2(LI)=XC2(LI)/REFL
      XC3(LI)=XC3(LI)/REFL
      XC4(LI)=XC4(LI)/REFL
      YC1(LI)=YC1(LI)/REFL
      YC2(LI)=YC2(LI)/REFL
      YC3(LI)=YC3(LI)/REFL
      YC4(LI)=YC4(LI)/REFL
      ELSE IF (NTRAN .EQ. 2 .OR. NTRAN .EQ. 3) THEN
C
C---- INPUT MID-SIDE NODE COORDINATES. ( NTRAN = 2,3 )
C
      READ(5,*) XC5(LI),XC6(LI),XC7(LI),XC8(LI)
      READ(5,*) YC5(LI),YC6(LI),YC7(LI),YC8(LI)
C
      IF(NTRAN.EQ.3 .AND. LI.EQ.1) THEN
      WRITE(6,9402)
      ELSE
      WRITE(6,9404)
      ENDIF
      WRITE(6,9551) XC5(LI),XC6(LI),XC7(LI),XC8(LI)
      IF(NTRAN.EQ.3 .AND. LI.EQ.1) THEN
      WRITE(6,9406)
      ELSE
      WRITE(6,9408)
      ENDIF
      WRITE(6,9551) YC5(LI),YC6(LI),YC7(LI),YC8(LI)
      IF(NTRAN.EQ.3 .AND. LI.EQ.1 .AND. NPRA1.EQ.1) THEN
```

```
      READ(5,*)X5,X6,X7,X8
      WRITE(6,9404)
      WRITE(6,9551) X5,X6,X7,X8
      READ(5,*)Y5,Y6,Y7,Y8
      WRITE(6,9408)
      WRITE(6,9551) Y5,Y6,Y7,Y8
      ENDIF
C       WRITE(6,9404)
C       WRITE(6,9551) XC5(LI),XC6(LI),XC7(LI),XC8(LI)
C       WRITE(6,9406)
C       WRITE(6,9551) YC5(LI),YC6(LI),YC7(LI),YC8(LI)
      XC5(LI)=XC5(LI)/REFL
      XC6(LI)=XC6(LI)/REFL
      XC7(LI)=XC7(LI)/REFL
      XC8(LI)=XC8(LI)/REFL
      YC5(LI)=YC5(LI)/REFL
      YC6(LI)=YC6(LI)/REFL
      YC7(LI)=YC7(LI)/REFL
      YC8(LI)=YC8(LI)/REFL
      ENDIF
  420 CONTINUE
C       ENDIF
      CALL GRIDIT

      IF (NBUG(4) .GE. 0 .AND. NBUG(4) .NE. 2) THEN
      WRITE(6,9900)
      CALL PRNTIT(0,XG)
      IF (N .NE. 1) THEN
      WRITE(6,9902)
      CALL PRNTIT(0,YG)
      ENDIF
      ENDIF
      IF (INTEG .GT. 0) THEN
      CALL IPRINT(1)
      END IF


      ELSE IF (NAME .EQ. NAMES(5)) THEN
C*********************************************************************
C              5. READ MATERIAL DIFFUSION DATA, AND ASSIGN
C                 MATERIAL DATA SPECIFICATION  FOR
C                 MACROELEMENT DISCRETIZATION (MATL)
C*********************************************************************


      IF (N .EQ. 1) THEN
      WRITE(6,9560) NAME
      DO 103 I=1,NPRA(1)
C             IF (MSW .EQ. 0) THEN
      READ(5,*)  NKPRM,AK(I*2-1),AK(I*2)
      WRITE(6,9541) NKPRM,AK(I*2-1),AK(I*2)
C             ELSE
C                READ(5,*)  NKPRM,AK(I*2-1),AK(I*2),ZSTAR
C                WRITE(6,9541) NKPRM,AK(I*2-1),AK(I*2),ZSTAR
C             END IF
 103  CONTINUE


      ELSE IF (N .EQ. 2) THEN
      WRITE(6,9565) NAME
      DO 104 I=1,NPRA(1)
      DO 104 J=1,NPRA(2)
      READ(5,*)  NKPRM,AK(NKPRM)
      WRITE(6,9541) NKPRM,AK(NKPRM)
 104  CONTINUE
      END IF
```

```

      CALL MATLIT
      IF (NBUG(3) .GT. 0 ) THEN
      WRITE(6,*) '           AK:'
      CALL PRNTIT(0,AK)
      ENDIF

      ELSE IF (NAME .EQ. NAMES(6)) THEN
C***********************************************************************
C               6. DEFINE DIRICHLET B.C. SPECIFICATIONS.(DIRI).
C                  READ NUMBER OF DIRICHLET B.C. NODES IN DISCRETIZATION
C***********************************************************************

      READ(5,*)NFIX,KNUM,ISOR
      WRITE(6,9780) NAME,NFIX,KNUM,ISOR

C          KNUM = 0, INPUT DIRICHLET DATA FOR EVERY NODE
C                 1, INPUT DIRICHLET DATA FOR A STRING OF NODES

      IF (NFIX .NE. 0) THEN

C                  READ NODE NUMBER AND VALUE OF DEPENDENT VARIABLE AT
C                  DIRICHLET NODES.  THE FIXED DATA
C                  ARE LOADED DIRECTLY INTO F(JFIX).
      IF (ISOR .EQ. 0) THEN
      IF (KNUM.EQ.0) THEN
      READ(5,*)  (JFIX(I),I=1,NFIX)
      WRITE(6,9690) (JFIX(I),I=1,NFIX)
      READ(5,*) (F(JFIX(I)),I=1,NFIX)


      ELSE IF (KNUM.NE.0) THEN
      I2=0
      READ(5,*)  (JFIX(I),I=1,NFIX)
      WRITE(6,9690) (JFIX(I),I=1,NFIX)
C                    READ(5,*)  (F(JFIX(I)),I=1,NFIX)
      DO 101 I=1,NFIX
      READ(5,*) ILOC,DRDT
      DO 106 I1=1,ILOC
      I2=I2+1
      F(JFIX(I2))=DRDT
  106 CONTINUE
      IF(I2.GE.NFIX) GO TO 23
  101 CONTINUE
   23 END IF
      WRITE(6,9551) (F(JFIX(I)),I=1,NFIX)
      END IF

C                        READ SIDE NUMBER AND VALUE OF TEMPERATURE AT
C                        DIRICHLET NODES.  THE FIXED TEMPERATURE DATA
C                        ARE LOADED DIRECTLY INTO F(JFIX).

C          ISOR = 0, DEFAULT SPECIFICATION
C               = 1-4, (BYPASS PROCEDURE)
C                       SIDE NUMBER CONVENTION FOR
C                       A GENERAL FOUR SIDED DOMAIN
C
C                            3
C                   -----------------
C                  |                 |
C                  |                 |
C               4  |                 |  2
C                  |                 |
C                  |                 |
C                  |                 |
C                   -----------------
```

```
C                              1
C
      ELSE
      DO 107 IS=1,ISOR
      READ(5,*) ISIDE
      CALL SDIR3(ISIDE)
 107  CONTINUE
      ENDIF

      ELSE IF (NAME.EQ.NAMES(7)) THEN
C**********************************************************************
      NO = 1
C                 7. DEFINE ROBIN BOUNDARY CONDITION SPECIFICATIONS.(ROBN)
C                    READ NUMBER OF ROBIN B.C. NODES IN DISCRETIZATION.
C**********************************************************************

      IF(MSW.EQ.0)THEN
      READ(5,*)INNEU,IROB,JK
      WRITE(6,9650)NAME,INNEU
      ELSE
      READ(5,*)INNEU,IROB,JK,ZSTAR
      WRITE(6,9650)NAME,INNEU
      WRITE(6,9651)ZSTAR
      END IF

C                   READ NODE NUMBER AND VALUE OF CONSTRAINTS
C                   ( CONVECTION COEFF. H AND T REFERENCE)
C                   AT ROBIN B.C. NODES.

      IF(INNEU.GT.0) THEN
      IF(N.EQ.1 .AND. JK.EQ.0) THEN
      NNEU(1)=INNEU
      JK=1
      ELSE
      READ(5,*)  (NNEU(J),J=1,JK)
      END IF
      NNEU(0)=0
      IIP=0
      DO 49 JO=1,JK
      READ(5,*)  (JNEU(I),I=NNEU(JO-1)+1,NNEU(JO))
      WRITE(6,9760)  (JNEU(I),I=NNEU(JO-1)+1,NNEU(JO))
      WRITE(6,*)'                           H          TREF'
      IF(IROB.EQ.0) THEN
      DO 40 IK=1,NNEU(JO)-NNEU(JO-1)
      IIP=IIP+1
      READ(5,*) (ANUMAN(J,IIP),J=1,2)
      WRITE(6,9551)  (ANUMAN(J,IIP),J=1,2)
 40   CONTINUE
      ELSE IF (IROB .GT. 0) THEN
      DO 42 I=1,NNEU(JO)-NNEU(JO-1)
      READ(5,*) M,DROB1,DROB2
      DO 43 J=1,M
      IIP=IIP+1
      ANUMAN(1,IIP)=DROB1
      ANUMAN(2,IIP)=DROB2
      WRITE(6,9551)  (ANUMAN(IK,IIP),IK=1,2)
      IF (IIP.GE.NNEU(JO)) GO TO 44
 43   CONTINUE
 42   CONTINUE
 44   CONTINUE
      END IF
 49   CONTINUE
      END IF

      ELSE IF (NAME.EQ.NAMES(8)) THEN
C**********************************************************************
```

```
      NO = 1
C               8. DEFINE NEUMANN BOUNDARY CONDITION SPECIFICATIONS.(NEU
C                   READ NUMBER OF NEUMANN B.C. NODES IN DISCRETIZATION.
C*********************************************************************

      READ(5,*)INNEU,IROB,JK
      WRITE(6,9650)NAME,INNEU

C                  READ NODE NUMBER AND VALUE OF CONSTRAINT
C                  (FIXED FLUX) AT NEUMANN NODES

      IF(INNEU.GT.0) THEN
      READ(5,*)  (NNEU(J),J=1,JK)
      NNEU(0)=0
      IIP=0
      DO 407 II=1,JK
      READ(5,*)   (JNEU(I),I=NNEU(II-1)+1,NNEU(II))
      WRITE(6,9760)  (JNEU(I),I=NNEU(II-1)+1,NNEU(II))
      WRITE(6,*)'                    FIXED FLUX'
C                 IFNNEU=1
      IF(IROB.EQ.0) THEN
      DO 41 IJ=1,NNEU(II)
      IIP=IIP+1
      READ(5,*) ANUMAN(2,IIP)
      WRITE(6,9551) ANUMAN(2,IIP)
      ANUMAN(1,IIP)=0.0
 41   CONTINUE
      ELSE IF (IROB .GT. 0) THEN
      DO 47 IJ=1,NNEU(II)
      READ(5,*) M,DROB1
      DO 45 J=1,M
      JJ=JJ+1
      ANUMAN(2,JJ)=DROB1
      ANUMAN(1,JJ)=0.0
      WRITE(6,9551) ANUMAN(2,JJ)
      IF (JJ.GE.NNEU(II)) GO TO 46
 45   CONTINUE
 47   CONTINUE
 46   END IF
 407  CONTINUE
      END IF


      ELSE IF (NAME .EQ. NAMES(9)) THEN
C*******************************************************************
C         9. FORM ELEMENT CONTRIBUTIONS AND ASSEMBLE SMAT AND F.(FORM).
C*******************************************************************

      DO 121 I=1,NDOFS
      Q(I)=F(I)
 121  CONTINUE
      TSTEP=1.
      DELTAT=1.
      TRNFLG=0.
      THETA=1.
      CALL FORMIT(0)
      IF (NBUG7 .EQ. 1) THEN
      WRITE(6,*)'  FORM:  SYSTEM MATRIX  [K + H]; COLUMN FORMAT'
      CALL PRNTIS
      WRITE(6,9906)
      CALL PRNTIT(0,F)
      END IF
      IF (NBUG(6) .EQ. 1) THEN
      WRITE(6,9904) NAME
      CALL PRNTIS
      WRITE(6,9906)
```

```
      CALL PRNTIT(0,F)
C             ELSE
C                 WRITE(6,9908) NAME
      ENDIF


      ELSE IF (NAME .EQ. NAMES(10)) THEN
C***************************************************************
C              10. INPUT SOURCE DISTRIBUTION. (SORC)
C                 READ NUMBER OF NODES WITH SOURCE IMPOSED
C***************************************************************
      READ(5,*)NSORC,IFUNC
C                PRINT*, IFUNC

C        DEFINITION KEYED TO IFUNC
C        IFUNC =0= READ NODES AND CORRESPONDING SORC(*)
C            =1= SINE WAVE,
C             =2= GAUSSIAN

      IF(IFUNC.EQ.0) THEN
      WRITE(6,9686) NAME,NSORC

      ELSE IF(IFUNC.EQ.1) THEN
      WRITE(6,9687) NAME,NSORC

      ELSE IF(IFUNC.EQ.2) THEN
      WRITE(6,9688) NAME,NSORC

      ELSE IF(IFUNC.EQ.3) THEN
      WRITE(6,9689) NAME,NSORC

      END IF

      IF(NSORC .GT. 0) THEN
      CALL SORCR(IFUNC)
      ENDIF

      ELSE IF (NAME .EQ. NAMES(11)) THEN
C***************************************************************
C        11.  STEADY STATE DIRECT SOLUTION (SOLV)
C             SET TRANSIENT ITERATIVE PARAMETERS TO DEFAULT
C***************************************************************



      ISOLV=1
      IF (NBAND .EQ. 3) THEN
      CALL SOLVT(NDOFS)
      ELSE
      CALL SOLVR(NDOFS,NBAND,SMAT,F)
      END IF
      IF (NBUG(4) .GE. 0) THEN
      WRITE(6,*)
      WRITE(6,*)'  SOLV:  NODAL SOLUTION (REAL VARIABLES)'
      CALL PRNTIT(0,F)
      ENDIF

      IF (INTEG .GT. 0) THEN
      CALL IPRINT(3)
      END IF

      DO 122 I=1,NDOFS
      Q(I)=F(I)
  122 CONTINUE

      ELSE IF (NAME .EQ. NAMES(12)) THEN
```

```
C*************************************************************
C         12.  TRANSIENT SOLUTION
C              8.  FORM  CONTRIBUTIONS AND ASSEMBLE SPMAT AND FSPLIT.
C              9.  SOLVE EQUATION SYSTEM : SPMAT*DQ=FSPLIT .(INTE).
C*************************************************************
      CALL INTGRN

      ELSE IF (NAME .EQ. NAMES(13)) THEN
C*************************************************************
C        13. COMPUTE APPROXIMATE SOLUTION ENERGY NORM.(ENEH).
C*************************************************************
      IENERG=1
      CALL FORMIT(IENERG)
      WRITE(6,9201) ENORMH

      ELSE IF (NAME .EQ. NAMES(14)) THEN
C****************************************************
C         14. STEADY-STATE SOLUTION USING
C             GAUSS-SEIDEL ITERATIVE PROCEDURE (GAUS)
C****************************************************
      ISOLV=1
      NBNDTH = NBAND/2+1
C          SEE COMMENTS IN MAIN
      ISOLVR=5450/NBAND
      READ(5,*)TOL
      IF (TOL.EQ.0.)TOL=0.001
      TOLE=TOL
      CALL GAUS(TOL,NDOFS,NBAND,NBNDTH,IFLAG,IO,ISOLVR,SMAT,F,XL)
C            END IF
      IF (NBUG(4) .GE. 0) THEN
      WRITE(6,*)
      WRITE(6,*)'  GAUS:  NODAL SOLUTION (REAL VARIABLES),
     .TOL = ',TOLE
      CALL PRNTIT(0,F)
      ENDIF

      IF (INTEG .GT. 0) THEN
      CALL IPRINT(3)
      END IF




      DO 123 I=1,NDOFS
      Q(I)=F(I)
 123  CONTINUE


      ELSE IF (NAME .EQ. NAMES(15)) THEN
C****************************************************
C           15. READ INITIAL CONDITIONS. (INIT)
C               INITIALIZE NODAL VELOCITY AU(.) ARRAY
C****************************************************
      READ(5,*) NINI,LAMBDA,INT
      WRITE(6,9780) NAME,NINI

      IF (NINI .GT. 0) THEN

C               READ NODE NUMBER AND INITIAL VALUE OF TEMPERATURE AT
C               ALL NODES.  THE FIXED TEMPERATURE DATA
C               ARE ALSO LOADED DIRECTLY INTO F(JINI).
C               WHICH WILL BE SET EQUAL TO Q(JINI)

      IF (INT.EQ.0) THEN

      READ(5,*)   (JINI(I),I=1,NINI)
```

```
      WRITE(6,9690) (JINI(I),I=1,NINI)
      READ(5,*) (F(JINI(I)),I=1,NINI)
      WRITE(6,9551) (F(JINI(I)),I=1,NINI)

      ELSE IF (INT.EQ.1) THEN

      DO 29 I=1,NNODES
      JINI(I)=I
 29   CONTINUE

      WRITE(6,9690) (JINI(I),I=1,NNODES)
      READ(5,*) (F(JINI(I)),I=1,NNODES)
      WRITE(6,9551) (F(JINI(I)),I=1,NNODES)

      ELSE IF (INT.EQ.2) THEN
      READ(5,*) DAIN
      DO 39 I=1,NNODES
      JINI(I)=I
      F(JINI(I))=DAIN
 39   CONTINUE

      WRITE(6,9690) (JINI(I),I=1,NNODES)
C                       READ(5,*) (F(JINI(I)),I=1,NNODES)
      WRITE(6,9551) (F(JINI(I)),I=1,NNODES)

      ELSE IF (INT.GE.3) THEN
      II1=0
      DO 9 J=1,(INT-1)
      INDENT=0
      DAINI=0.
      READ(5,*) INDENT,DAINI
      DO 9 I=1,INDENT
      II1=II1+1
      JINI(I)=II1
      F(JINI(I))=DAINI
  9   CONTINUE

      END IF

C              FOR ROTATING CONE PROBLEM WITH PARAMETER (LAMBDA)
C              IF LAMBDA>0, INITIALIZE Q WITH COSINE HILL
C              FROM 0.75 TO 1.75 LAMBDA

      ELSE IF (LAMBDA .GT. 0) THEN
      NINI=0
      IF (N .EQ. 2) THEN
      DO 911 I=1,NNODES
      RADIUS=SQRT(((XG(I)-1.25*FLOAT(LAMBDA))**2)+
     *((YG(I)-1.25*FLOAT(LAMBDA))**2))
      IF (RADIUS .LE. 1.0) THEN
      NINI=NINI+1
      F(I)=1.+COS(3.141592654*(2.+RADIUS))
      ENDIF
 911  CONTINUE
      ENDIF
      CALL PRNTIT(0,F)
      ENDIF


      ELSE IF (NAME .EQ. NAMES(16)) THEN
C***************************************************************
C         16. CREATE CONVECTION VELOCITY FIELD (UVEL)
C             READ VELOCITY CONSTANT 'UCON' , PARAMETER 'NUPRM'
C             NUPRM=0 : U(NODE)=UCON,V(NODE)=0
C             NUPRM=1 : U(NODE)=UCON,V(NODE)=UCON
C             NUPRM=2 : U(NODE)=0,V(NODE)=UCON
```

```
C              NUPRM=3 : (U,V)=RADIUS * OMEGA
C              OMEGA=UCON, READ (X,Y) CENTER
C              INITIALIZE AU(.) ARRAY
C              INITIALIZE AV(.) ARRAY
C*****************************************************************

      READ(5,*) NUPRM,UCON
C STORE NUPRM IN UNUSED COMMON VARIABLE IFNNEU
      IFNNEU=NUPRM
      PE=0.
      IF (AK(1) .GT. 0.) PE=UCON*(XF(1,NPRA(1))-XI(1))/AK(1)
      WRITE(6,*) '  UVEL:  UCON,  SELECTION,  PECLET'
      WRITE(6,9562) UCON,NUPRM,PE
      UMUL=0.
      VMUL=0.
      IF (NUPRM .LE. 1) UMUL=1.
      IF (NUPRM .EQ.  0  .OR. NUPRM .EQ. 2) VMUL=1.
      IF (NUPRM .EQ. 3) THEN
      READ(5,*) XCEN,YCEN
      WRITE(6,9551) XCEN,YCEN
      DO 910 I=1,NNODES
      AU(I)=-(YG(I)-YCEN)*UCON*2.
      AV(I)=(XG(I)-XCEN)*UCON*2.
 910  CONTINUE
      ELSE
      DO 909 I=1,NDOFS
      AU(I)=UCON*UMUL
      AV(I)=UCON*VMUL
 909  CONTINUE
      ENDIF
      IF (NBUG(4) .GE. 0) THEN
      WRITE(6,9563)
      CALL PRNTIT(0,AU)
      WRITE(6,9564)
      CALL PRNTIT(0,AV)
      ENDIF

         ELSE IF (NAME .EQ. NAMES(17)) THEN
C************************************************************
C         17. GENERATE THE DATA FILE FOR PLOT HARDCOPY. (GRAF)
C************************************************************
          READ (5,*)(NSWITCH(I),I=1,5)
          CALL PLOTIT (NSWITCH)
               IPLOT=1

      ELSE IF (NAME .EQ. NAMES(18)) THEN
C**************************************************
C         18. NEW PROBLEM DEFINITION.(EXIT).
C**************************************************
      WRITE(6,9510) NAME
      IGOHON=1

      ELSE IF (NAME .EQ. NAMES(19)) THEN
C**************************************************
C         19. END OF JOB.(STOP).
C**************************************************
      WRITE(6,9510) NAME
      IGOHON=0

C**************************************************
C               ILLEGAL-SITUATION HANDLING
C**************************************************
      ELSE

      WRITE(6,9600) NAME
      IFAIL=IFAIL+1
```

```
      IF (IFAIL .EQ. 18) THEN
      WRITE(6,9600) NAME
      NAME=NAMES(19)
      IGOHON=0
      ENDIF
      ENDIF
      GOTO 102
      ENDIF
      RETURN

 9901 FORMAT(12I6)
 9201 FORMAT(/3X,'NORM:  F.E. SOLN. ENERGY SEMI-NORM:   E(QH)=', E15.8)
 9395 FORMAT(/,10X,'DIMENSIONAL VARIABLE')
 9401 FORMAT(10X,'NTRAN: ',I3)
 9400 FORMAT(4X,3HNPR,3X)
 9402 FORMAT(17X,'X1',9X,'X2',9X,'X3',9X,'X4',3X)
 9404 FORMAT(17X,'X5',9X,'X6',9X,'X7',9X,'X8',3X)
 9406 FORMAT(17X,'Y1',9X,'Y2',9X,'Y3',9X,'Y4',3X)
 9408 FORMAT(17X,'Y5',9X,'Y6',9X,'Y7',9X,'Y8',3X)
 9520 FORMAT(' ')
 9420 FORMAT(10X,'FINITE ELEMENTS  1-2-3  CODE: LEARN FINITE ELEMENTS ')
 9500 FORMAT(A4)
 9510 FORMAT(/,3X,A4)
 9530 FORMAT(10X,14I5)
 9531 FORMAT(10X,14I5)
 9532 FORMAT(10X,11I6)
 9535 FORMAT(10X,2I3,I8,F6.1,2I6,I7,2I6)
 9533 FORMAT(10X,2I3,I8,F6.1,I6,I7,2I6)
 9534 FORMAT( 8X,6I7)
C 9572 FORMAT(' WARNING: NODE ', I5 , ' BOTH DIRI AND INIT. VALUE=' ,
C     *  E12.5)
 9572 FORMAT(/)
 9540 FORMAT(10X,I5,5X,8E12.5)
 9541 FORMAT(10X,I5,1X,E12.5,1X,E12.5,1X,E12.5)
C 9541 FORMAT(10X,I5,1X,6(F8.2,1X))
 9550 FORMAT(10X,6E12.5)
 9551 FORMAT(10X,6F11.5)
 9552 FORMAT(10X,7F9.4)
 9556 FORMAT(10X,6E12.5,I5)
 9557 FORMAT(10X,6E12.5,I5)
 9558 FORMAT(10X,'TIME=', F12.5, ' STEP=', I5, ' #RES=',I5 ,
     *' NEW DELTAT=' ,F12.5)
 9559 FORMAT(10X,'TIME=', F12.5, '  DIR=', I5, ' #SWP=',I5 )
 9560 FORMAT(/,3X,A4,':',3X,'MACRO NODAL CONDUCTIVITIES')
 9561 FORMAT(10X,F12.5,I5,F12.5)
 9562 FORMAT(/10X,F12.5,I5,F12.5)
 9563 FORMAT(/,10X,'U VELOCITY')
 9564 FORMAT(/,10X,'V VELOCITY')
 9565 FORMAT(/,3X,A4,':',3X,'MACRO ELEMENT CONDUCTIVITIES')
 9570 FORMAT(I5,1X,E12.5,1X,E12.5)
 9571 FORMAT('          TIME      # STEPS',/,1X,F12.5,I12)
 9580 FORMAT(10X,I5,4X,F12.4,4X,F12.4)
 9600 FORMAT(/3X,4HNAME,2X,A4,2X,9HNOT LEGAL/)
 9620 FORMAT(/,3X,A4,':',2X,1HK,5X,1HN,2X,6HNNODEL,2X,3HNPR,
     *1X,5HNTRAN,2X,4HNAXI)
 9641 FORMAT(4X,'MACREL:',5I5,3X)
 9650 FORMAT(/,3X,A4,':',2X,5HNROB:,4I5)
 9651 FORMAT(/,10X,'ZSTAR:',F5.2)
 9686 FORMAT(/,3X,A4,':',2X,6HNSORC:,I5,3X,'USER SPECIFICATION'/)
 9687 FORMAT(/,3X,A4,':',2X,6HNSORC:,I5,3X,'SINE FUNCTION'/)
 9688 FORMAT(/,3X,A4,':',2X,6HNSORC:,I5,3X,'GAUSSIAN DISTRIBUTION'/)
 9689 FORMAT(/,3X,A4,':',2X,6HNSORC:,I5,3X,'LINE SOURCE'/)
 9760 FORMAT(10X,5HJROB:,15I5)
 9761 FORMAT(10X,5HJNEU:,15I5)
 9670 FORMAT(18X,1HH,8X,4HTREF,3X)
 9780 FORMAT(/,3X,A4,':',2X,5HNFIX:,I5,2X,5HKNUM:,I3,2X,5HISOR:,I2)
```

```
9781 FORMAT(/,3X,A4,':',2X,6HNSORC:,I5)
9690 FORMAT(10X,5HJFIX:,10I5)
9700 FORMAT(/,3X,A4,':',2X,A80)
9720 FORMAT(10X,2I5)
9850 FORMAT(/,3X,A4,':',2X,'NBUG(1)',3X,'(2)',3X,'(3)',3X,'(4)')
9900 FORMAT(/,10X,'NODE COORDINATES (X REAL VARIABLES)')
9902 FORMAT(/,10X,'NODE COORDINATES (Y REAL VARIABLES)')
9904 FORMAT(/,3X,A4,':',2X,'SYSTEM MATRIX  [K + H]  COOLUMN FORMAT')
9906 FORMAT(/,10X,'DATA MATRIX  {b}  ROW FORMAT')
9908 FORMAT(/,3X,A4)
9911 FORMAT(/,10X,'NODE COORDINATES (X REAL VARIABLES)'
    ./,10X,7F8.2)
     END
```

```
      SUBROUTINE FORMIT(IENERG)
C
C     FORM THE ALGORITHM MATRIX STATEMENT
C**************************************************

      IMPLICIT REAL(A-H,O-Z)

      COMMON/BIGMT/SMAT(5450),F(1090),Q(1090),JINI(1090),
     *XG(1090),YG(1090),AK(1090),AU(1090),AV(1090),SORC(1090),
     *XL(1090),JFIX(1090),MEL(4800)
      COMMON/CONST/K,N,NAXI,NNODEL,NBAND,NDOFS,NDOFN,MSW,NTRAN,INTEG,
     *NELES,NNODES,NSORC,MO,NPR,NFIX,INNEU,JK,IFNNEU,JNEU(68),
     *NNOD(3),NELE(3),NBUG(6),NEM(2,3),NPRA(2),NNEU(0:4),
     *DELTAT,TRNFLG,THETA,PREMAC,ENORMH,ZSTAR,
     *ANUMAN(2,68),REFL,XI(2),XF(2,3),PR(2,3)
      COMMON/MATX/A3000L(2,2,2),A3011L(2,2,2),A3000Q(3,3,3),
     *A3011Q(3,3,3),A3011C(2,4,4),A3011H(2,4,4),
     *A3001L(2,2,2),A3001Q(3,3,3),A200C(4,4),
     *B211B(4,4),B222B(4,4),B212B(4,4),B221B(4,4),
     *B200L(3,3),B200B(4,4)
      COMMON/COORDN/XC1(9),XC2(9),XC3(9),XC4(9),XC5(9),XC6(9),
     *XC7(9),XC8(9),YC1(9),YC2(9),YC3(9),YC4(9),
     *YC5(9),YC6(9),YC7(9),YC8(9)

      REAL A3000(3,4,4),A3001(3,4,4),A3011(3,4,4)
      REAL R(4),U(4),ENGY(4),ZETA(3,2)
      REAL EMAT(8,8),CMAT(8,8)

C*****************************************************************
C     INITIALIZE
C*****************************************************************

      NDOFE=NNODEL*NDOFN
      NTSPL=0
C     TIMESPLIT 2-D
      IF(NNODEL .EQ. 2 .AND. N .EQ. 2) THEN
      N=1
C       DIR=1
      NTSPL=1
C       DIR=2
      IF(JINI(2) .NE. 2) NTSPL=2
      ELSE IF(NNODEL .EQ. 3 .AND. N .EQ. 2 .AND. K .EQ. 2) THEN
      N=1
C       DIR=1
      NTSPL=1
C       DIR=2
      IF(JINI(2) .NE. 2) NTSPL=2
      ENDIF
      DO 1 I=1,NDOFE
      ENGY(I)=0.
      U(I)=0.
      R(I)=0.
  1   CONTINUE
      KINT=2
      KN1 = K +(1-NDOFN)
      KN = KN1+1
      IF (N .EQ. 1 .AND. K .EQ. 2) THEN
      KN1=1
      KINT=3
      ENDIF
      ENORMH=0.
      NBNDTH = NBAND/2 +1

C***************
C STORE 1-D MATRICES TO AVOID IF STATEMENTS
C IF FIRST PASS (TRNFLG .LE. 1, IENERG .EQ. 0)
```

```
C***************

      IF (TRNFLG .LE. 1. .AND. IENERG .EQ. 0) THEN
      IF (K .EQ. 1) THEN
      DO 341 I=1,KINT
      DO 341 J=1,K+1
      DO 341 L=1,K+1
      A3000(I,J,L)=A3000L(I,J,L)
      A3001(I,J,L)=A3001L(I,J,L)
      A3011(I,J,L)=A3011L(I,J,L)
  341 CONTINUE
      ELSE IF (K .EQ. 2) THEN
      DO 342 I=1,KINT
      DO 342 J=1,K+1
      DO 342 L=1,K+1
      A3000(I,J,L)=A3000Q(I,J,L)
      A3001(I,J,L)=A3001Q(I,J,L)
      A3011(I,J,L)=A3011Q(I,J,L)
  342 CONTINUE
      ELSE IF (K .EQ. 3) THEN
      DO 347 I=1,K-1
      DO 347 J=1,K+1
      DO 347 L=1,K+1
      A3011(I,J,L)=A3011C(I,J,L)
  347 CONTINUE
      ENDIF
      ENDIF
C*************
C         PARAMETER IUNFRM = 1 INDICATES UNIFORM GRID, NON-AXISYMMETRIC
C         AND/OR UNIFORM CONDUCTIVITY.  EMAT AND CMAT ARE THUS CONSTANT,
C         HENCE WILL NOT BE CREATED FOR EACH ELEMENT.
C*************
      IUNFRM=1
      NPR=MAX(NPRA(1),NPRA(2))
      IF (NAXI .EQ. 1 .OR. NPR .GT. 1 .OR. NTRAN .NE. 0) IUNFRM=0
      IF (PR(1,1) .NE. 1.0 .OR. PR(N,1) .NE. 1.0) IUNFRM=0
      IF (IENERG .EQ. 1) IUNFRM=0
C***************
C     ZERO F AND SMAT ARRAYS
C***************
      DO 108 I=1,NDOFS
      F(I)=0.
      IF (AU(I) .NE. AU(1) .OR. AK(I) .NE. AK(1)) IUNFRM=0
  108 CONTINUE
      DO 109 ISMAT=1,NBAND*NDOFS
      SMAT(ISMAT)=0.
  109 CONTINUE

C
C
C*******************************************************************
C     1*) SOURCE : ELEMENT LOOP AND ASSEMBLE INTO F
C*******************************************************************
      IF (NSORC .GE. 1 .AND. IENERG .EQ. 0 .AND. NTSPL .LT. 2) THEN
      NEL=NELES*(1-IUNFRM)+IUNFRM
      JEL=0
      DO 310 IEL=1,NEL
      JEL=JEL+1
      ML1=MEL(1+(JEL-1)*NNODEL)
      ML2=MEL(2+(JEL-1)*NNODEL)
      ML3=MEL(3+(JEL-1)*NNODEL)
      ML4=MEL(4+(JEL-1)*NNODEL)
C**************
C           1-D
C**************
      IF (N .EQ. 1) THEN
```

```
      DET=XG(MEL(KN+(JEL-1)*NNODEL))-XG(ML1)
      I1=0
      DO 311 L=1,NDOFE,KN1
      I1=I1+1
      KROW=MEL(L+(JEL-1)*NNODEL)
      R(I1)=(XG(KROW)-1.)*FLOAT(NAXI)+1.
      IF (MSW.GE.1)R(I1)=R(I1)*ZSTAR
 311  CONTINUE
      DO 320 L=1,NDOFE
      DO 321 J=1,NDOFE
      CMAT(L,J) = 0.E0
      DO 322 I=1,KINT
      CMAT(L,J)=CMAT(L,J)+R(I)*A3000(I,L,J)*DET
 322  CONTINUE
 321  CONTINUE
 320  CONTINUE
C***************
C           2-D
C***************
      ELSE IF (N .EQ. 2) THEN
C              TRIANGLES
      IF (NDOFE .LE. 3) THEN
      DET= (XG(ML2)*YG(ML3)+XG(ML1)*
     *YG(ML2)+
     *XG(ML3)*YG(ML1)-XG(ML2)*
     *YG(ML1)-
     *XG(ML3)*YG(ML2)-XG(ML1)*
     *YG(ML3))
      AREA=DET/2.
      DO 323 L=1,NDOFE
      KROW = MEL(L+(JEL-1)*NNODEL)
      DO 324 J=1,NDOFE
      CMAT(L,J)=B200L(L,J)*AREA
 324  CONTINUE
 323  CONTINUE
C               QUADS
      ELSE IF (NDOFE .GE. 4) THEN
      ETA11 = ((YG(ML4)-YG(ML1)) + (YG(ML3)-YG(ML2)))/4.E0
      ETA21 = ((XG(ML1)-XG(ML4)) + (XG(ML2)-XG(ML3)))/4.E0
      ETA12 = ((YG(ML1)-YG(ML2)) + (YG(ML4)-YG(ML3)))/4.E0
      ETA22 = ((XG(ML2)-XG(ML1)) + (XG(ML3)-XG(ML4)))/4.E0
      DET=(ETA11*ETA22-ETA12*ETA21)
      DO 325 L=1,NDOFE
      KROW = MEL(L+(JEL-1)*NNODEL)
      DO 326 J=1,NDOFE
      CMAT(L,J)=B200B(L,J)*DET
 326  CONTINUE
 325  CONTINUE
      ENDIF
      ENDIF
C*********************************************************
C     ASSEMBLE GLOBAL CMAT USING SMAT WORKSPACE
C*********************************************************
      NEL2=NELES*IUNFRM+1-IUNFRM
      JEL=JEL-IUNFRM
      DO 312 IEL2=1,NEL2
      JEL=JEL+IUNFRM
      DO 332 L=1,NDOFE
      KROW=MEL(L+(JEL-1)*NNODEL)
      DO 331 J=1,NDOFE
      LOC = MEL(J+NNODEL*(JEL-1)) - KROW + NBNDTH
      LOC=LOC*(1-(LOC/(NDOFS+1)))
      LOC=MAX(LOC,0)
      ISMAT=LOC+NBAND*(KROW-1) +
     *(NDOFS/(LOC+NDOFS))*(1+NBAND*(NDOFS-KROW+1))
      SMAT(ISMAT) = SMAT(ISMAT) + CMAT(J,L)
```

```
  331  CONTINUE
  332  CONTINUE
  312  CONTINUE
  310  CONTINUE

       IF (IENERG .EQ. 0 .AND. NBUG(6) .EQ. 1) THEN
       WRITE(6,*)'  INTE:  SMAT- SOURCE'
       CALL PRNTIS
       ENDIF
C**********************************************
C         PLACE SOURCE CONTRIBUTION INTO F
C**********************************************

       DO 360 KROW=1,NDOFS
       KCMN=MAX(1,1+NBNDTH-KROW)
       KCMX=MIN(NBAND,NDOFS+NBNDTH-KROW)
       DO 370 KCOL=KCMN,KCMX
       LOC = KROW + KCOL - NBNDTH
       ISMAT=KCOL+NBAND*(KROW-1)
       F(KROW)= F(KROW)+SMAT(ISMAT)* SORC(LOC)
  370  CONTINUE
  360  CONTINUE
       IF (IENERG .EQ. 0 .AND. NBUG(6) .EQ. 1) THEN
       WRITE(6,*)'  INTE:  F    -SOURCE'
       CALL PRNTIT(0,F)
       ENDIF
C****************************************
C         ZERO SMAT ARRAY
C****************************************
       DO 390 I=1,NDOFS
       DO 395 J=1,NBAND
       ISMAT=J+NBAND*(I-1)
       SMAT(ISMAT) = 0.
  395  CONTINUE
  390  CONTINUE
       ENDIF
C
C*************************************************************
C          END SOURCE
C      1*) BEGIN CONDUCTION MATRIX : ELEMENT LOOP
C*************************************************************
C
       JEL=0
       NEL=NELES*(1-IUNFRM)+IUNFRM
       DO 199 IEL=1,NEL
       JEL=JEL+1
       ML1=MEL(1+(JEL-1)*NNODEL)
       ML2=MEL(2+(JEL-1)*NNODEL)
       ML3=MEL(3+(JEL-1)*NNODEL)
       ML4=MEL(4+(JEL-1)*NNODEL)
C
C
C************************************
C   BEGIN 1-D
C************************************
C
C
C         2*)   1-D PROBLEMS.
C         ELEMENTS AVAILABLE:
C         LINEAR,QUADRATIC,
C         CUBIC LAGRANGE,
C         CUBIC HERMITE.
       IF (N .EQ. 1) THEN
C           2.1)       INITIALIZE CUBIC HERMITE STANDARD MATRIX
C            IF (K .EQ. 4) THEN
C               DO 113 I=1,NDOFE,2
```

```
C                 DO 112 L=1,2
C                    DO 111 J=1,NDOFE,2
C                       A3011H(L,I+1,J+1)=A3011H(L,I+1,J+1)*DET
C                       A3011H(L,I,J)=A3011H(L,I,J)/DET
C 111                CONTINUE
C 112             CONTINUE
C 113          CONTINUE
C           ENDIF
C***************
C          RADIUS*CONDUCTIVITY FOR AXISYMM AND NON-AXISYMM
C***************
      I1=0
      DO 385 L=1,NDOFE,KN1
      I1=I1+1
      KROW=MEL(L+(JEL-1)*NNODEL)
C GAUSSIAN PLUME: IFNNEU=1 INDICATES UNIFORM VELOCITY FIELD
C                 SET DIFFUSION TO 0 IN VEL. DIRECTION
      IF(IFNNEU .GT. 0 .AND. NTSPL .EQ. IFNNEU) THEN
      R(I1)=0.
      ELSE
      R(I1)=AK(KROW)*(XG(KROW)*FLOAT(NAXI)+FLOAT(1-NAXI))
      ENDIF
      IF (MSW.GE.1) R(I1)=R(I1)*ZSTAR
      U(I1)=AU(KROW)
      IF(IENERG .EQ. 1) U(I1)=0.
 385  CONTINUE
C***************
C          2.3) FORM ELEMENT CONDUCTIVITY MATRIX.
C***************
      DET=XG(MEL(KN+(JEL-1)*NNODEL))-XG(ML1)
      DO 100 L=1,NDOFE
      DO 110 J=1,NDOFE
      EMAT(L,J) = 0.E0
      DO 120 I=1,KINT
      EMAT(L,J)=EMAT(L,J)+R(I)*A3011(I,L,J)/DET +  U(I)*A3001(I,L,J)
 120  CONTINUE
 110  CONTINUE
 100  CONTINUE
C          FOR CUBIC HERMITE, RETURN STANDARD MATRIX.
C           IF (K .EQ. 4) THEN
C              DO 130 I=1,NDOFE,2
C                 DO 140 L=1,2
C                    DO 150 J=1,NDOFE,2
C                       A3011H(L,I+1,J+1)=A3011H(L,I+1,J+1)/DET
C                       A3011H(L,I,J)=A3011H(L,I,J)*DET
C 150                CONTINUE
C 140             CONTINUE
C 130          CONTINUE
C           ENDIF
C
C***************
C  END 1-D, BEGIN 2-D
C***************

      ELSE IF (N .EQ. 2) THEN
C***************
C          ELEMENT AVERAGE CONDUCTIVITY AK(ML1)
C***************
C          3*) 2-D PROBLEMS.
C       ELEMENTS AVAILABLE:
C       LINEAR TRIANGLES,
C       LINEAR QUADRILATERALS
      IF (NDOFE .LE. 3) THEN
C***************
C             T R I A N G L E S
C***************
```

```
C             3.1) LINEAR TRIANGLES
C             SET UP ZETAS
      ZETA(1,1)=YG(ML2)-YG(ML3)
      ZETA(2,1)=YG(ML3)-YG(ML1)
      ZETA(3,1)=YG(ML1)-YG(ML2)
      ZETA(1,2)=XG(ML3)-XG(ML2)
      ZETA(2,2)=XG(ML1)-XG(ML3)
      ZETA(3,2)=XG(ML2)-XG(ML1)
      DET=(XG(ML2)*YG(ML3)+XG(ML1)*
     *YG(ML2)+
     *XG(ML3)*YG(ML1)-XG(ML2)*
     *YG(ML1)-
     *XG(ML3)*YG(ML2)-XG(ML1)*
     *YG(ML3))
      AREA=DET/2.
C                 ELEMENT AVERAGE CONDUCTIVITY
      COND= (AK(ML1) + AK(ML2) + AK(ML3))/3.
C                 SET UP DETERMINANT 2*AREA(ELEMENT)
C                 SET UP ELEMENT D FFUSION MATRIX
      DO 160 I=1,NDOFE
      DO 170 J=1,NDOFE
      EMAT(I,J)=COND*(ZETA(I,1)*ZETA(J,1)+
     *ZETA(I,2)*ZETA(J,2))/(4.*AREA)
  170 CONTINUE
  160 CONTINUE
C
C             3.2) QUADRATIC TRIANGLES
C                  IMPLEMENTATION PROCEDURES GO HERE
C
C***************************************
C       Q U A D R I L A T E R A L S
C***************************************
C
      ELSE IF (NDOFE .GE. 4) THEN
C             3.3) BILINEAR QUADRILATERALS
C                  DETERMINANT OF JACOBIAN
C                  EVALUATED AT ELEMENT CENTROID
      ETA11 = ((YG(ML4)-YG(ML1)) + (YG(ML3)-YG(ML2)))/4.E0
      ETA21 = ((XG(ML1)-XG(ML4)) + (XG(ML2)-XG(ML3)))/4.E0
      ETA12 = ((YG(ML1)-YG(ML2)) + (YG(ML4)-YG(ML3)))/4.E0
      ETA22 = ((XG(ML2)-XG(ML1)) + (XG(ML3)-XG(ML4)))/4.E0
      DET=4.*(ETA11*ETA22-ETA12*ETA21)
      COND= (AK(ML1) + AK(ML2) + AK(ML3) + AK(ML4))/4.
C                  ELEMENT CONDUCTIVITY MATRIX
      DO 175 I=1,NDOFE
      DO 176 J=1,NDOFE
      EMAT(I,J) = B211B(I,J)*(ETA11**2+ETA21**2)
     *+(B212B(I,J)+B221B(I,J))
     **(ETA11*ETA12)
     *+(B221B(I,J)+B212B(I,J))
     **(ETA21*ETA22)
     *+B222B(I,J)*(ETA22**2+ETA12**2)
      EMAT(I,J)= EMAT(I,J)*COND*4./DET

  176 CONTINUE
  175 CONTINUE
C             3.4) QUADRATIC QUADRILATERALS
C                  IMPLEMENTATION SEQUENCE GOES HERE
      ENDIF
      ENDIF
      IF (IENERG .EQ. 0 .AND. NBUG(6) .EQ. 1) THEN
      WRITE(6,*)' INTE:  EMAT, EL# ',JEL
      DO 101 J=1,NDOFE
      WRITE(6,9900) J,(EMAT(I,J),I=1,NDOFE)
  101 CONTINUE
      ENDIF
```

```
C
C
C**********************************************
C   END 2-D
C   BEGIN ASSEMBLY MATRIX LOOP 1-D AND/OR 2-D
C**********************************************
C
C
C         4*)  ASSEMBLE SMAT CONDUCTIVITY
C         MATRIX FOR 1D AND 2D PROBLEMS
C         STORED IN 'SMAT(LOC,KROW)'

      IF (IENERG .EQ. 0) THEN
      NEL2=NELES*IUNFRM+1-IUNFRM
      JEL=JEL-IUNFRM
      DO 396 IEL2=1,NEL2
      JEL=JEL+IUNFRM
      DO 383 L=1,NDOFE
      KROW = MEL(L+(JEL-1)*NNODEL)
      DO 384 J=1,NDOFE
      LOC = MEL(J+NNODEL*(JEL-1)) - KROW + NBNDTH
      LOC=LOC*(1-(LOC/(NDOFS+1)))
      LOC=MAX(LOC,0)
      ISMAT=LOC+NBAND*(KROW-1) +
     *(NDOFS/(LOC+NDOFS))*(1+NBAND*(NDOFS-KROW+1))
      SMAT(ISMAT) = SMAT(ISMAT) +  EMAT(J,L)
 384  CONTINUE
 383  CONTINUE
 396  CONTINUE

      ENDIF
      IF(NELE(1).EQ.1.AND.N.EQ.1)THEN
      DO 222 J=1,K
      DO 222 I=1,K-J+1
      SMAT(NBAND*J-K+I)=SMAT(NBAND*(I+J-1)+K-I+1)
 222  CONTINUE
      ENDIF


C*********************************************
C            5*)    ENERGY NORM PREPARATION
C*********************************************
      DO 220 I=1,NDOFE
      ENGY(I)=0.E0
      DO 210 J=1,NDOFE
      I2=MEL(J+(JEL-1)*NNODEL)
      ENGY(I)=ENGY(I)+EMAT(I,J)*Q(I2)
 210  CONTINUE
      I2=MEL(I+(JEL-1)*NNODEL)
      ENORMH=ENORMH+ENGY(I)*Q(I2)
 220  CONTINUE
C 197        CONTINUE
C            ENDIF
 199  CONTINUE
C         DO 309 ISMAT=1,NBAND*NDOFS
C            IF(SMAT(ISMAT) .LT. 1.0E-06) SMAT(ISMAT)=0.
C309      CONTINUE
C
C
C**************************************************************
C     END ELEMENT LOOP:CONDUCTIVITY TERM
C**************************************************************
      IF (IENERG .EQ. 0 .AND. NBUG(6) .EQ. 1) THEN
      WRITE(6,*)'  INTE:  SMAT- ASSEMBLED'
      CALL PRNTIS
      ENDIF
```

```
C
C
C*****************************************************************
C 6.)  BEGIN   ASSEMBLY: ROBIN/NEUMANN BC ADDITION TO SMAT AND F
C*****************************************************************
C
C**************
C     BEGIN 1-D
C**************
      IF (INNEU .GT. 0 .AND. NTSPL .LT. 2) THEN
      IF (N .EQ. 1 ) THEN
      I=0
      DO 231 J=1,JK
      DO 231 II=1,NNEU(J)
      I=I+1
      I3=JNEU(I)
      X1=((XG(I3)-1.)*FLOAT(NAXI)+1.)*ANUMAN(1,I)
      ISMAT=NBNDTH+NBAND*(I3-1)
      SMAT(ISMAT) = SMAT(ISMAT) + X1
      ENORMH = ENORMH + (Q(I3)**2) * X1
      F(I3) = F(I3) + ANUMAN(2,I) * X1
 231  CONTINUE
      ELSE
C
C****************
C     END 1-D, BEGIN 2-D
C****************
C     7*)  FORM 1D BOUNDARY ELEMENTS
C     FOR 2D PROBLEM FOR ASSEMBLY
C     INTO SMAT OR  ENERGY NORM
C     EVALUATION.
C
      I1=0
      DO 240 JJ=1,JK
      DO 241 IOP=1,NNEU(JJ)-NNEU(JJ-1)-K
      I1=I1+1
      N1 = JNEU(I1)
      N2 = JNEU(I1+K)
      DET = ((XG(N2)-XG(N1))**2) + ((YG(N2)-YG(N1))**2)
      DET = SQRT(DET)
      DO 250 L=1,K+1
      KROW = JNEU(I1-1+L)
      DO 260 J=1,K+1
      JN = (I1-1+J)
      CMAT(J,L) = 0.E0
      ENGY(J)=0.
      DO 270 I=1,K+1
      IN=I1-1+I
      CMAT(J,L) =CMAT(J,L)+ANUMAN(1,IN)*A3000(I,J,L)
      ION= JNEU(IN)
      ENGY(J)=ENGY(J)+CMAT(J,L)*Q(ION)
C                        ENGY(J)=ENGY(J)+EMAT(J,L)*Q(IN)
 270  CONTINUE
      LOC = JNEU(JN) - KROW + NBNDTH
C                 IF (LOC .GT. NDOFS) LOC=0
      LOC=LOC*(1-(LOC/(NDOFS+1)))
      LOC=MAX(LOC,0)
C                 IF (LOC .EQ. 0) ISMAT=NBAND*NDOFS+1
      ISMAT=LOC+NBAND*(KROW-1) +
     *(NDOFS/(LOC+NDOFS))*(1+NBAND*(NDOFS-KROW+1))
      SMAT(ISMAT) = SMAT(ISMAT) + DET * CMAT(J,L)
      F(KROW) = F(KROW) + DET * CMAT(J,L)*ANUMAN(2,JN)
 260  CONTINUE
      ENORMH=ENORMH+ENGY(L)*Q(KROW)*DET
 250  CONTINUE
 241  CONTINUE
```

```
      I1=I1+1
 240  CONTINUE
      ENDIF

      IF (IENERG .EQ. 0 .AND. NBUG(6) .EQ. 1) THEN
      WRITE(6,*)'  INTE:  SMAT -NEUMANN'
      CALL PRNTIS
      WRITE(6,*)'  INTE:  F    -NEUMANN'
      CALL PRNTIT(0,F)
      ENDIF
      ENDIF
C
C
C******************************************************************
C6.)     END ROBIN/NEUMANN BC ADDITION TO SMAT AND F
C******************************************************************
C
C******************************************************************
C7.)     BEGIN TRANSIENT: COMPUTE RHS: FSPLIT=F-SMAT*QOLD*DELTAT
C******************************************************************
      IF (TRNFLG .GE. 1. .AND. IENERG .EQ. 0) THEN
      DO 181 KROW=1,NDOFS
C           FKSUM=0.
      KCMN=MAX(1,1+NBNDTH-KROW)
      KCMX=MIN(NBAND,NDOFS+NBNDTH-KROW)
      DO 191 KCOL=KCMN,KCMX
      LOC = KROW + KCOL - NBNDTH
      ISMAT=KCOL+NBAND*(KROW-1)
      SMATI=SMAT(ISMAT)*Q(LOC)
      F(KROW)= F(KROW)-SMATI
      SMAT(ISMAT)=SMAT(ISMAT)*THETA*DELTAT
      IF(NBUG(6) .EQ. 1) THEN
      WRITE(6,9919) KROW,KCOL,LOC,ISMAT,Q(LOC),SMATI,F(KROW),SMAT(ISMAT)
      ENDIF
 9919 FORMAT(4I5,4E15.8)
C             SMATI=ABS(SMATI)*DELTAT
C             FKSUM=FKSUM+SMATI
 191  CONTINUE
      F(KROW)= DELTAT* F(KROW)
C           IF (FKSUM .GT. 1.0E-10) THEN
C             FKSUM=F(KROW)/FKSUM
C             IF (ABS(FKSUM) .LE. 1.0E-10) F(KROW)=0.
C           ENDIF
 181  CONTINUE
C***************
C     1*) BEGIN ELEMENT LOOP: MASS MATRIX (1-D)
C***************
      NEL=NELES*(1-IUNFRM)+IUNFRM
      JEL=0
      DO 193 IEL=1,NEL
      JEL=JEL+1
      DET=XG(MEL(KN+(JEL-1)*NNODEL))
     *-XG(MEL(1+(JEL-1)*NNODEL))
      DO 182 L=1,NDOFE
      KROW = MEL(L+(JEL-1)*NNODEL)
      DO 192 J=1,NDOFE
      CMAT(L,J) = 0.E0
      DO 195 I=1,KINT
      CMAT(L,J)=CMAT(L,J)+A3000(I,L,J)*DET
 195  CONTINUE
 192  CONTINUE
 182  CONTINUE

      IF (IENERG .EQ. 0 .AND. NBUG(6) .EQ. 1) THEN
      WRITE(6,*)'  INTE:  CMAT, EL# ',JEL
      DO 194 J=1,NDOFE
```

```
      WRITE(6,9900) J,(CMAT(I,J),I=1,NDOFE)
  194 CONTINUE
      ENDIF

      NEL2=NELES*IUNFRM+1-IUNFRM
      JEL=JEL-IUNFRM
      DO 196 IEL2=1,NEL2
      JEL=JEL+IUNFRM
      DO 183 L=1,NDOFE
      KROW = MEL(L+(JEL-1)*NNODEL)
      DO 184 J=1,NDOFE
      LOC = MEL(J+NNODEL*(JEL-1)) - KROW + NBNDTH
C                    IF (LOC .GT. NDOFS) LOC=0
      LOC=LOC*(1-(LOC/(NDOFS+1)))
      LOC=MAX(LOC,0)
C                    IF (LOC .EQ. 0) ISMAT=NBAND*NDOFS+1
      ISMAT=LOC+NBAND*(KROW-1) +
     *(NDOFS/(LOC+NDOFS))*(1+NBAND*(NDOFS-KROW+1))
      SMAT(ISMAT) = SMAT(ISMAT) +  CMAT(J,L)
  184 CONTINUE
  183 CONTINUE
  196 CONTINUE
  193 CONTINUE


C***************
C     END ELEMENT LOOP
C***************
      IF (IENERG .EQ. 0 .AND. NBUG(6) .EQ. 1) THEN
      WRITE(6,*)'  INTE:  SMAT -TRANSIENT'
      CALL PRNTIS
      WRITE(6,*)'  INTE:  F    -TRANSIENT'
      CALL PRNTIT(0,F)
      ENDIF
      ENDIF
C*******************************************************************
C     END TRANSIENT
C*******************************************************************
C
C
C
C*******************************************************************
C     8.)    BEGIN DIRICHLET DATA LOOP
C*******************************************************************
      IF (NFIX .GT. 0 .AND. IENERG .EQ. 0) THEN
      TRNTMP=MIN(TRNFLG,1.)
      DO 280 I=1,NFIX
      JNB = JFIX(I)
      DO 290 J=1,NBAND
      ISMAT=J+NBAND*(JNB-1)
      SMAT(ISMAT) = 0.
  290 CONTINUE
      ISMAT=NBNDTH+NBAND*(JNB-1)
      SMAT(ISMAT)=1.0E0 +(TRNTMP/PREMAC/PREMAC)
      F(JNB)= Q(JNB)* (1.0 - TRNTMP)
  280 CONTINUE
C        8.1)      MODIFY SMAT AND F FOR CUBIC HERMITE
C         IF (K .EQ. 4) THEN
C            DO 252 I=1,NFIX
C               JNB = JFIX(I)
C               DO 253 IA=1,NDOFS
C                  IF (IA .NE. JNB) THEN
C                     IB=JNB+NBNDTH-IA
C                     IF (IB .GT. 0 .AND. IB .LE. NBAND) THEN
C                        F(IA)=F(IA)-SMAT(IB,IA)*F(JNB)
C                        SMAT(IB,IA)=0.E0
```

```
C                       ENDIF
C                    ENDIF
C 253             CONTINUE
C 252          CONTINUE
C              DO 255 IC=1,NBAND
C                 SMAT(IC,2)=0.E0
C 255          CONTINUE
C              SMAT(NBNDTH,2)=1.E0
C              SMAT(NBNDTH-1,2)=-2.E0
C              F(2)=-2.E0*ANUMAN(2,1)
C           ENDIF
      IF (IENERG .EQ. 0 .AND. NBUG(6) .EQ. 1) THEN
      WRITE(6,*)'  INTE:  SMAT -DIRICHLET'
      CALL PRNTIS
      WRITE(6,*)'  INTE:  F    -DIRICHLET'
      CALL PRNTIT(0,F)
      ENDIF
      ENDIF
      ENORMH=ENORMH*0.5
C*****************************************************************
C     8.1)    END DIRICHLET DATA LOOP
C*****************************************************************
 9900 FORMAT(10X,I6,9F8.3)
 9901 FORMAT(10X,12I6)
 9902 FORMAT(10X,6E12.5)
 9011 FORMAT(10X,3I6,3F12.5)
      RETURN
      END
```

```
      BLOCK DATA

      IMPLICIT REAL(A-H,O-Z)

      COMMON/BIGMT/SMAT(5450),F(1090),Q(1090),JINI(1090),
     *XG(1090),YG(1090),AK(1090),AU(1090),AV(1090),SORC(1090),
     *XL(1090),JFIX(1090),MEL(4800)
      COMMON/CONST/K,N,NAXI,NNODEL,NBAND,NDOFS,NDOFN,MSW,NTRAN,INTEG,
     *NELES,NNODES,NSORC,MO,NPR,NFIX,INNEU,JK,IFNNEU,JNEU(68),
     *NNOD(3),NELE(3),NBUG(6),NEM(2,3),NPRA(2),NNEU(0:4),
     *DELTAT,TRNFLG,THETA,PREMAC,ENORMH,ZSTAR,
     *ANUMAN(2,68),REFL,XI(2),XF(2,3),PR(2,3)
      COMMON/MATX/A3000L(2,2,2),A3011L(2,2,2),A3000Q(3,3,3),
     *A3011Q(3,3,3),A3011C(2,4,4),A3011H(2,4,4),
     *A3001L(2,2,2),A3001Q(3,3,3),A200C(4,4),
     *B211B(4,4),B222B(4,4),B212B(4,4),B221B(4,4),
     *B200L(3,3),B200B(4,4)
      COMMON/COORDN/XC1(9),XC2(9),XC3(9),XC4(9),XC5(9),XC6(9),
     *XC7(9),XC8(9),YC1(9),YC2(9),YC3(9),YC4(9),
     *YC5(9),YC6(9),YC7(9),YC8(9)

C     FOLLOWING TO BE DIMENSIONED /MAX. # OF NODES ( 315)/
      DATA AK/1090*0./
      DATA AU/1090*0./
      DATA F/1090*0./
      DATA Q/1090*0./
      DATA SORC/1090*0./
      DATA XG/1090*0./
      DATA YG/1090*0./
      DATA XL/1090*0./
C     FOLLOWING TO BE DIMENSIONED
C     FOR MAXIMUM 2-D SOLVE
      DATA SMAT/5450*0./
C     FOLLOWING TO BE DIMENSIONED /(2 OR 1) *MAX # OF NEUMANN BC'S(68)/
      DATA ANUMAN/136*0./,JNEU/68*0/
C     FOLLOWING TO BE DIMENSIONED /MAX # OF DIRICHLET BC'S(68)/
      DATA JFIX/1090*0/
C     FOLLOWING TO BE DIMENSIONED/NDIMENSIONS(2)/
      DATA XI/2*0./,NNOD/3*0/,NELE/3*0/
C     FOLLOWING TO BE DIMENSIONED/NDIMENSIONS*NPR(2*3)/
      DATA XF/6*0./,PR/6*0./,NEM/6*0/,NPRA/2*0/
C     FOLLOWING TO BE DIMENSIONED/NPR**NDIMENSIONS(3**2)
      DATA XC1,XC2,XC3,XC4,XC5,XC6,XC7,XC8/72*0./
      DATA YC1,YC2,YC3,YC4,YC5,YC6,YC7,YC8/72*0./
C     FOLLOWING TO BE DIMENSIONED/8* MAX # OF ELEMENTS(8*280)/
      DATA MEL/4800*0/
C     FOLLOWING NOT DEPENDENT ON PROBLEM SIZE
      DATA ENORMH/1*0./
      DATA NFIX/1*0/
      DATA JK/1*0/
      DATA MSW/1*0/
      DATA ZSTAR/1*0./
      DATA NNEU/5*0/
      DATA NSORC/0/
      DATA K,N,NAXI,NNODEL,NBAND,NDOFS,
     *NDOFN,NTRAN,NELES,NNODES/10*0/
      DATA DELTAT,TRNFLG,THETA,PREMAC/3*0.,1.0E-07/
      DATA NBUG/6*0/,NPR/0/



      DATA A3011L /
     11.0E 00    ,  -1.0E 00
     2,            1.0E 00    ,  -1.0E 00
     3,           -1.0E 00    ,   1.0E 00
     4,           -1.0E 00    ,   1.0E 00
```

```
     5/
C
      DATA A3000L /
     13.0E 00    ,    1.0E 00
     2,              1.0E 00    ,   1.0E 00
     3,              1.0E 00    ,   1.0E 00
     4,              1.0E 00    ,   3.0E 00
     5/
C
      DATA A3001L /
     1-2.0E 00    ,    2.0E 00
     2,           -1.0E 00    ,   1.0E 00
     3,           -1.0E 00    ,   1.0E 00
     4,           -2.0E 00    ,   2.0E 00
     5/
C
C     DATA A3001L /
C    1                        -2.0E 00    ,  -1.0E 00
C    2              ,         -1.0E 00    ,  -2.0E 00
C    3              ,          2.0E 00    ,   1.0E 00
C    4              ,          1.0E 00    ,   2.0E 00
C    5              /
      DATA A3000Q /
     139.0E 00  ,    20.0E 00   ,    -3.0E 00
     2,      20.0E 00  ,   16.0E 00  ,    -8.0E 00
     3,      -3.0E 00  ,   -8.0E 00  ,    -3.0E 00
     4,      20.0E 00  ,   16.0E 00  ,    -8.0E 00
     5,      16.0E 00  ,  192.0E 00  ,    16.0E 00
     6,      -8.0E 00  ,   16.0E 00  ,    20.0E 00
     7,      -3.0E 00  ,   -8.0E 00  ,    -3.0E 00
     8,      -8.0E 00  ,   16.0E 00  ,    20.0E 00
     9,      -3.0E 00  ,   20.0E 00  ,    39.0E 00
     1/
C
      DATA A3011Q /
     137.0E 00   ,   -44.0E 00   ,     7.0E 00
     2,  36.0E 00   ,  -32.0E 00   ,    -4.0E 00
     3,  -3.0E 00   ,   -4.0E 00   ,     7.0E 00
     4, -44.0E 00   ,   48.0E 00   ,    -4.0E 00
     5, -32.0E 00   ,   64.0E 00   ,   -32.0E 00
     6,  -4.0E 00   ,   48.0E 00   ,   -44.0E 00
     7,   7.0E 00   ,   -4.0E 00   ,    -3.0E 00
     8,  -4.0E 00   ,  -32.0E 00   ,    36.0E 00
     9,   7.0E 00   ,  -44.0E 00   ,    37.0E 00
     1/
C
      DATA A3001Q /
     1-30.0E 00   ,    36.0E 00   ,    -6.0E 00
     2,     -18.0E 00  ,   24.0E 00  ,    -6.0E 00
     3,       3.0E 00  ,    0.0E 00  ,    -3.0E 00
     4,     -18.0E 00  ,   24.0E 00  ,    -6.0E 00
     5,     -48.0E 00  ,    0.0E 00  ,    48.0E 00
     6,       6.0E 00  ,  -24.0E 00  ,    18.0E 00
     7,       3.0E 00  ,    0.0E 00  ,    -3.0E 00
     8,       6.0E 00  ,  -24.0E 00  ,    18.0E 00
     9,       6.0E 00  ,  -36.0E 00  ,    30.0E 00
     1/
C
      DATA A3011C /
     1262.0E 00  ,    -327.0E 00   ,    78.0E 00   ,    -13.0E 00
     2,     34.0E 00   ,     -51.0E 00   ,    30.0E 00   ,     -13.0E 00
     3,   -327.0E 00   ,     594.0E 00   , -297.0E 00   ,      30.0E 00
     4,    -51.0E 00   ,     270.0E 00   , -297.0E 00   ,      78.0E 00
     5,     78.0E 00   ,    -297.0E 00   ,  270.0E 00   ,     -51.0E 00
     6,     30.0E 00   ,    -297.0E 00   ,  594.0E 00   ,    -327.0E 00
     7,    -13.0E 00   ,      30.0E 00   ,  -51.0E 00   ,      34.0E 00
```

```
     8,  -13.0E 00  ,     78.0E 00  , -327.0E 00  ,    262.0E 00
     9/
C
      DATA A200C /
     1128.0E 00  ,    99.0E 00  ,    -36.0E 00  ,  19.0E 00
     2,   99.0E 00  , 648.0E 00  ,     -81.0E 00  , -36.0E 00
     3,  -36.0E 00  , -81.0E 00  ,     648.0E 00  ,  99.0E 00
     4,   19.0E 00  , -36.0E 00  ,      99.0E 00  , 128.0E 00
     5/

C
      DATA A3011H /
     136.0E 00  ,    0.0E 00  ,    -36.0E 00  ,   6.0E 00
     2,   36.0E 00  ,     6.0E 00  ,     -36.0E 00  ,    0.0E 00
     3,    0.0E 00  ,     6.0E 00  ,       0.0E 00  ,   -1.0E 00
     4,    6.0E 00  ,     2.0E 00  ,      -6.0E 00  ,   -1.0E 00
     5,  -36.0E 00  ,     0.0E 00  ,      36.0E 00  ,   -6.0E 00
     6,  -36.0E 00  ,    -6.0E 00  ,      36.0E 00  ,    0.0E 00
     7,    6.0E 00  ,    -1.0E 00  ,      -6.0E 00  ,    2.0E 00
     8,    0.0E 00  ,    -1.0E 00  ,       0.0E 00  ,    6.0E 00
     9/
C
      DATA B200L /
     12.0E 00  ,    1.0E 00  ,    1.0E 00
     2,    1.0E 00  ,     2.0E 00  ,   1.0E 00
     3,    1.0E 00  ,     1.0E 00  ,   2.0E 00
     4/
C
      DATA B200B/
     14.0E 00  ,  2.0E 00  ,  1.0E 00  ,  2.0E 00
     2,    2.0E 00  ,  4.0E 00  ,  2.0E 00  ,  1.0E 00
     3,    1.0E 00  ,  2.0E 00  ,  4.0E 00  ,  2.0E 00
     4,    2.0E 00  ,  1.0E 00  ,  2.0E 00  ,  4.0E 00
     5/
C
      DATA B211B/
     12.0E 00  , -2.0E 00  , -1.0E 00  ,  1.0E 00
     2,   -2.0E 00  ,  2.0E 00  ,  1.0E 00  , -1.0E 00
     3,   -1.0E 00  ,  1.0E 00  ,  2.0E 00  , -2.0E 00
     4,    1.0E 00  , -1.0E 00  , -2.0E 00  ,  2.0E 00
     5/

C
      DATA B222B /
     12.0E 00  ,  1.0E 00  , -1.0E 00  , -2.0E 00
     2,    1.0E 00  ,  2.0E 00  , -2.0E 00  , -1.0E 00
     3,   -1.0E 00  , -2.0E 00  ,  2.0E 00  ,  1.0E 00
     4,   -2.0E 00  , -1.0E 00  ,  1.0E 00  ,  2.0E 00
     5/
C
      DATA B212B /
     11.0E 00  ,  1.0E 00  , -1.0E 00  , -1.0E 00
     2,   -1.0E 00  , -1.0E 00  ,  1.0E 00  ,  1.0E 00
     3,   -1.0E 00  , -1.0E 00  ,  1.0E 00  ,  1.0E 00
     4,    1.0E 00  ,  1.0E 00  , -1.0E 00  , -1.0E 00
     5/

      DATA B221B /
     11.0E 00  , -1.0E 00  , -1.0E 00  ,  1.0E 00
     2,    1.0E 00  , -1.0E 00  , -1.0E 00  ,  1.0E 00
     3,   -1.0E 00  ,  1.0E 00  ,  1.0E 00  , -1.0E 00
     4,   -1.0E 00  ,  1.0E 00  ,  1.0E 00  , -1.0E 00
     5/

      END
```

```
      SUBROUTINE ELMAT
C     MASTER MATRIX LIBRARY
C*****************************************************

      IMPLICIT REAL(A-H,O-Z)
      COMMON/MATX/A3000L(2,2,2),A3011L(2,2,2),A3000Q(3,3,3),
     *A3011Q(3,3,3),A3011C(2,4,4),A3011H(2,4,4),
     *A3001L(2,2,2),A3001Q(3,3,3),A200C(4,4),
     *B211B(4,4),B222B(4,4),B212B(4,4),B221B(4,4),
     *B200L(3,3),B200B(4,4)

C*********************************************
C       1.1   ONE DIMENSIONAL, LINEAR
C*********************************************
      CALL TRANS(2,2,2,8,A3000L)
      CALL TRANS(2,2,2,8,A3001L)
      CALL TRANS(2,2,2,8,A3011L)

      ICOUNT=0
      DO 490 L=1,2
      DO 500 I=1,2
      ICOUNT=ICOUNT+1
      DO 510 J=1,2
      A3000L(L,I,J) = A3000L(L,I,J)/12.E0
      A3001L(L,I,J) = A3001L(L,I,J)/6.E0
      A3011L(L,I,J) = A3011L(L,I,J)/2.E0
510   CONTINUE
500   CONTINUE
490   CONTINUE

C*********************************************
C       1.2   ONE DIMENSIONAL, QUADRATIC
C*********************************************
      CALL TRANS(3,3,3,27,A3000Q)
      CALL TRANS(3,3,3,27,A3001Q)
      CALL TRANS(3,3,3,27,A3011Q)

C       DO 11210 L=1,2
C       USE FULL INTERPOLATION
      DO 11210 L=1,3
      DO 11220 I=1,3
      DO 11230 J=1,3
      A3000Q(L,I,J) = A3000Q(L,I,J)/420.E0
      A3001Q(L,I,J) = A3001Q(L,I,J)/90.E0
      A3011Q(L,I,J) = A3011Q(L,I,J)/30.E0
11230 CONTINUE
11220 CONTINUE
11210 CONTINUE

C*********************************************
C       1.3  ONE DIMENSIONAL,CUBIC LAGRANGE
C*********************************************
      CALL TRANS(2,4,4,32,A3011C)

      DO 11240 L=1,2
      DO 11250 I=1,4
      DO 11260 J=1,4
      A3011C(L,I,J)=A3011C(L,I,J)/80.E0
11260 CONTINUE
11250 CONTINUE
11240 CONTINUE

      CALL TRANS(1,4,4,16,A200C)

      DO 1125 I=1,4
      DO 1126 J=1,4
```

```
      A200C(I,J)=A200C(I,J)/1680.E0
 1126 CONTINUE
 1125 CONTINUE

C************************************************
C        1.4  ONE DIMENSIONAL,CUBIC HERMITE
C************************************************
      CALL TRANS(2,4,4,32,A3011H)

      DO 11270 L=1,2
      DO 11280 I=1,4
      DO 11290 J=1,4
      A3011H(L,I,J)=A3011H(L,I,J)/60.E0
11290 CONTINUE
11280 CONTINUE
11270 CONTINUE

C************************************************
C        2.1  TWO DIMENSIONAL LINEAR TRIANGLE
C************************************************

      DO 420 L=1,3
      DO 440 J=1,3
      B200L(L,J) = B200L(J,L)/12.E0
 440  CONTINUE
 420  CONTINUE


C************************************************
C        2.2   TWO DIMENSIONAL, BILINEAR QUADRILATERAL
C************************************************
      DO 520 L=1,4
      DO 530 J=1,4
      B200B(L,J) = B200B(J,L)
      B211B(L,J) = B211B(J,L)
      B222B(L,J) = B222B(J,L)
      B221B(L,J) = B221B(J,L)
      B212B(L,J) = B212B(J,L)
530   CONTINUE
520   CONTINUE

      DO 540 L=1,4
      DO 550 J=1,4
      B200B(L,J) = B200B(L,J)/9.E0
      B211B(L,J) = B211B(L,J)/6.E0
      B222B(L,J) = B222B(L,J)/6.E0
      B221B(L,J) = B221B(L,J)/4.E0
      B212B(L,J) = B212B(L,J)/4.E0
550   CONTINUE
540   CONTINUE

      RETURN
      END
```

```
      SUBROUTINE INTGRN

      COMMON/BIGMT/SMAT(5450),F(1090),Q(1090),JINI(1090),
     *XG(1090),YG(1090),AK(1090),AU(1090),AV(1090),SORC(1090),
     *XL(1090),JFIX(1090),MEL(4800)
      COMMON/CONST/K,N,NAXI,NNODEL,NBAND,NDOFS,NDOFN,MSW,NTRAN,INTEG,
     *NELES,NNODES,NSORC,MO,NPR,NFIX,INNEU,JK,IFNNEU,JNEU(68),
     *NNOD(3),NELE(3),NBUG(6),NEM(2,3),NPRA(2),NNEU(0:4),
     *DELTAT,TRNFLG,THETA,PREMAC,ENORMH,ZSTAR,
     *ANUMAN(2,68),REFL,XI(2),XF(2,3),PR(2,3)
      DIMENSION AQQ(1090)

      READ(5,*) TIME,TSTOP,TSTEP,THETA,EPSILN,QDMAX,TSTEPI,NTSMAX
      ISTEPN=0
      IF(TSTEP .EQ. TSTEPI) THEN
      ISTEPN=1
      ENDIF
      DO 99 I=1,NNODES
      Q(I)=F(I)
      AQQ(I)=Q(I)+SORC(I)
 99   CONTINUE
      NTS=0
      DELTAT=0.

      WRITE(6,*) ' '
      WRITE(6,*) '  INTE:'
      WRITE(6,*) '           TZERO     TSTOP      TSTEP      THETA',
     1'    DQMIN       QDMAX     TSTEPI NTSM'
      WRITE(6,9557) TIME,TSTOP,TSTEP,THETA,EPSILN,QDMAX,TSTEPI,NTSMAX
      IF (NBUG(5) .GT. 0) THEN
      IF (N.EQ.1) WRITE(6,9568) TIME
      IF(N.EQ.1)WRITE(6,9564)(AQQ(I),I=1,NNODES)
      IF (N.GT.1) WRITE(6,9558)
      IF(N.GT.1) WRITE(6,9559) TIME,NTS,ICOUNT,DELTAT
      CALL PRNTIT(0,AQQ)
      ENDIF


      TSTEP=MAX(TSTEP,PREMAC)
      IF (TSTEPI .LT. TSTEP*PREMAC .OR. TSTEPI .GT. TSTEP) TSTEPI=
     1TSTEP*1.E-05
      DELTAT=TSTEPI
      THETA=MAX(THETA,0.)
      EPSILN=MAX(EPSILN,PREMAC)

C     TRANSIENT SOLUTION LOOP


      DQMAX=EPSILN+EPSILN
C      DO 299 WHILE
C     1   (NTS.LE.NTSMAX.AND.TIME.LE.TSTOP.AND.DQMAX.GE.EPSILN)
      DO 299 IDP=1,NTSMAX+1
      IF(TIME.GT.TSTOP.OR.DQMAX.LT.EPSILN)GOTO 299
      NTS=IDP

      ICOUNT=0
      ILCOUN=0
      DQMAX=EPSILN+QDMAX
C      DO 499 WHILE (DQMAX .GE. QDMAX .AND. ICOUNT .LE. 5)
      DO 499 IKL=1,6
      ICOUNT=IKL
      IF (DQMAX .LT. QDMAX ) GOTO 499

      TRNFLG=1.
      IF (N .GT. 1) THEN
      CALL TSPLIT(NTS,ICOUNT)
```

```
      ELSE
      CALL FORMIT(0)
      IF (NBAND .EQ. 3) THEN
      CALL SOLVT(NDOFS)
      ELSE
      CALL SOLVR(NDOFS,NBAND,SMAT,F)
      ENDIF
      ENDIF

      DQMAX=0.
      DO 234 I=1,NDOFS
      DQMAX=MAX(DQMAX,ABS(F(I)))
  234 CONTINUE
      IF (ISTEPN .EQ. 0 ) THEN
      TSTEPN=DELTAT*QDMAX/DQMAX/FLOAT(ICOUNT)
      IF (ICOUNT .GT. 3) TSTEPN=TSTEPN*DQMAX/QDMAX
      TSTEPN=MIN(TSTEP,TSTEPN)
      TSTEPN=MIN(TSTOP-TIME-DELTAT,TSTEPN)
      TSTEPN=MAX(TSTEPN,TSTEPI)
      ELSE
      TSTEPN=TSTEP
      ENDIF
C         ENDIF
  499 CONTINUE

      IF (ICOUNT .GT. 1) THEN
C            WRITE(6,9558) TIME,NTS,ICOUNT,DELTAT
C            WRITE(6,*) '  INTE: MAX # OF RESTARTS- TERMINATED'
      IF(NBUG(5).GT.0) THEN
C            IF  (N.EQ.1) WRITE(6,9568) TIME
C            IF(N.EQ.1)WRITE(6,9564)(Q(I),I=1,NNODES)
      IF (N.GT.1) WRITE(6,9558)
      IF(N.GT.1)WRITE(6,9559) TIME,NTS,ILCOUN,DELTAT
      ENDIF
      ENDIF
      TIME=TIME+DELTAT
      DO 236 I=1,NDOFS
      Q(I)=Q(I)+F(I)
  236 CONTINUE
C       PRINT SOLUTION
      IF (NBUG(5) .GT. 0 .AND. TIME.LT.TSTOP) THEN
      IF (FLOAT(NTS/NBUG(5)) .EQ.
     1FLOAT(NTS)/FLOAT(NBUG(5))) THEN
      IF (N.EQ.1) WRITE(6,9568) TIME
      IF(N.EQ.1.AND.NBUG(4).EQ.2)WRITE(6,9564)(Q(I),I=1,NNODES)
      IF (N.GT.1) WRITE(6,9558)
      IF(N.GT.1) WRITE(6,9559) TIME,NTS,ICOUNT,DELTAT
      CALL PRNTIT(0,Q)
      ENDIF
      ENDIF
      DELTAT=TSTEPN
C       ENDIF
  299 CONTINUE

      IF (TIME .GE. TSTOP) THEN
      TIME=TIME-DELTAT
      WRITE(6,*)
     1'  INTE: REACHED STOP TIME BEFORE STEADY STATE'

      ELSE IF ( DQMAX .LE. EPSILN) THEN
      WRITE(6,*) '  INTE: REACHED STEADY STATE'

      ELSE IF (NTS .GE. NTSMAX) THEN
      WRITE(6,*) '  INTE: EXCEEDED MAXIMUM NUMBER OF TIMESTEPS'
      ENDIF
```

```
C        WRITE(6,9558) TIME,NTS,ICOUNT,DELTAT
      NTS1=NTS+1
      WRITE(6,9560) (TIME+DELTAT),NTS1,ICOUNT,DELTAT
      WRITE(6,*)
      WRITE(6,*)'  INTE:  NODAL SOLUTION'
      CALL PRNTIT(0,Q)
      IF(N.EQ.1.AND.NBUG(4).EQ.2)WRITE(6,9564)(Q(I),I=1,NNODES)
C     STORE TIME VALUE IN COMMON VARIABLE DELTAT FOR SUB PLOTIT
      DELTAT=TIME

      RETURN
 9564 FORMAT(10X,9F7.2)
 9557 FORMAT(10X,3F8.4,7X,F4.2,X,3F8.4,X,I5)
C 9558 FORMAT(3X,'TIME=', F12.5, ' STEP=', I5, ' #RES=',I5 ,
C     1 ' NEW DELTAT=' ,F12.5)
 9558 FORMAT(//10X,'TIME=', 7X, ' STEP=', 7X, ' #TSTPCHG=', 4X,
     1' NEW DELTAT=')
 9568 FORMAT(//10X,'TIME=',F8.4)
 9559 FORMAT(10X, F7.3,5X,I5,8X,I5,8X,F7.4)
 9560 FORMAT(/10X,'TIME=', F12.5, ' STEP=', I5, ' #TSTPCHG=',I5 ,
     1' FINAL DELTAT=' ,F12.5)
 9111 FORMAT(/,10X,A4,':',2X,'SYSTEM MATRIX')
      END
```

```
      SUBROUTINE TSPLIT(NTS,ICOUNT)

      IMPLICIT REAL(A-H,O-Z)

      COMMON/BIGMT/SMAT(5,1090),F(1090),Q(1090),JINI(1090),
     *XG(1090),YG(1090),AK(1090),AU(1090),AV(1090),SORC(1090),
     *XL(1090),JFIX(1090),MEL(4800)
      COMMON/CONST/K,N,NAXI,NNODEL,NBAND,NDOFS,NDOFN,MSW,NTRAN,INTEG,
     *NELES,NNODES,NSORC,MO,NPR,NFIX,INNEU,JK,IFNNEU,JNEU(68),
     *NNOD(3),NELE(3),NBUG(6),NEM(2,3),NPRA(2),NNEU(0:4),
     *DELTAT,TRNFLG,THETA,PREMAC,ENORMH,ZSTAR,
     *ANUMAN(2,68),REFL,XI(2),XF(2,3),PR(2,3)
      INTEGER ISV(6)
      REAL RSV(2)
      REAL QSV(1090)

      IF (NTS .EQ. 1 .AND. ICOUNT .EQ. 1) ND1=2
      DO 350 NOD=1,NNODES
      QSV(NOD)=Q(NOD)
  350 CONTINUE

      DO 210 NSPLIT=1,N
C        REEORDER
      ND1=ND1+1
      IF (ND1 .LT. 1 .OR. ND1 .GT. N) ND1=1
      IF (ND1 .EQ. 1) THEN
      ND2=2
      ELSE IF (ND1 .EQ. 2) THEN
      ND2=1
      ENDIF
      IF (ND1 .EQ. 1) THEN
      DO 340 NOD=1,NNODES
      JINI(NOD)=NOD
  340 CONTINUE
      ELSE IF (ND1 .EQ. 2) THEN
      NOD=0
      DO 320 NC1=1,NNOD(ND1)
      DO 320 NC2=1,NNOD(ND2)
      NOD=NOD+1
      NPD=NC1+(NC2-1)*NNOD(ND1)
      SMAT(1,NOD)=XG(NOD)
      SMAT(2,NOD)=YG(NOD)
      SMAT(3,NOD)=AK(NOD)
      JINI(NOD)=NPD
  320 CONTINUE
      ENDIF
      IF (ND1 .EQ. 1) THEN
      CONTINUE
      ELSE
C           COPY BACK TO ORIGINAL MATRICES
      DO 341 NOD=1,NNODES
      NPD=JINI(NOD)
C              SWAP XG,YG
      YG(NPD)=SMAT(1,NOD)
      XG(NPD)=SMAT(2,NOD)
      AK(NPD)=SMAT(3,NOD)
      SMAT(1,NOD)=F(NOD)
      SMAT(2,NOD)=SORC(NOD)
      SMAT(3,NOD)=Q(NOD)
  341 CONTINUE
      DO 342 NOD=1,NNODES
      NPD=JINI(NOD)
      F(NPD)=SMAT(1,NOD)
      SORC(NPD)=SMAT(2,NOD)
      Q(NPD)=SMAT(3,NOD)
      SMAT(2,NOD)=AU(NOD)
```

```
      SMAT(3,NOD)=AV(NOD)
  342 CONTINUE
      DO 343 NOD=1,NNODES
      NPD=JINI(NOD)
C              SWAP AU,AV
      AV(NPD)=SMAT(2,NOD)
      AU(NPD)=SMAT(3,NOD)
  343 CONTINUE
      IF (NFIX .GT. 0) THEN
      DO 344 NFI=1,NFIX
      KFIX=JFIX(NFI)
      JFIX(NFI)=JINI(KFIX)
  344 CONTINUE
      ENDIF
      IF (INNEU .GT. 0) THEN
      DO 345 NFI=1,NNEU(JK)
      KNEU=JNEU(NFI)
      JNEU(NFI)=JINI(KNEU)
  345 CONTINUE
      ENDIF
      ENDIF

      RSV(1)=DELTAT
      ISV(1)=N
      ISV(2)=NNOD(1)
      ISV(3)=NNOD(2)
      ISV(4)=NELES
      ISV(5)=NAXI
      ISV(6)=NNODEL
      NELES=NELE(ND1)*(K*NELE(ND2)+1)
      NBAND=2*K+1
      NNODEL=K+1
      DELTAT=DELTAT*0.5
      IF(ND1 .NE. 1) NAXI=0

C        OVERWRITE MEL
      JEL=0
      DO 346 JEL1=1,K*NELE(ND2)+1
      DO 356 JEL2=1,NELE(ND1)
      JEL=JEL+1
      DO 347 LNOD=1,K+1
      LMEL=NNODEL*(JEL-1)+LNOD
      NOD=(JEL1-1)*(K*NELE(ND1)+1)+(JEL2-1)*K+LNOD
      MEL(LMEL)=NOD
  347 CONTINUE
  356 CONTINUE
  346 CONTINUE

      IF (NBUG(6) .EQ. 1) THEN
      WRITE(6,*) ' XG DIR=',ND1
      WRITE(6,9115) (XG(I1),I1=1,NNODES)
      WRITE(6,*) ' YG DIR=',ND1
      WRITE(6,9115) (YG(I1),I1=1,NNODES)
      WRITE(6,*) ' AK DIR=',ND1
      WRITE(6,9115) (AK(I1),I1=1,NNODES)
      WRITE(6,*) '  Q DIR=',ND1
      WRITE(6,9115) ( Q(I1),I1=1,NNODES)
      WRITE(6,*) ' AU DIR=',ND1
      WRITE(6,9115) (AU(I1),I1=1,NNODES)
      WRITE(6,*) ' AV DIR=',ND1
      WRITE(6,9115) (AV(I1),I1=1,NNODES)
      WRITE(6,*) ' ML DIR=',ND1
      WRITE(6,9116) (MEL(I1),I1=1,NELES*(K+1))
      ENDIF

      CALL FORMIT(0)
```

```
C      IF (NBUG(6) .EQ. 1) THEN
C         WRITE(6,*) ' DIR=',ND1
C         IF (ND1 .EQ. 2) THEN
C         KK=1
C         IND1=16
C         ELSE
C         IND1=42
C         KK=10
C         ENDIF
C         WRITE(6,9119) F(IND1),F(IND1+KK),F(IND1+2*KK)
C         WRITE(6,9119) Q(IND1),Q(IND1+KK),Q(IND1+2*KK)
C         WRITE(6,9119) SMAT(1,IND1),SMAT(2,IND1),SMAT(3,IND1)
C         WRITE(6,9119) SMAT(1,IND1+KK),SMAT(2,IND1+KK),SMAT(3,IND1+KK)
C      WRITE(6,9119)SMAT(1,IND1+2*KK),SMAT(2,IND1+2*KK),SMAT(3,IND1+2*KK)
C      ENDIF

       IF (NBAND .EQ. 3) THEN
       CALL SOLVT(NDOFS)
       ELSE
       CALL SOLVR(NNODES,NBAND,SMAT,F)
       END IF

       DELTAT=RSV(1)
       N=ISV(1)
       NNOD(1)=ISV(2)
       NNOD(2)=ISV(3)
       NELES =ISV(4)
       NAXI  =ISV(5)
       NNODEL=ISV(6)

       NOD=0
C      COUNT IN CURRENT ORDER
       IF (ND1 .EQ. 1) THEN
       DO 410 NC2=1,NNOD(ND2)
       DO 410 NC1=1,NNOD(ND1)
       NOD=NOD+1
       NPD=NC1+(NC2-1)*NNOD(ND1)
       JINI(NOD)=NPD
C UPDATE Q
       Q(NOD)=Q(NOD)+F(NOD)
 410   CONTINUE
       ELSE IF (ND1 .EQ. 2) THEN
       DO 451 NC2=1,NNOD(ND2)
       DO 451 NC1=1,NNOD(ND1)
       NOD=NOD+1
       NPD=NC2+(NC1-1)*NNOD(ND2)
C              REEORDER INDEX TO JINI
       SMAT(1,NOD)=XG(NOD)
       SMAT(2,NOD)=YG(NOD)
       SMAT(3,NOD)=AK(NOD)
       JINI(NOD)=NPD
 451   CONTINUE
       DO 442 NOD=1,NNODES
       NPD=JINI(NOD)
C               SWAP XG,YG
       YG(NPD)=SMAT(1,NOD)
       XG(NPD)=SMAT(2,NOD)
       AK(NPD)=SMAT(3,NOD)
       SMAT(1,NOD)=F(NOD)
       SMAT(2,NOD)=SORC(NOD)
       SMAT(3,NOD)=Q(NOD)
 442   CONTINUE
       DO 443 NOD=1,NNODES
       NPD=JINI(NOD)
       F(NPD)=SMAT(1,NOD)
```

```
      SORC(NPD)=SMAT(2,NOD)
C UPDATE Q
      Q(NPD)=SMAT(3,NOD)+SMAT(1,NOD)
      SMAT(2,NOD)=AU(NOD)
      SMAT(3,NOD)=AV(NOD)
  443 CONTINUE
      DO 444 NOD=1,NNODES
      NPD=JINI(NOD)
C                SWAP AU,AV
      AV(NPD)=SMAT(2,NOD)
      AU(NPD)=SMAT(3,NOD)
  444 CONTINUE
      IF (NFIX .GT. 0) THEN
      DO 445 NFI=1,NFIX
      KFIX=JFIX(NFI)
      JFIX(NFI)=JINI(KFIX)
  445 CONTINUE
      ENDIF
      IF (INNEU .GT. 0) THEN
      DO 447 NFI=1,NNEU(JK)
      KNEU=JNEU(NFI)
      JNEU(NFI)=JINI(KNEU)
  447 CONTINUE
      ENDIF
      ENDIF

C     IF (NBUG(6) .EQ. 1) THEN
C        WRITE(6,*) ' XG DIR=',ND1
C        WRITE(6,9115) (XG(I1),I1=1,NNODES)
C        WRITE(6,*) ' YG DIR=',ND1
C        WRITE(6,9115) (YG(I1),I1=1,NNODES)
C     ENDIF
      IF (NBUG(6) .EQ. 1) THEN
      WRITE(6,*) ' NTS=', NTS,' DIR=',ND1
      CALL PRNTIT(0,F)
      CALL PRNTIT(0,Q)
      ENDIF
  210 CONTINUE


      DO 449 NOD=1,NNODES
      F(NOD)=Q(NOD)-QSV(NOD)
      Q(NOD)=QSV(NOD)
  449 CONTINUE

 9113 FORMAT(1X,8F9.3)
 9115 FORMAT(1X,9F8.3)
 9116 FORMAT(16I5)
 9119 FORMAT(3E20.12)

      RETURN
      END
```

```
      SUBROUTINE SORCR(IFUNC)

      IMPLICIT REAL(A-H,O-Z)

      COMMON/BIGMT/SMAT(5450),F(1090),Q(1090),JINI(1090),
     *XG(1090),YG(1090),AK(1090),AU(1090),AV(1090),SORC(1090),
     *XL(1090),JFIX(1090),MEL(4800)
      COMMON/CONST/K,N,NAXI,NNODEL,NBAND,NDOFS,NDOFN,MSW,NTRAN,INTEG,
     *NELES,NNODES,NSORC,MO,NPR,NFIX,INNEU,JK,IFNNEU,JNEU(68),
     *NNOD(3),NELE(3),NBUG(6),NEM(2,3),NPRA(2),NNEU(0:4),
     *DELTAT,TRNFLG,THETA,PREMAC,ENORMH,ZSTAR,
     *ANUMAN(2,68),REFL,XI(2),XF(2,3),PR(2,3)
      COMMON/MATX/A3000L(2,2,2),A3011L(2,2,2),A3000Q(3,3,3),
     *A3011Q(3,3,3),A3011C(2,4,4),A3011H(2,4,4),
     *A3001L(2,2,2),A3001Q(3,3,3),A200C(4,4),
     *B211B(4,4),B222B(4,4),B212B(4,4),B221B(4,4),
     *B200L(3,3),B200B(4,4)
      COMMON/COORDN/XC1(9),XC2(9),XC3(9),XC4(9),XC5(9),XC6(9),
     *XC7(9),XC8(9),YC1(9),YC2(9),YC3(9),YC4(9),
     *YC5(9),YC6(9),YC7(9),YC8(9)
      REAL XSORC(3,40)


C       XL(*) : SOURCE AMPLITUDE
C       SMAT(*) : HALF-WIDTH
C       XSORC(*,*) : CENTROIDAL CO-ORDINATES OF SOURCE

      DO 121 I=1,NDOFS
      SORC(I)=0.0
 121  CONTINUE
      IF (IFUNC .EQ. 0) GO TO 1100

      DO 301 I = 1,N
      READ(5,*) (XSORC(I,J),J=1,NSORC)
 301  CONTINUE
      DO 10 I1 = 1,NSORC
      READ(5,*) XL(I1),SMAT(I1)
      WRITE(6,59) XL(I1),SMAT(I1),
     .(XSORC(ID,I1),ID=1,N)
 10   CONTINUE

      NEL=NNODES
      DO 102 IL = 1,NSORC
      IF (XL(IL).GT.0. .AND. SMAT(IL).GT.0.) THEN
      IF (IFUNC.EQ.1) THEN
      DO 103 IJ = 1,NEL
      DIST = SQRT((XG(IJ)-XSORC(1,IL))**2. +
     .(YG(IJ)-XSORC(2,IL))**2.)

      Z = DIST/SMAT(IL)
      IF (Z.LT.1.) THEN
      SORC(IJ) = SORC(IJ) + XL(IL)*COS(Z*3.141593/2.)
C          PRINT*, 'IL, IJ, SORC(IJ) = ',IL,IJ,SORC(IJ)
      END IF
 103  CONTINUE
      ELSE IF (IFUNC.EQ.2) THEN
      DO 101 IK = 1,NEL
      DIST = SQRT((XG(IK)-XSORC(1,IL))**2. +
     .(YG(IK)-XSORC(2,IL))**2.)
      Z = DIST/SMAT(IL)
      IF (Z.LT.1.) THEN
      SORC(IK) = SORC(IK) + XL(IL)*EXP(-Z*Z/2.)
      PRINT*, 'IL, IK, SORC(IK) = ',IL,IK,SORC(IK)
      END IF
 101  CONTINUE
      END IF
```

```
      END IF
 102  CONTINUE
      WRITE(6,54)  (SORC(JJ),JJ=1,NNODE)

 1100 IF (IFUNC.EQ.0) THEN

C        JINI(*) : NODE NUMBER
C        SORC(*) : SOURCE DISTRIBUTION

      READ(5,*)   (JINI(I),I=1,NSORC)
      WRITE(6,68)  (JINI(I),I=1,NSORC)
      READ(5,*)  (SORC(JINI(I)),I=1,NSORC)
      WRITE(6,97)  (SORC(JINI(I)),I=1,NSORC)
      END IF
      IF (INTEG .GT. 0) THEN
      CALL IPRINT(2)
      ELSE
      CALL PRNTIT(0,SORC)
      ENDIF

 59   FORMAT(10X,F6.2,7X,F6.2,7X,F6.2,7X,F6.2)
 50   FORMAT(10X,'AMPLITUDE',5X,'HALF-WIDTH',5X,'X COORD',5X,'Y COORD')
 54   FORMAT(10X,8F6.2)
 55   FORMAT(25F6.2)
 56   FORMAT(10X,'FINITE LINE SOURCE LOCATED ON Z AXIS'/
     .13X,'ZA',8X,'ZB',8X,'ZP',8X,'R'/
     .10X,F6.2,4X,F6.2,4X,F6.2,4X,F6.2)
 57   FORMAT(10X,2F6.2)
 68   FORMAT(10X,6HJSORC:,10I5)
 95   FORMAT(5E12.5)
 97   FORMAT(10X,5F12.2)

      RETURN
      END
```

# LIST OF SYMBOLS

|  |  |
|---|---|
| $a$ | expansion coefficient; measure |
| $A$ | plane area; one-dimensional master matrix prefix |
| $[A]$ | factored global matrix |
| $b$ | coefficient; boundary condition subscript; body force component; measure |
| $\{b\}$ | global data (load) matrix |
| $B$ | two-dimensional master matrix prefix |
| $\mathbf{B}$ | body force |
| $c$ | coefficient; specific heat |
| $C$ | Hooke's law matrix; three-dimensional master matrix prefix |
| $d$ | coefficient; master matrix indicator |
| $D$ | diagonal matrix |
| $e$ | element-dependent subscript; unit vector component |
| $e(\cdot)$ | error |
| $E$ | energy seminorm subscript; Young's modulus |
| $g$ | dependent variable |
| $G$ | elasticity shear modulus |
| $h$ | discretization superscript; convective heat transfer coefficient |
| $H$ | macro domain superscript; Gauss quadrature weight |
| $[H]$ | global convection matrix |
| $i$ | summation index |
| $\mathbf{i}$ | unit vector parallel to $x$ |
| $I$ | moment of inertia; element matrix summation index |
| $j$ | summation index |
| $\mathbf{j}$ | unit vector parallel to $y$ |
| $J$ | element matrix summation index |
| $k_{ij}$ | element of a square matrix |
| $k$ | thermal conductivity; basis degree; summation index; diffusion coefficient |
| $\{K\}$ | discretized conductivity nodal array |
| $[K]$ | global conduction matrix |
| $K$ | element matrix summation index |
| $l$ | one-dimensional element length; summation index |
| $l(\cdot)$ | differential operator on $\partial\Omega$ |
| $L$ | one-dimensional domain span |
| $L(\cdot)$ | differential operator on $\Omega$ |
| $M$ | total elements in discretization $\Omega^h$; moment; matrix name prefix |
| $n$ | summation matrix; normal subscript; dimension of domain $\Omega$ |
| $\mathbf{n}$ | outward pointing unit vector normal to $\partial\Omega$ |
| $N$ | last summation term index value; approximation superscript |

| | |
|---|---|
| $\{N_k\}$ | finite element basis of degree $k$ |
| $p$ | macro grid progression ratio; load (data) |
| $P$ | discretized nodal load; Gauss quadrature order |
| $\{P\}$ | global intermediate column matrix |
| $q$ | flux on $\partial\Omega$; generalized dependent variable |
| $Q$ | discretized dependent variable nodal value |
| $r$ | reference state subscript; radius |
| $R$ | discretized radial coordinate; Euclidean space |
| $\{R\}$ | global matrix statement residual |
| $s$ | source term on $\Omega$ |
| $\mathbf{s}$ | unit vector tangent to $\partial\Omega$ |
| $S$ | finite element assembly operator; discretized source nodal value |
| $t$ | time |
| $T$ | temperature |
| $\mathbf{T}$ | surface traction |
| $\mathbf{u}$ | displacement vector; velocity vector |
| $u$ | cartesian $x$ component |
| $U$ | discretized nodal value |
| $[U]$ | global fluid convection matrix |
| $v$ | cartesian $y$ component |
| $V$ | shear; volume; discretized nodal value |
| $w$ | weight function; fin thickness; cartesian $z$ component |
| $W$ | weight |
| $WS$ | Galerkin weak statement |
| $x$, $x_i$ | cartesian coordinate (system $1 \le i \le n$) |
| $\bar{x}$ | local coordinate |
| $X$ | discretized cartesian coordinate |
| $y$ | displacement; cartesian coordinate |
| $Y$ | discretized cartesian coordinate |
| $z$ | cartesian coordinate |
| $Z$ | thickness ratio; discretized cartesian coordinate |
| $\{\cdot\}$ | column matrix |
| $\{\cdot\}^T$ | row matrix (column transpose) |
| $[\cdot]$ | square matrix |
| $\bigcup$ | union |
| $\bigcap$ | intersection |
| det $[\cdot]$ | matrix determinant |
| sym | symmetric |
| $\beta$ | coefficient |
| $\delta_{ij}$ | Kronecker delta |
| $\Delta$ | discrete increment |
| $\varepsilon$ | normal strain |
| $\phi(\cdot)$ | trial space function; potential function |
| $\gamma$ | shear strain |
| $\eta_i$ | tensor product coordinate system |
| $\kappa_{\alpha\beta}$ | element of a square matrix |
| $\lambda$ | Lamé parameter |
| $\mu$ | Lamé parameter |
| $\nu$ | Poisson ratio |
| $O(\cdot)$ | order of truncation error |
| $\pi$ | pi (3.1415926 ...) |
| $\theta$ | time integration implicitness factor |
| $\rho$ | density |
| $d\sigma$ | differential element on $\partial\Omega$ |
| $\tau$ | normal stress |
| $\sum$ | summation symbol |
| $d\tau$ | differential element on $\Omega$ |
| $\tau$ | shear stress |
| $\omega$ | frequency |
| $\Omega^h$ | discretization of $\Omega$ |
| $\Omega$ | domain of differential equation |
| $\partial\Omega$ | boundary of domain $\Omega$ |
| $\zeta_i$ | natural coordinate system |
| $d(\cdot)/dx$ | ordinary derivative |
| $\partial(\cdot)/\partial x$ | partial derivative |
| $\nabla$ | vector derivative |

# INDEX